W. de Leeuw
R. van Liere (eds.)

Data Visualization 2000

Proceedings of the Joint
EUROGRAPHICS and IEEE TCVG
Symposium on Visualization
in Amsterdam, The Netherlands,
May 29–31, 2000

Eurographics

SpringerWienNewYork

Dr. ir. Willem Cornelis de Leeuw
ir. Robert van Liere
Center for Mathematics and Computer Science,
Amsterdam, The Netherlands

© 2000 Springer-Verlag/Wien
Reprint of the original edition 2000

Typesetting: Camera-ready by authors

Graphic design: Ecke Bonk

Printed on acid-free and chlorine-free bleached paper

SPIN: 10768503

With 166 partly coloured Figures

ISSN 0946-2767
ISBN-13: 978-3-211-83515-9   e-ISBN-13: 978-3-7091-6783-0
DOI: 10.1007/978-3-7091-6783-0

Preface

This book contains the papers presented at VisSym '00, the Second Joint Visualization Symposium organized by the Eurographics Association and the IEEE Computer Society Technical Committee on Visualization and Graphics (TCVG). The event took place from May 28 to May 31, 2000, in Amsterdam. We hope that these papers will be valuable, not only for visualization researchers, but also for practitioners developing or using visualization applications.

We are glad to report that the visiblility of the symposium continues to increase and that visualization researchers and practitioners from all over the world have submitted papers. This year, 66 papers and case studies were submitted of which 27 were accepted. In addition, we are glad to see that the focus of the symposium is also expanding. Topics are shifting from scientific data visualization (eg. flow and volume visualization) towards new areas in visualization. We accepted 7 research papers on information visualization and there was a broad range of other topics.

We would like to thank all those involved in organizing the symposium. In particular, special thanks to Wim de Leeuw and Guy Melançon who managed the electronic paper submission and review process. Also, many thanks to Mieke Brune who was in charge of the local organization. We want to thank the international program committee for their excellent, yet laborious, job in reviewing all submitted papers. The high quality of the symposium is a reflection of the quality of the submitted papers and the quality of the reviewing process.

Symposium co-chairs:
Ivan Herman
William Ribarsky
Robert van Liere

Paper co-chairs:
Frits Post
Jarke van Wijk

Chairs, IPC, and Reviewers

Symposium Co-Chairs

Robert van Liere, CWI, Amsterdam, Netherlands
Ivan Herman, CWI, Amsterdam, Netherlands
William Ribarsky, Georgia Institute of Technology, Atlanta, Georgia

Paper Co-Chairs

Frits Post, Delft Universiy of Technology, Netherlands
Jarke van Wijk, Technical University Eindhoven, Netherlands

Organizing Co-Chairs

Wim de Leeuw, CWI Amsterdam, Netherlands
Guy Melançon, CWI Amsterdam, Netherlands

International Program Committee

D. Bartz,	H. Hagen,	H.-G. Pagendarm,
K. Brodlie,	H. Hauser,	H. Pfister,
S. Card,	I. Herman,	F. Post,
D. Cohen-Or,	A. Kaufman,	W. Ribarsky,
S. Coquillart,	D. Keim,	M. Rumpf,
R. Crawfis,	D. Kenwright,	G. Sakas,
S. Eick,	W. de Leeuw,	R. Scopigno,
T. Ertl,	R. van Liere,	D. Silver,
S. Gibson,	W. Lefer,	P. Slavik,
M. Grave,	N. Max,	R. Spence,
E. Gröller,	R. Moorhead,	H. Spoelder,
M. Gross,	G. Nielson,	J. van Wijk

Additional reviewers

C. Bajaj,	C. Johnson,	G. Melançon
G. Di Battista,	M. van Kreveld,	J. Mulder,
D. Bergeron,	T. Kuipers,	T. Munzner,
S. Gumhold,	K.-L. Ma,	A. Pang,

VIII

K. Polthier,
A. Sadarjoen,
D. Saupe,
H. Schumann,

J. Smit,
P.J. Stappers,
R. Tamassia,
L. Treinish,

R. Veltkamp,
M. Ward,
R. Westermann,
H. van de Wetering

Table of Contents

Volume and Flow Visualization

Visualization Systems

Applications and Case Studies

Invited Speaker

Colin Ware

Data Visualization Research Lab,
Coastal and Ocean Mapping Center,
University of New Hampshire

Visualization as Applied Perception

Information visualization employs the high bandwidth of the visual system to enable people to perceive complex patterns in data. But it is important that the information is presented in a way that it is compatible with the pattern finding mechanisms of the brain. A poorly implemented visualization will mislead rather than enlighten. From this perspective visualization can be considered as a discipline of applied perception science and we examine how this science can help us invent and evaluate better visualization techniques. Examples are given relating to effective network visualization in 3D and 2D, the use of stereo views in geographic information systems, and the application of principles of pattern perception to the representation of abstract data structures.

Biography

Dr. Colin Ware takes the "visual" in visualization very seriously. He has advanced degrees in both computer science (MMath, Waterloo) and the psychology of perception (PhD. Toronto). He has published over ninety articles in scientific and technical journals and at leading conferences, many of which relate to the use of color, texture, motion and 3D information visualization and his book "Information Visualization: Perception for Design" appeared in January, 2000. In addition to more formal research, Professor Ware also builds innovative visualization software systems. He has been involve in developing 3D interactive visualization software (Fledermaus) for of ocean mapping for over 12 years, and he directed the initial development of the NestedVision3D system for visualizing very large networks of information. Both of these projects led to commercial spin-offs. Ware is Director of the Data Visualization Research Lab and Professor of Computer Science at the University of New Hampshire.

DAG Drawing from an Information Visualization Perspective

G. Melançon and I. Herman

Centrum voor Wiskunde en Informatica (CWI)
P.O. Box 94079
1090 GB Amsterdam, The Netherlands
{Guy.Melancon,Ivan.Herman,}@cwi.nl

Abstract. When dealing with a graph, any visualization strategy must rely on a layout procedure at least to initiate the process. Because the visualization process evolves within an interactive environment the choice of this layout procedure is critical and will often be based on efficiency.

This paper compares two popular layout strategies, one based on the extraction of a spanning tree, the other based on edge crossing minimization of directed acyclic graphs. The comparison is made based on a large number of experimental evidence gathered through random graph generation. The main conclusion of these experiments is that, contrary to the popular belief, usage of edge crossing minimization algorithms may be extremely useful and advantageous, even under the heavy requirements of information visualization.

1   Introduction

Graph visualization has emerged lately as a sub–field of information visualization, specializing in visualizing data that comes with inherent relations. In that case, the data to be visualized is interpreted as a graph, and applications usually offer different strategies to view, to navigate in, and to interact with the graph. This has many areas of application in biology, chemistry, computer science, web navigation, or document management systems. See [6].

Graph visualization can rely on the rich body of knowledge developed over the years by the graph drawing community, which grew around the yearly Symposia on Graph Drawing. A large number of layout strategies are at disposal, which have been collected recently in the book of di Battista *et al.*[1]. However, it is not always easy to decide which layout algorithm to use for a given application and for a specific class of graphs: the practical requirements, concerns, assumptions, etc., are often different. The graph drawing community usually assumes graphs to be sparse and to contain only a small number of nodes. A few hundred nodes, containing only about two times more edges than the number of nodes, is a common measure. In contrast, the information visualization community has to deal with graphs often containing thousands of nodes. Of course, large graphs are often clustered, yielding much smaller graphs; nevertheless, the real size and density numbers are still different. This issue will be examined in more details later in the paper.

Trees occur quite often in information visualization, whenever the data form some kind of hierarchy. It is also generally admitted that tree layout algorithms have the lowest complexity. Trees are sparse graphs (trees with N nodes only have $N-1$ edges) and they can be efficiently traversed. Moreover, trees are planar and most algorithms will produce an "aesthetically pleasant" layout for a tree. The running time of a tree layout algorithm is usually low (see [1, Chap. 10]). As a consequence, tree visualization have a wide variety of usage in information visualization. The advantages of relying on tree drawing algorithms has motivated many authors to base their visualization strategy on the extraction and layout of a spanning tree [10, 11, 16] to gain the efficiency of the tree layout. Also, interacting with the graph often translates into the necessity of redrawing it a large number of times over very small time intervals; under those conditions, relying on the efficiency of a tree layout algorithms is sometimes viewed as a mandatory choice. One of the goals of this paper is to put this claim in a better perspective, and to show when it is justified and when it is not.

Directed acyclic graphs ("dag"–s) can be seen as a natural generalization of trees. They have no cycles and their nodes can be assigned to layers so that all edges point in the same direction, usually downwards. Many layout methods have been developed for these graphs, too. The core of most of these algorithms (usually referred to as the Sugiyama methods) mainly focus on minimizing the number of edge crossings on the generated layout. The reason is that a dag is generally not planar, and the usual assumption is that an "aesthetically pleasant" drawing of a dag is the one with a minimum number of edge crossings. The problem is that minimizing the number of crossings is difficult. The computation of the optimal solution compares to a classical sorting problem and is thus well above linear time complexity. However, some heuristics have been developed that produce a layout approaching the optimal solution.

Dag–s form a good intermediary class between trees and general graphs. Efficient methods exist which first extract a directed acyclic graph from a general graph (see [1, 3], for example), which makes them generally usable. As a consequence, instead of relying on spanning trees, information visualization systems could also decide to use the Sugiyama approach: extract a "spanning dag" first, layout the dag, and add the missing edges. The claims of "aesthetics", or "better readability" of the graph, which focuses on edge crossings may make this choice a good alternative.

The problem, which researchers in information visualization face, is that there aren't any systematic comparisons available in the literature, which would make it clear when the spanning tree approach is preferable over the spanning dag. As a consequence, most dag based algorithms are disregarded altogether, being viewed as too complicated anyway. Offering such a comparison is the main goal of this paper. The question addressed here is "under what limits is it reasonable to rely on a layout technique for directed acyclic graphs when visualizing a graph in an interactive environment?". The answer can be very valuable for information visualization: improving the overall readability of a graph layout is certainly desirable although this has to be balanced against response time and interactivity. Having a better knowledge on the limits of crossing minimization algorithms can only help to take the appropriate decision.

2 Algorithmic details

As we have said in the introduction, the goal is to compare two layout approaches: the first is based on spanning dag–s, whereas the second relies on the usage of spanning trees. Although both layout strategies concentrate on a special classes of graphs (dag–s and trees, respectively), their applicability is much wider. Many different algorithms extracting a spanning tree can be found in the literature[8] which makes the spanning tree approach fairly general. As for dag–s, several techniques exist, too, to extract an spanning acyclic subgraph. The reader should consult [1] for further details. Consequently, the methods being compared are both of a fairly general use.

2.1 Layout of dag–s

Global minimization of the number of crossings in a graph is a very difficult problem. Instead, a common approach is to look first at the more restricted problem of a graph consisting of two layers of nodes, where edges connect nodes in different layers only (also referred to as *bipartite* graphs). If the solution to this restricted problem is found, it can then be extended to the full graph by sweeping through the graph layer by layer. However, simply sweeping through the layers is not enough; one has to make use of the so–called *dummy nodes*, too. Indeed, in order to apply the Sugiyama algorithm, each edge has to be transformed into a sequence of appropriate edges only crossing neighbour levels. The dummy nodes (which are not visible on the final layout) are inserted between nodes which are not on neighbouring layers. Dummies are inserted as necessary to yield a series of truly bipartite graphs when sweeping through the layers.

Even the solution of the bipartite case has proven to be difficult, more exactly, to be NP-complete[4, 2]. Consequently, various heuristics have been explored to produce a sub–optimal solution (see, e.g., di Battista *et al.*[1, Section 9.2]). The layout strategy chosen for the purposes of this study is the so–called "barycentre" heuristic. The underlying idea is rather simple and is based on the intuition that a node should be positioned according to its neighbours' positions. Hence, given a node in a layer, the x coordinate of a node can be computed as being the barycentre of the coordinates of its neighbours. That process can be iterated to refine the final positions of nodes. Among all possible crossing minimization heuristics or algorithms, the barycentre heuristic possesses many attractive qualities that were stressed in a recent paper by Juenger and Mutzel[7]. Maybe the most important feature is that the complexity of the technique is *linear* in the number of nodes. Also, it provides a solution that compares very well to the optimal solution with respect to the number of crossings.

The linear complexity of the barycentre heuristics makes it a good competitor to the tree layout algorithm. The actual implementation developed for this study also minimizes the number of iterations further: assuming layer i has been assigned positions, we first compute the next layer of nodes at level $i + 1$. In doing this, we need to introduce new dummy nodes for every edge crossing level $i + 1$. The nodes on layer $i + 1$ are then positioned at the barycentric coordinate of their predecessors on layer i. The nodes on layer i are then revisited and positioned at the barycentric coordinate of *all their neighbours*, this time including those on layer $i + 1$. The implementation takes care of "throwing away" the dummy nodes which are no longer in use.

2.2   The spanning tree strategy

As said before, the book of Jungnickel[8] is an excellent source to find spanning tree extraction algorithms. A simple extraction satisfying no particular condition can be achieved within $|E|$ steps, where E denotes the set of edges in the graph. For example, a breath–first-search traversal of the graph with an appropriate selection criteria for edges will output a spanning tree of a graph. The extraction process can sometimes be bound to an optimization problem, too. Although it may be tempting to follow this connection in applications, the tree satisfying an optimization condition might have undesirable geometric properties, like being very poorly balanced. In other words, the extraction strategy must be carefully selected. The experiences leading to this paper have shown that iteratively building the spanning tree, selecting at each step a node reachable from the set of already selected nodes, is quite appropriate. The selection of the next node is made by minimizing the distance to the root (or starting point of the search). In most cases, this will result in a fairly balanced spanning trees.

The tree layout we used is the classical algorithm due to Reingold and Tilford [14, 15]. It runs in two steps, first traversing the tree in post–order and computing temporary coordinates and modifier fields values for each node. A second traversal, this time visiting the nodes in pre–order, computes the actual coordinates by cumulating the modifier values of nodes from the root to the leaves of the tree. This algorithm is linear in the number of nodes, in spite of the fact it also relies on the necessity of visiting and comparing coordinates of nodes on the right and left border of neighbor subtrees.

Care should be taken, however, when applying this algorithm to the spanning trees of dag–s. Indeed, the traditional Reingold and Tilford algorithm will position nodes at a specific depth in the tree on the same layer. However, the depth of a node in the spanning tree might not coincide with its level in the dag (and applications might require to keep this visual relation). One solution is to follow the same approach most implementations of the Sugiyama methods use, i.e., to introduce dummy nodes into the tree. This, however, may have very negative consequences on the memory and time requirements of the algorithm. Instead, in our case, the y-coordinate of a node is assigned according to its level in the dag, while preserving the x-coordinate as computed by the Reingold–Tilford algorithm. As a consequence, the load of introducing dummy nodes into the graph is avoided in this case.

Figure 4 in the appendix shows a tree with 200 nodes and about 400 edges laid out using the two different layout methods. We will comment on those in later sections.

3   The graph sample

The graphs, on which the experiments were based, were randomly generated using a Markov chain process. It would go beyond the scope of this paper to describe the details of how this Markov chain works, which will be published elsewhere [9]. Suffices it to say that, with no restriction, the chain produces a set of uniformly random dag–s (all dag–s appear with equal probability). It can be proven that the average edge density is of $N^2/4$ (note that this is higher than what the graph drawing community usually considers!). The interesting point is that the state space of the chain can be controlled by imposing an upper bound on either the total number of edges in the dag, or the maximal degree of nodes. It appears that when restricting nodes to have a degree of at most k, the randomly generated graphs will have $kN/2$ edges on average. (The Markov chain works by systematically adding edges, on condition that a new edge does not create a cycle in the graph. The chain will tend to add edges until it reaches the maximum number of edges it can add. If all nodes reach degree k, the total number of edges must be $kN/2$.)

This Markov chain generation has proven to be a very useful tool because it offered a full control over the edge densities in the samples. With $k = 2$ (node degree at most 2), the average number of edges becomes N. Graphs with edge density N are obviously sparse since this condition will impose a large number of isolated vertices. When $k = 4$, the resulting sample consists of graphs which are considered as "dense" by the graph drawing community. However, in our view, $k = 4$ should not be considered as really dense for information visualization, and actually constitutes the lowest reasonable density to be examined (density $k = 4$ means that each data node has, on average, "relations" with at most 4 other data nodes, which is an unrealistic condition). Higher density cases have to be examined: graphs with edge density varying from $3N$ up to $5N$ (that is with k varying from 6 to 10) were also considered (although we only report here the cases with $k \leq 6$, see section 4).

For each value of $k = 2, \ldots, 10$, graphs of different sizes varying from 10 to 1000 nodes were generated. Exactly 20 graphs per category were processed 30 times through the layout algorithms, measuring the time it took to perform the layout itself (disregarding the time needed to display or read in the graph). Our conclusions are based on the average value of the results computed out of those 600 layout trials for each size category.

4   Results of the experimentation

We ran the experiment based on a Java implementation of the layout methods[1]; it can be expected that C implementations, for instance, would result in more efficient computing time. However, most of our conclusions are based on the relative efficiency measures of the two layout methods, i.e. our results remain valid in general, too. In the cases when we do refer to absolute timing results, our conclusions could only be made stronger by a faster implementation, as we will see in what follows.

4.1   Computing time complexity

As can be expected, time complexity is not a real concern for graphs with a few nodes and a low density (see Figures 1 & 2). The barycentre method is actually linear in the number of nodes (dummies *and* real nodes). With low density graphs, the number of dummies introduced in the layout process is low, as shown on Figure 3. For $k = 2, 3$ the number of dummies is at most equal to 4 or 5 times the number of real nodes in the graph. The curves in Figures 1 & 2 show that both layout techniques are equivalent with

---

[1] Our Java implementation is based on a general framework for graph visualization (see [5]).

8

respect to time in these cases; actually, the barycentre technique proves to be slightly superior when $k = 2$. Note also that for $k = 2$, even with graphs of size up to 1000 nodes, both algorithms have running time below half a second, which is acceptable when the layout occurs in an interactive environment. The same conclusion holds when $k = 3$, for graphs with up to 500 nodes.

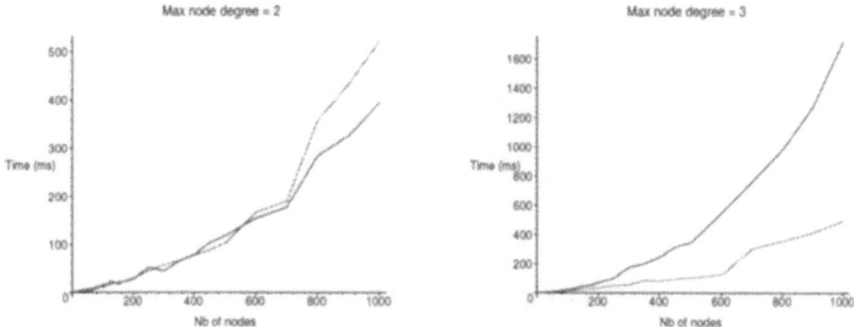

Fig. 1. Computing time for spanning tree and barycentre layouts. The gray colored curves show the spanning tree results, whereas the darker curves stand for the barycentre layout.

For graphs with higher densities, the number of inserted dummy nodes increases significantly. For a specific graph size, the number of dummy nodes appears to increase linearly with respect to edge density. Figure 3 shows that for a graph of size 250 the total number of nodes (dummies and real) varies from 5 to 30 times the actual number of real nodes, as k goes from 3 to 5. For graph of size 500, the same measure will vary from 5 to a little less than 50 times the number of nodes in the graph, over the same interval for edge density. These dummies create a significant load on the algorithm, because the weight of memory management will overshadow any theoretical speed improvement of the algorithm. Although the Java virtual machine might not be the best example for a good memory management, it is still significant that our implementation ran into serious memory fault problems at graphs with a few thousands nodes already. Hence, the barycentre approach seems to loose at higher densities; the extremely high number of dummies proves to be the Achilles' heel of the method.[2]

All these numbers seem to suggest that the barycentre method — or any other Sugiyama like layout method — is inappropriate for information visualization in real applications. However, one should not jump on conclusions too hastily. As Figures 1 & 2 show, for example, the barycentre method still scores well *in absolute times* for lower density graphs of size less than 400. When $k = 4$ (which we regard as the lowest

---

[2] This is, as a matter of fact, also an important conclusion of our measurements: researchers in graph drawing tend to concentrate on various ways of improving the edge crossing optimization step, and the problems incurred by the dummy nodes is often considered as less important.

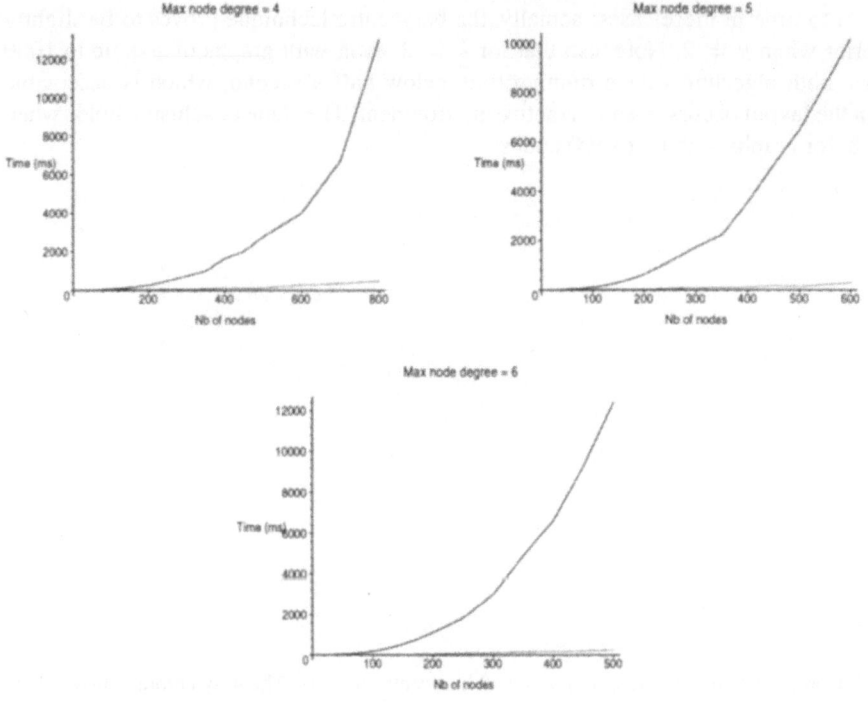

Fig. 2. Computing time for spanning tree and barycentre layouts (continued).

complexity level for information visualization) both the tree based and the barycentre strategies are comparable for graphs of a size up to about 250 nodes, since the barycentre method still lays out the graph in less than half a second. For higher densities, that is when $k = 5$ or 6, the same conclusion holds although the maximum graph size then has to be lowered down to remain under a 500 ms computing time (to 200 for $k = 5$ and to 150 for $k = 6$).[3] Although graphs with much more than 250 nodes frequently appear in applications, displaying all elements is often not a real option anyway: even the fastest and best layout strategy may produce a image barely understandable to humans. As a consequence, applications often offer an overview diagram of the graph (based, for example, on a suitable clustering of the original graph), and interaction occurs on this overview rather than the original data. Such an overview graph may have a size and edge density falling under the limits listed above, in which case it could be laid out using a Sugiyama method for improved readability.

---

[3] The reader should remember that these numbers stem from a Java implementation which could be improved, if necessary, by choosing a faster implementation environment.

10

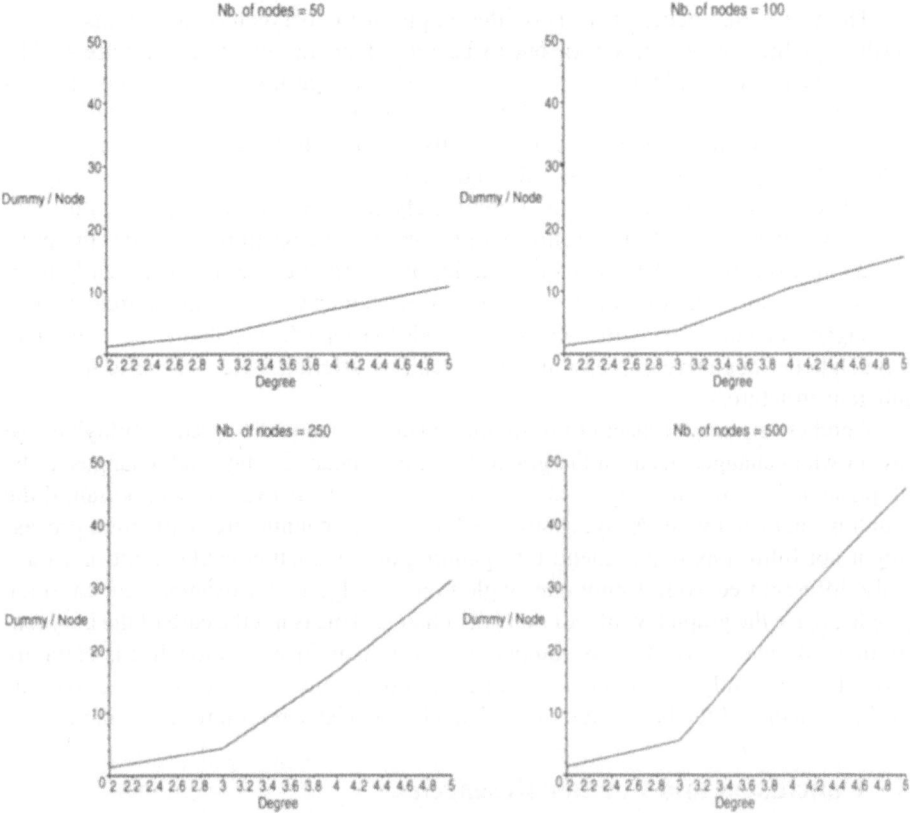

Fig. 3. Number of dummies introduced during barycentre layout (ratio per node)

4.2 Aesthetics

The usual reason for implementing the barycentre heuristics, or any other crossing min-imization heuristics, is to produce a layout with as few edge crossings as possible. This aesthetics has been proven to have a measurable effect on the "readability" of the graph [12, 13], hence it may play a major role in the user interface aspect of the information visualization application. This, added to the fact that the barycentre method works in reasonable time when the size of the graph does not exceed some bounds, seems to make the barycentre method a real option.

There is a counter argument, though. One can show that as edge density increases, any crossing minimization algorithm will produce a solution that approaches the opti-mal solution (see [1, Section 9.2.5]). In other words, if the graph is dense, the number of unavoidable crossings will be high and any approach is susceptible to provide an ac-ceptable solution — which will incidentally contain a large number of crossings. This argument supports the conclusion that the spanning tree strategy is as good as any other method, as far as edge crossings are concerned. So why bother?

The fact is that better analysis on the output of the barycentre algorithms reveals further qualities which may turn out to be important. Let us look at Figure 4. The barycentre layout results in less edge crossings, although this might not be immediately noticeable. However, the fact that the method positions a node at the barycentric x-coordinates of its neighbours explains why the overall picture has a more or less "slender" shape in the middle. The Reingold and Tilford layout is not affected in this way. The spanning tree extraction is completely separated from the positioning process and thus leads to a "square" image of the graph. The result is that nodes might be spread all over the displayable area, resulting in a large number of long, nearly horizontal edges: these are the edges which were disregarded by the spanning tree process. We suggest that the barycentre view, which tends to keep adjacent nodes together, gives a much better cognitive impression of the graph by offering a view closer to its "real" inherent structure.

Another important aspect in information visualization is the "predictability" of the layout when changes occur in the graph. What this means is that small changes in the graph should not result in significant layout changes; this is extremely important if the graph is interacted with. A layout strategy based on a spanning tree extraction process might not fulfill this requirement: the spanning tree extraction could result in a radically different tree even if only one single node or edge is, for example, added to the graph. Hence the graph layout will radically change. This is not the case of the barycentre method, where any effect of interaction will remain local. Although it is certainly possible to carefully design the tree extraction process and preserve predictability, this intrinsic quality of the barycentre method can be a decisive reason for choosing it.

5 Conclusion and Further Research

Our experiments with the two layout strategies have shown that the popular belief telling us that information visualization should rely on spanning tree algorithms is not really correct. Edge crossing minimization methods for dag–s should not be disregarded, as they often are: with very reasonable restrictions on the size and the densities of the graphs, these methods do not only compete with tree based layouts in terms of time, but the quality of the output they produce may be superior.

Edge crossing minimization is not the only graph layout strategy which has been developed over the years. We think that a systematic overview and comparison of other methods (like the force–directed ones, for example), confronted with the requirements of information visualization, is necessary. Such comparisons would help the information visualization community to choose the right approach when implementing a particular application. Care should be taken to base these comparisons on real data and implementations, rather than on theoretical results only; our measurement results on the explosion on dummy nodes is a good example for the problems which can be revealed that way.

References

1. di Battista, G., Eades, P., Tamassia, R., and Tollis, I.G. *Graph Drawing: Algorithms for the Visualisation of Graphs*. Prentice Hall, 1999.

12

2. Eades, P. and Whitesides, S.H. Drawing Graphs in Two Layers. *Theoretical Computer Science*, 131(2):361–374, 1994.

3. Gansner, E.R., North, S.C., and Vo, K.P. DAG - A Program that Draws Directed Graphs. *Software: Practice and Experience*, 18(11):1047–1062, 1988.

4. Garey, M.R. and Johnson, D.S. Crossing number is NP-complete. *SIAM Journal of Algebraic and Discrete Methods*, 4(3):312–316, 1983.

5. Herman I., Marshall M. S., and Melançon G. An Object-Oriented Design for Graph Visualization. Technical Report INS-0001, Centrum voor Wiskunde en Informatica (CWI), 2000. See ftp://ftp/cwi/nl/pub/CWIreports/INS.

6. Herman I., Marshall M. S., and Melançon G. Graph Visualization and Navigation in Information Visualization. *IEEE Transactions on Visualization and Computer Graphics*, 6, 2000 (to appear).

7. Juenger, M. and Mutzel, P. 2-layer straightline crossing minimization: Performance of exact and heuristic algorithms. *Journal of Graph Algorithms and Applications*, 1(25):33–59, 1997.

8. Jungnickel, D. *Graphs, Networks and Algorithms*. Springer, 1999.

9. Melançon G., Dutour I., and Bousquet-Mélou M. Random generation of Dags for Graph Drawing. Technical Report INS-0005, Centrum voor Wiskunde en Informatica (CWI), 2000. See ftp://ftp/cwi/nl/pub/CWIreports/INS.

10. Munzner, T. H3: Laying out Large Directed Graphs in 3D Hyperbolic Space. In *IEEE Symposium on Information Visualization (InfoVis '97)*, pages 2–10. IEEE CS Press, 1997.

11. Munzner, T. Drawing Large Graphs with H3Viewer and Site Manager. In *Symposium on Graph Drawing GD '98*, Lecture Notes in Computer Science, pages 384–393. Springer-Verlag, 1998.

12. Purchase, H., Cohen, R.F., and James, M. Validating Graph Drawing Aesthetics. In *Symposium Graph Drawing GD'95*, volume 1027 of *Lectures Notes in Computer Science*, pages 435–446. Springer-Verlag, 1995.

13. Purchase, H.C. Which aesthetic has the greatest effect on human understanding? In *Symposium on Graph Drawing GD '97*, Lecture Notes in Computer Science, pages 248–261. Springer-Verlag, 1998.

14. Reingold, E.M. and Tilford, J.S. Tidier Drawing of Trees. *IEEE Transactions on Software Engineering*, 7(2):223–228, 1981.

15. Walker, J.Q. A Node-positioning Algorithm for General Trees. *Software: Practice and Experience*, 20(7):685–705, 1990.

16. Wills, G.J. Niche Works - Interactive Visualization of Very Large Graphs. In *Symposium on Graph Drawing GD'97*, volume 1353 of *Lectures Notes in Computer Science*, pages 403–414. Springer, 1997.

Editors' Note: see Appendix, p. 281 for colored figure of this paper

Contextual Visualization
of Actor Status in Social Networks

Ulrik Brandes* and Dorothea Wagner

University of Konstanz, Department of Computer and Information Science,
Box D 188, 78457 Konstanz, Germany
{Ulrik.Brandes,Dorothea.Wagner}@uni-konstanz.de

Abstract. We propose a novel information visualization approach for an analytical method applied in the social sciences. In social network analysis, social structures are formally represented as graphs, and structural properties of these graphs are assumed to be useful in the explanation of social phenomena. A particularly important such property is the relative status of actors in a given network.

Since operationalizations of status are aggregate indices of vertices, researchers are not only interested in status, but also in the context leading to these values, i.e. the underlying social network. We therefore visualize the network in a layered fashion, mapping status scores to vertical coordinates. The resulting problem of determining horizontal positions of vertices such that the overall layout is readable, is algorithmically difficult, yet well-studied in the literature on graph drawing. We outline a customized approach that routinely produces satisfactory pictures at interactive speed.

1 Introduction

Different from categorical data analysis, aggregate indices of relational data are typically insufficient to fully appreciate and understand the information contained in the data. In any kind of network analysis, it is therefore desirable to always provide a representation of the actual network as well.

Most types of networks are traditionally visualized using point-and-line representations [4]. If the network has no underlying spatial layout (unlike, e.g., data associated with geographic networks [3]), a layout has to be computed explicitly. But in addition to the inherent difficulty of laying out an abstract network in a readable way [8], this raises the problem of trust in its analysis. Who is going to comfortably interpret complex aggregate data, when it is difficult to relate it to the base data? As a potential remedy for this problem we propose contextual visualization, i.e. the simultaneous representation of base and derived data in a single diagram that is based on some express principles. Simple examples of

---
* Part of this research was done while with the Department of Computer Science at Brown University. I gratefully acknowledge the *German Academic Exchange Service* (DAAD/Hochschulsonderprogramm III) for financial support.

this principle are found, e.g., in bar charts with an additional line indicating the mean value. We here pursue this idea in an application from the social sciences.

Social network analysis is a subdiscipline of the social sciences, using graph-theoretic concepts to understand and explain social phenomena. A social network consists of a set of actors, who may be arbitrary entities like persons or organizations, and one or more types of relations between them. For a comprehensive overview of methods and applications see [29]. We here confine ourselves to networks of a single, directed relation.

The concept is illustrated by an example group of 14 employees, the internal auditing staff of a larger company. This group is analyzed in [17], where its formal organization is compared to an informal relation called "advice", i.e. who does an actor turn to for help or advice at work about work-related questions or problems. Organizational and advice relation data are given in Fig. 1.

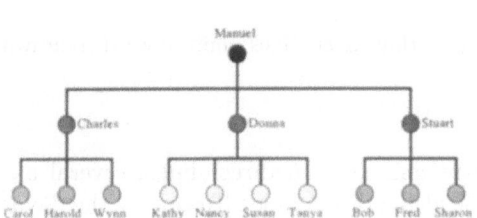

```
Manuel  0|0 0 0|0 0 0 0 0 0|0 1 0 0   manager
Charles 1|0 0 1|0 0 0 0 0 0|0 1 0 0
Donna   1|0 0 0|0 0 0 0 0 0|0 1 0 0   supervisors
Stuart  1|1 0 0|0 0 0 0 0 0|0 1 0 0
Bob     0|0 0 1|0 0 1 0 1 0|0 0 0 0
Carol   0|1 0 0|0 0 0 0 0 0|0 0 0 0
Fred    0|0 0 1|0 0 0 0 0 0|0 0 0 0
Harold  0|1 0 0|0 0 0 0 0 0|0 0 0 0   auditors
Sharon  0|0 0 1|0 0 0 0 0 0|0 0 0 0
Wynn    0|1 0 0|0 0 0 0 0 0|0 0 0 0
Kathy   0|0 1 0|0 0 0 0 0 0|0 1 0 1
Nancy   0|0 1 0|0 0 0 0 0 0|0 0 0 0   secretaries
Susan   0|0 1 0|0 0 0 0 0 0|1 0 0 1
Tanya   0|0 1 0|0 0 0 0 0 0|1 1 0 0
```

Fig. 1. Formal organizational chart and adjacency matrix of advice relationship. If a matrix entry is 1, the row actor turns to the column actor for advice

The advice relation largely resembles the organizational hierarchy with one notable exception, as illustrated in Fig. 2. In a tiresome and error-prone[1] process, vertices were manually arranged such that most edges point in upward direction, thus depicting an informal status in the advice network. Based on this graphical support, the conclusion of [17] is that changes the manager introduced to increase through-put may have been ineffective because he had not made sure that the secretary presiding the informal hierarchy of advice was backing them.

Qualitative results like this can be supported routinely using the formal apparatus of social network analysis. Networks of relationships are conveniently modeled by graphs $G = (V, E)$, where vertex set V represents the set of actors, and the set $E \subseteq V \times V$ of directed edges represents the relation under study, i.e., in our example, $(u, v) \in E$, if and only if the actor represented by u turns to the actor represented by v for advice. For convenience, we usually omit the distinction between actors and vertices, or edges and relations.

A simple, yet crude, quantitative measure of an actor's network status is its indegree, defined as the number of edges directed to the vertex. Since this

---

[1] This is by necessity. See the paragraph on layer assignment in Section 3.

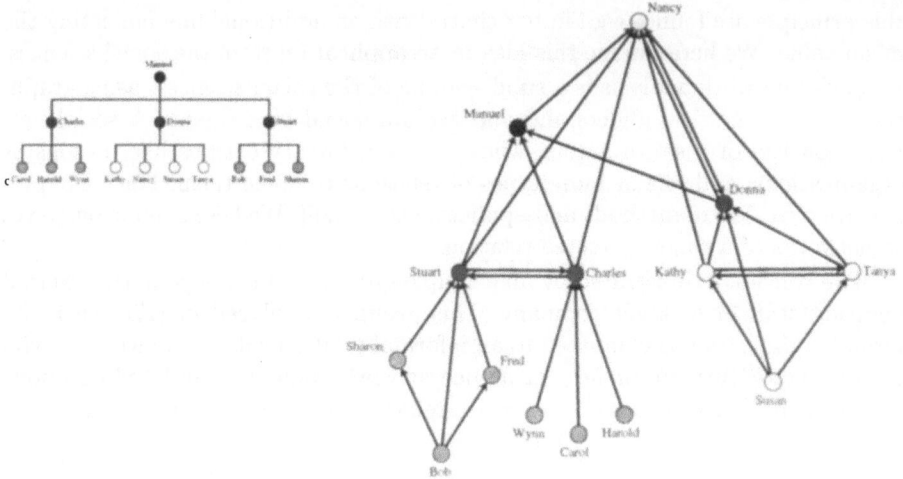

Fig. 2. Advice network, *manually* arranged so that most edges point upward (redrawn from [17])

definition takes into account only status gained from direct links, several approaches have been developed to include also indirect links. To convey the flavor of these approaches, a commonly used definition of status is presented. Introduced in [16], it rests on the assumption that links from actors that have high status themselves contribute more to a receiving actor's status than links from others. This recursive definition leads to the following equilibrium equation. Let $a < 1$ be an attenuation factor indicating the decrease of status passed along edges in the graph. If A denotes the adjacency matrix of the graph, then a solution $s = (s_v)_{v \in V}$ of

$$\left(\frac{1}{a} \cdot I - A^T \right) \cdot s = d^-,$$

where I is the unit matrix, A^T the transpose of A, and d^- the vector of indegrees, describes the relative status according to the above model. For ease of comparison, each entry in s is divided by the maximum entry. Status results for the example network are given in the next section. We refer the reader to Chapter 5 of [29] for references on sociological interpretation and other models of status.

The goal of this work is to provide automatic visual support for status analysis in social networks. Such work has three main aspects [5]: the substance to be visualized, a graphical design, and an algorithm realizing it. We have already described the substance we are interested in. The remainder of this paper is therefore organized as follows. In Sect. 2, we develop a graphical design for contextual visualization of networks and status therein, and give algorithms to produce such drawings in Sect. 3. We conclude in Sect. 4 with visualizations from a sociological study using a prototype implementation of our approach.

2 Contextual Visualization

In this section, we develop a graphical design for visualizations that contextualize status values with the underlying network. Currently available tools for network visualization[2] do not achieve this, because they essentially produce general purpose visualizations focusing on the ease of perceiving connectedness information, i.e. the presence or absence of edges between pairs of vertices. Figure 3 shows the type of visualization thus typically encountered. The network is shown separate from the result of the analysis (here, status according to the measure described above). Though the image is very readable, it does not convey the interesting status information. Its design is inherently undirected (the picture would be the same even if all edge directions are reversed), and it is next to impossible to relate the status values to the picture. Such visualizations are typical in the work of sociologists, and others applying their concepts (see, e.g., [13]).

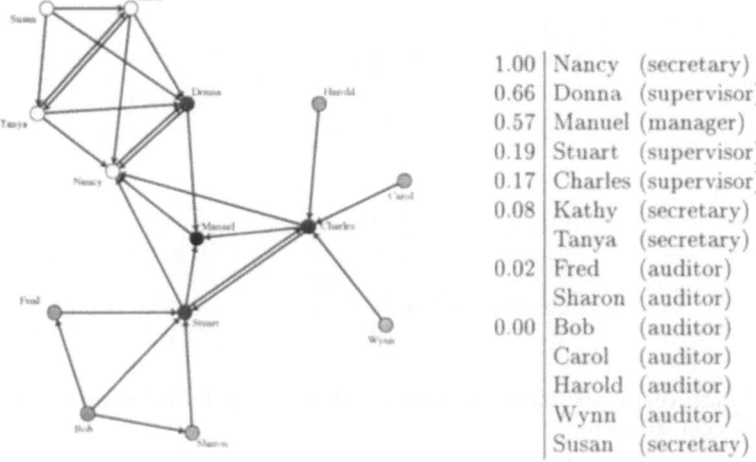

1.00	Nancy	(secretary)
0.66	Donna	(supervisor)
0.57	Manuel	(manager)
0.19	Stuart	(supervisor)
0.17	Charles	(supervisor)
0.08	Kathy	(secretary)
	Tanya	(secretary)
0.02	Fred	(auditor)
	Sharon	(auditor)
0.00	Bob	(auditor)
	Carol	(auditor)
	Harold	(auditor)
	Wynn	(auditor)
	Susan	(secretary)

Fig. 3. Non-contextual automatic visualization of status and advice network (spring embedder type layout and stem-and-leaf diagram)

Empirical evidence suggests that network layout affects not only the ease of reading [25], but has an influence on the understanding and interpretation of substantive content as well [21]. Building on the familiar everyday notion of "higher" and "lower" status, it seems natural to graphically represent status through vertical positioning. Instead of using bar charts to depict status, the placement of vertices themselves can be restricted to levels signaling their status. The idea is illustrated in Fig. 4, where the status index described in the

---

[2] Best known are KrackPlot [18], Pajek [2], and MultiNet [27]. They mainly offer layouts based on variants of the spring embedder [9], multidimensional scaling, and layouts based on eigenvectors of the adjacency or Laplacian matrix of the graph.

introduction is used to assign, to each vertex, a y-coordinate proportional to its status.

Note how the vertical ordering differs from that in Fig. 2. While the stem-and-leaf diagram of Fig. 3 makes this obvious as well, this visualization also explains the reason why: the measure of status used assigns values according to the values of sending neighbors, and Donna is the only actor that Nancy asks for advice.

While, in principle, any definition of status is applicable, we will see in the next section that, e.g., the requirement of a maximum number of upward pointing edges is not suitable, since it leads to computational difficulties that make interpretation infeasible.

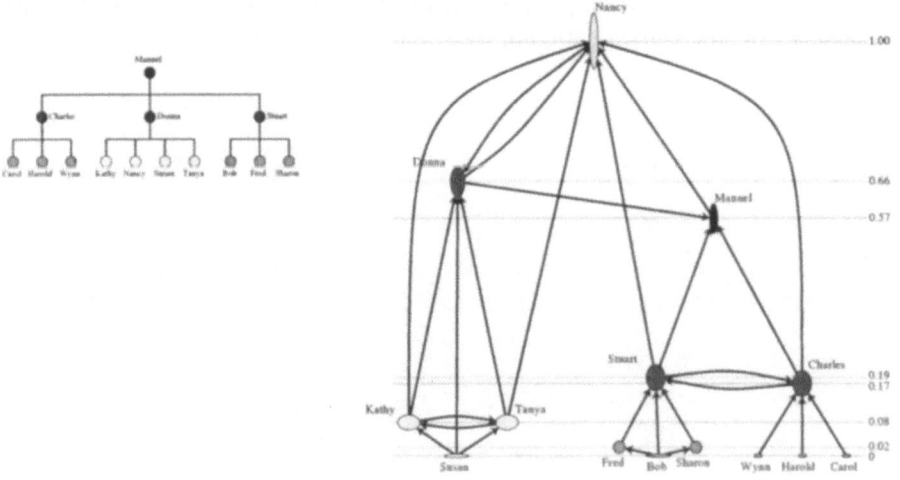

Fig. 4. Contextual visualization of status in advice network (organizational hierarchy shown for comparison)

We have integrated additional information in Fig. 4 by depicting vertices as ellipses rather than circles. This way, the ratio of incoming and outgoing edges is incorporated into the drawing without changing the layout. Let $d_G^-(v)$ and $d_G^+(v)$ denote the in- and outdegree of vertex v. Then, a horizontal radius $r_h(v)$ and a vertical radius $r_v(v)$ for the ellipse are chosen to satisfy

$$\frac{r_v(v)}{r_h(v)} = \frac{d_G^-(v)}{d_G^+(v)},$$

$$r_v(v) + r_h(v) = \pi \cdot d_G^-(v) \cdot d_G^+(v),$$

so that the ratio of in- and outdegrees is visually represented by the ratio of height and width, and the sum of the degrees is represented by the area of a vertex feature. A minimum height and width is used for zero in- and outdegree, and simple adjustments of the second equation account for vertex shapes other than ellipses (rectangles, rhombs).

Other than substantive, there are ergonomic criteria a visualization should satisfy. For example, a large number of crossing edges makes a drawing difficult to read [25]. Visualizations like the one in Fig. 4 are therefore more difficult to produce than, e.g., bar charts, because we can not just place vertices at specified y-coordinates. Algorithms to generate readable drawings under the above substantive constraint are described in the next section.

3 Automatic Layout

To automatically generate layered visualizations of social networks, we have to provide algorithms to compute x-coordinates for vertices and bend points of edges in the graph. This is a special case of a *graph drawing* problem. See [8] for an overview of the field.

The most commonly used framework for horizontally layered drawings of graphs is presented in [28]. It consists of the following generic steps:

1. determine a layer for each vertex,
2. introduce an edge bend point for each layer an edge spans and determine a relative ordering of vertices and bend points on the same layer, and finally
3. assign x- and y-coordinates to each vertex and bend point.

Steps 2 and 3 are separated to enable the use of combinatorial methods in the second step, which serves to reduce the number of crossing edges. Note that crossings severely affect the readability of a drawing [24], and that the number of crossings between two adjacent layers is determined by the relative ordering of vertices and bend points, independent of the actual coordinates (hence the introduction of bend points, see Fig. 5). A comprehensive overview of approaches to carry out the above steps is given in Chapter 9 of [8]. Though there is a whole range of implementations, most notably [11], our specific needs in the first step rule out their usage.

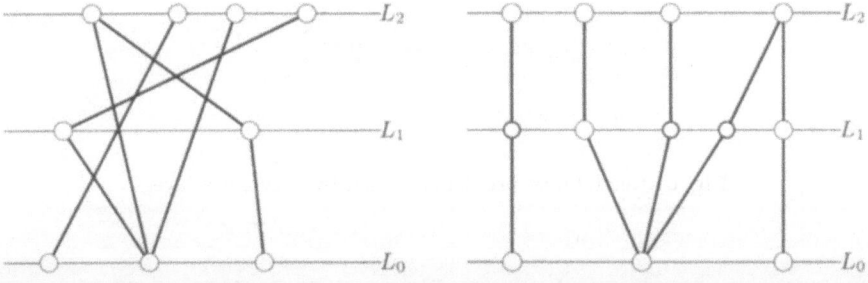

Fig. 5. A three-layer graph with many crossings, and the same graph with reordered vertices and dummy vertices

Layer assignment. We first argue, why the criterion of a maximum number of upward pointing edges must not form the basis of automatically generated status visualizations. A fairly common approach to layering is to break all directed cycles, if any, by temporarily reversing some edges, and assign vertices to layers by topological sorting. Reversing the minimum number of edges nicely corresponds to finding a layering with a maximum number of upward pointing edges.

There are three substantive reasons against this approach. First of all, the implicit definition of status (directed edges imply that the receiver has a higher status than the sender) yields only a partial ordering. Secondly, a minimum cardinality set of cycle breaking edges need not be unique. And thirdly, a straightforward reduction from feedback arc set shows that the problem of determining such a set with minimum cardinality is $\mathcal{NP}$-hard [15]. Since all three of these difficulties introduce arbitrariness into the complete ordering of actor status that any computed layering implies, interpretation of relative status becomes unreliable, if not impossible.

Assuming that formal status indices have a sound theoretical basis (a discussion of the appropriateness of an interval scale measurement is beyond the scope of this paper), any such index can directly be used for the y-coordinate of each vertex (up to scaling). We don't know of other approaches dealing with y-coordinates that are already given by the context. Let $s = (s_v)_{v \in V}$ be a status vector, a trivial layer assignment then is a partition $L_0 = \{v_0\}, \ldots, L_{|V|-1} = \{v_{|V|-1}\}$ of V, such that $i < j$ implies $s_{v_i} \leq s_{v_j}$. Status values often differ only marginally, though, leading to very close layers that cause perceptual problems like, e.g., several crossing (or non-crossing?) edge segments running almost horizontally (Fig. 6). To avoid such problems, status values are clustered and all vertices with status values in the same cluster are assigned to the same layer. Though any clustering may be used, we apply an agglomerative clustering scheme starting with singletons and merging two clusters, if the minimum status difference between any pair of vertices in different clusters is below a fixed threshold $0 \leq \varepsilon < 1$.

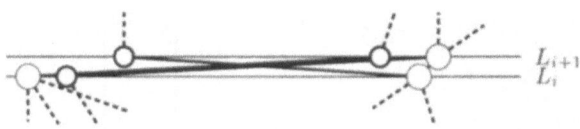

Fig. 6. Readability problems caused by very close layers

Crossing reduction. In this step, we are given a layering $L_0, \ldots, L_k$ of the vertices and our goal is to define a horizontal ordering of vertices in and edges spanning the same layer such that the number of edge crossings is small. An edge $(u, v) \in E$ is said to *span* a layer L_i, if $u \in L_{j_1}$, $v \in L_{j_2}$, and $j_1 < i < j_2$ or $j_2 < i < j_1$.

For each layer an edge spans, a dummy vertex representing a bend point is introduced, subdividing that edge and placed in the appropriate layer. We can now assume that we are given a layering such that no edge spans any layer. Note that the number of crossings is now dependent only on the ordering of vertices in each layer.

Finding an ordering that minimizes the number of edge crossings is another $\mathcal{NP}$-hard problem [12]. A common heuristic is the layer-by-layer sweep, in which the ordering in, say, L_0 is fixed and L_1 is reordered to reduce the number of crossings. Then, the order in L_1 is fixed, and L_2 is reordered, and so on. After reaching L_k, the process is reversed and repeated up and down the layering until no further improvement is made. Note however, that minimizing the number of crossings between adjacent layers, where the ordering in one layer is fixed, is $\mathcal{NP}$-hard [10] as well. Though in praxis this problem can be solved optimally for medium sized instances [14] using integer linear programming, the overall number of crossing will not be minimum. For simplicity, we use one of several heuristics (e.g., the median heuristic, placing a vertex at the median position of its neighbors in the adjacent layer) which are known to perform quite satisfactory.

Another heuristic, called global sifting [20], is used as a postprocessing step to the layer-by-layer sweep to reduce the number of remaining crossing. Roughly speaking, global sifting picks one vertex at a time and finds the locally optimal position within a layer by probing all of them. Our experiences are that this postprocessing is worth the additional effort.

Horizontal placement. Given y-coordinates, a layering, and an ordering of vertices and bend points within each layer, it remains to compute x-coordinates respecting the horizontal orderings. Currently, we are using an adoption of a fast heuristic provided in the AGD library [23], trying to straighten long edges and keep edge lengths small, but better strategies need to be explored for future use.

4 Example and Conclusion

One of several studies already applying our visualization approach is an analysis of the privatization processes of two industrial conglomerates in Eastern Germany after reunification [26]. Actors in these networks are political or corporate organizations, and different kinds of relations between them are investigated.

In this application, we represent the semantic attributes "sector" (government, political parties, unions and associations, corporations) by color and "level" (local, regional, federal) by shape. To reduce clutter due to bidirectional edges and arrow heads, non-downward pointing uni-directional edges are depicted in black, bidirectional edges in green, and downward pointing edges in red.

Figure 7 shows two relations between the same set of actors in the shipbuilding industry. On the left, edges indicate to what other actors an actor reports mandatorily, and on the right, edges indicate whose interests actors claim to have taken into account in important decisions. Even without any background knowledge, it is readily observed that fairly coordinated high-level governmental actors (blue rhombs) dominate the hierarchy of interest consideration.

Though our visualizations are considered very useful by those using them, we feel that several details – in particular regarding bend point placement – need further improvement. Moreover, we would like to provide automatic help for label placement, which has been refined manually for the above examples, and need to explore means of user interaction: what kind of improvements may a user make without running the risk of being suggestive? Similar work [6] is concerned with a structural index called centrality, but can we also provide automatic support for contextual visualizations of substance without an immediate geometric connotation?

Acknowledgments. Data sets are courtesy of David Krackhardt and Jörg Raab. Frank Müller implemented a prototype layout system in C++ using LEDA [22], AGD [23], and LAPACK [1]. This work would not have been possible without the fruitful discussions we had with Patrick Kenis, Jörg Raab, and Volker Schneider.

References

1. E. Anderson, Z. Bai, C. Bischof, S. Blackford, J. Demmel, J. Dongarra, J. Du Croz, A. Greenbaum, S. Hammarling, A. McKenney, and D. Sorensen. *LAPACK User's Guide.* Society for Industrial and Applied Mathematics, 3rd edition, 1999. See http://www.netlib.org/lapack/.
2. V. Batagelj and A. Mrvar. PAJEK – Program for large network analysis. *Connections,* 21(2):47–57, 1998. Project home page at http://vlado.mat.uni-lj.si/pub/networks/pajek/default.htm.
3. R. A. Becker, S. G. Eick, and A. R. Wilks. Visualizing network data. *IEEE Transactions on Visualization and Graphics,* 1(1):16–28, 1995.
4. J. Bertin. *Semiology of Graphics: Diagrams, Networks, Maps.* University of Wisconsin Press, 1983.
5. U. Brandes, P. Kenis, J. Raab, V. Schneider, and D. Wagner. Explorations into the visualization of policy networks. *Journal of Theoretical Politics,* 11(1):75–106, 1999.
6. U. Brandes, P. Kenis, and D. Wagner. Centrality in policy network drawings. In Kratochvíl [19], pages 250–258.
7. R. S. Burt. *Structure, Version 4.2.* Center for the Social Sciences, Columbia University, New York, 1991.
8. G. Di Battista, P. Eades, R. Tamassia, and I. G. Tollis. *Graph Drawing: Algorithms for the Visualization of Graphs.* Prentice Hall, 1999.
9. P. Eades. A heuristic for graph drawing. *Congressus Numerantium,* 42:149–160, 1984.
10. P. Eades and N. C. Wormald. Edge crossings in drawings of bipartite graphs. *Algorithmica,* 11:379–403, 1994.
11. E. R. Gansner, S. C. North, and K.-P. Vo. DAG – A program that draws directed graphs. *Software—Practice and Experience,* 17(1):1047–1062, 1988.
12. M. R. Garey and D. S. Johnson. Crossing number is $\mathcal{NP}$-complete. *SIAM Journal on Algebraic and Discrete Methods,* 4:312–316, 1983.

13. N. B. Harrison and J. O. Coplien. Patterns of productive software organziations. *Bell Labs Technical Journal*, pages 138–145, Summer 1996.
14. M. Jünger and P. Mutzel. 2-Layer straightline crossing minimization: Performance of exact and heuristic algorithms. *Journal on Graph Algorithms and Applications*, 1(1):1–25, 1997.
15. R. M. Karp. Reducibility among combinatorial problems. In R. Miller and J. Thatcher, editors, *Complexity of Computer Computations*, pages 85–103. Plenum Press, 1972.
16. L. Katz. A new status index derived from sociometric analysis. *Psychometrika*, 18:39–43, 1953.
17. D. Krackhardt. Social networks and the liability of newness for managers. In C. L. Cooper and D. M. Rousseau, editors, *Trends in Organizational Behavior*, volume 3, pages 159–173. John Wiley & Sons, 1996.
18. D. Krackhardt, J. Blythe, and C. McGrath. KrackPlot 3.0: An improved network drawing program. *Connections*, 17(2):53–55, 1994. Project home page at http://www.contrib.andrew.cmu.edu/~krack/.
19. J. Kratochvíl, editor. *Proceedings of the 7th International Symposium on Graph Drawing (GD '99)*, volume 1731 of *Lecture Notes in Computer Science*. Springer, 1999.
20. C. Matuszewski, R. Schönfeld, and P. Molitor. Using sifting for k-layer straightline crossing minimization. In Kratochvíl [19], pages 217–224.
21. C. McGrath, J. Blythe, and D. Krackhardt. The effect of spatial arrangement on judgments and errors in interpreting graphs. *Social Networks*, 19(3):223–242, 1997.
22. K. Mehlhorn and S. Näher. *The LEDA Platform of Combinatorial and Geometric Computing*. Cambridge University Press, 1999. Project home page at http://www.mpi-sb.mpg.de/LEDA/.
23. P. Mutzel, C. Gutwenger, R. Brockenauer, S. Fialko, G. W. Klau, M. Krüger, T. Ziegler, S. Näher, D. Alberts, D. Ambras, G. Koch, M. Jünger, C. Buchheim, and S. Leipert. A library of algorithms for graph drawing. In S. H. Whitesides, editor, *Proceedings of the 6th International Symposium on Graph Drawing (GD '98)*, volume 1547 of *Lecture Notes in Computer Science*, pages 456–457. Springer, 1998. Project home page at http://www.mpi-sb.mpg.de/AGD/.
24. H. C. Purchase. Which aesthetic has the greatest effect on human understanding? In G. Di Battista, editor, *Proceedings of the 5th International Symposium on Graph Drawing (GD '97)*, volume 1353 of *Lecture Notes in Computer Science*, pages 248–261. Springer, 1997.
25. H. C. Purchase, R. F. Cohen, and M. James. An experimental study of the basis for graph drawing algorithms. *ACM Journal of Experimental Algorithmics*, 2(4), 1997.
26. J. Raab. *Steuerungsstrukturen politisierter Großprivatisierungen in Ostdeutschland 1990–1994. Das Beispiel der Werft- und Stahlindustrie*. PhD thesis, University of Konstanz, 2000.
27. W. D. Richards. MultiNet. Software tool, see http://www.sfu.ca/~richards/Multinet/.
28. K. Sugiyama, S. Tagawa, and M. Toda. Methods for visual understanding of hierarchical system structures. *IEEE Transactions on Systems, Man and Cybernetics*, 11(2):109–125, February 1981.
29. S. Wasserman and K. Faust. *Social Network Analysis: Methods and Applications*. Cambridge University Press, 1994.

Editors' Note: see Appendix, p. 282 for colored figure of this paper

Improving Angular Resolution in Visualizations of Geographic Networks*

Ulrik Brandes[1], Galina Shubina[2], and Roberto Tamassia[2]

[1] University of Konstanz, Department of Computer and Information Science,
Box D 188, 78457 Konstanz, Germany, Ulrik.Brandes@uni-konstanz.de
[2] Brown University, Department of Computer Science, Providence,
Rhode Island 02912-1910, USA, {gs,rt}@cs.brown.edu

Abstract. In visualizations of large-scale transportation and communications networks, node coordinates are usually fixed to preserve the underlying geography, while links are represented as geodesics for simplicity. This often leads to severe readability problems due to poor angular resolution, i.e. small angles formed by lines converging in a node. We present a new method using automatically routed cubic curves that both preserves node coordinates and eliminates the resolution problem. The approach is applied to representations in the plane and on the sphere, showing European train connections and Internet traffic, respectively.

1 Introduction

Since nodes in large-scale transportation networks, such as airline flight plans, train connection maps, or extracts of the Internet, have a given geographic location, we call these networks *geographic networks*. Typical visualizations use given node coordinates either directly, or only apply an appropriate projection (e.g., from the surface of a sphere to a plane) to retain the viewers familiarity with the underlying geometry [3]. In general, the exact routing of connections is not important, so that links are often represented as geodesics (straight-lines or great circles). While computationally and visually simple, this approach does not take into account the perceptual organization of the resulting visualization.

Prior work on improving the visual quality has focused on moderate re-positioning of close or overlapping nodes [17, 10], and re-routing of edges cutting through nodes or other features [1, 8]. Since we address a different criterion of layout quality, our work can be potentially used in conjunction with these approaches.

Severe readability problems in visualizations of geographic networks stem from small angles formed by lines converging in a common node. Small angles cause viewers to perceive filled-in areas between the lines ◄, causing "blob"-effects and making it difficult to tell lines apart.

* Research partially supported by the U.S. National Science Foundation under grants CCR-9732327 and CDA-9703080, by the U.S. Army Research Office under grant DAAH04-96-1-0013, and by the German Academic Exchange Service (DAAD, Hochschulsonderprogramm III).

We present a new method that modifies a given visualization so that all angles formed by incident lines are of sufficient size. This is achieved by replacing straight-line and great-circle connections with cubic curves. While visually still simple, cubic curves allow to prescribe angles between incident curves at will. A computationally fast and simple method is introduced and demonstrated working on two real-world data sets.

The remainder of this paper is organized as follows. In Sect. 2, we provide some terminology and background on properties of angles in network visualizations. Our approach is described in Sect. 3 and some extensions are presented in Sect. 4. Finally, it is applied to visualizations in the plane and on the sphere in Sect. 5, using data from European train and ferry schedules, and the multicast backbone of the Internet, respectively.

2 Background on Angular Resolution

Networks are conveniently described as graphs $G = (V, E)$, where v is a set of *vertices* (nodes), and E is a set of *edges* (links). Without loss of generality we consider only undirected graphs without loops and multiple edges, so that every edge is an unordered pair of vertices.

A network visualization is a drawing of the graph, i.e. a mapping of vertices to points in the plane or in space together with a mapping of edges to curves connecting the points of their respective vertices. We confine ourselves to drawings that map vertices into the plane or the surface of a sphere.

Assume, we are given a drawing of a graph in the plane, such that every edge is represented as a straight line. The *local angular resolution* at vertex $v \in V$ is the minimum angle between a pair of edges incident to v. The minimum angle between any pair of edges incident to any $v \in V$ is called the *angular resolution* of the drawing and introduced in [9].

A trivial upper bound on the local angular resolution at vertex $v \in V$ is $\frac{2\pi}{d_G(v)}$, where $d_G(v)$ is the *degree* of v, defined as the number of edges incident to v. It is shown in [9] that every simple graph has a straight-line drawing with angular resolution $\Omega(\frac{1}{\Delta(G)})$, where $\Delta(G)$ is the maximum degree of any vertex in G. For planar graphs, i.e. graphs that can be drawn in the plane without crossing edges, it is shown how to construct drawings with asymptotically optimal angular resolution. However, these drawings are in general not planar.

Every planar graph has a planar straight-line drawing with angular resolution $\Omega(\frac{1}{\alpha^{\Delta(G)}})$ for some constant $\alpha > 1$ [18], but there are planar graphs for which the angular resolution in any planar straight-line drawing is bounded by $\mathcal{O}(\sqrt{\frac{\log \Delta(G)}{\Delta(G)^3}})$ [12]. Note that maximizing the angular resolution over all planar straight-line drawings of a planar graph is $\mathcal{NP}$-hard [11].

It is shown in [16] how to obtain planar drawings with asymptotically optimal angular resolution when edges may be represented as sequences of straight lines. This result is improved in [13] and [14], where it is shown that only two bends per edge are needed, so that edges can be drawn as cubic curves.

Our setting differs from the above in that the mapping of vertices to points is already fixed. We do not know of other work on this particular problem.

To alter the angular resolution, we clearly must use polyline or curved edges as well. We currently use cubic Bézier curves [4], partially for the pragmatic reason that they are built into PostScript and LEDA's [19] graph editor. They are defined by a sequence of four *control points*, $b_0, \ldots, b_3$ (and thus three *control segments* $\overline{b_0 b_1}$, $\overline{b_1 b_2}$, and $\overline{b_2 b_3}$), and have several advantages over other types of curves [5]. In particular, the tangents at b_0 and b_3 are collinear with the first and third control segment. The definitions of local and global angular resolution are hence easily generalized to drawings with cubic Bézier curves.

To replace a straight-line edge with a cubic Bézier curve, we only have to place the inner control points b_2 and b_3, since b_0 and b_3 must be assigned to the positions given for the incident vertices. The placement is then divided into

1. determining directions for the first and third control segment, and
2. determining the lengths of the first and third control segment.

Since the resulting angular resolution is fully determined by the first step, we do not consider the second step in the remainder of this paper. As a simple heuristic the length of the first and third control segment is chosen proportional to the length of the straight-line edge. Future work should guarantee further desirable properties [15] of, potentially higher-order, curves (e.g., to retain planarity of a drawing). Figure 1 shows the principal combinations of directions of the first and third control segment and resulting cubic Bézier curves. Note that all of them are easy to trace for the human eye. Moreover, we will see later that our approach avoids the four high-curvature cases on the right hand side whenever possible.

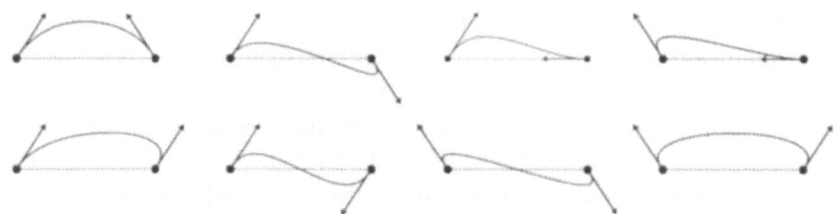

Fig. 1. Principal non-degenerate combinations of control segment orientations

3 Fixed Angular Resolution and Interesting Rotations

In this section, we show how to improve the angular resolution of a given straight-line drawing in the plane by choosing directions for the first and third control segment of Bézier curves replacing the straight lines.

On the one hand, these directions should yield good angular resolution. On the other hand, we also want them to resemble the straight-line directions as close as possible to avoid the high-curvature cases of Fig. 1, which also serves to keep the drawing simple. Note that we can treat control segments incident to one vertex independently from control segments incident to other vertices. Therefore, let $e_0, \ldots, e_{d_G(v)-1}$, be a counterclockwise ordering of the edges around some $v \in V$ in the given drawing (with ties broken arbitrarily), and denote by α_i, $i = 0, \ldots, d_G(v) - 1$, the angle between e_i and its counterclockwise neighbor.

Accordingly, we define $c_0, \ldots, c_{d_G(v)-1}$ to be the corresponding ordering of control segments incident to v. The angles between neighboring control segments are fixed to be $\frac{2\pi}{d_G(v)}$, thus ensuring optimal local, and hence global, angular resolution. Because of the simplicity requirement, we want to penalize the deviation of control segments from straight-line edges. Denote by x_i the angle between e_i and c_i, $i = 0, \ldots, d_G(v) - 1$, where $x_i > 0$, if e_i comes before, and $x_i < 0$, if e_i comes after c_i in the counterclockwise order around v. We call these quantities the *angular differences*. See Fig. 2 for an illustration and note that $x_3 < 0$.

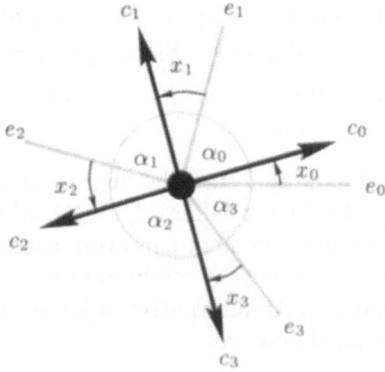

Fig. 2. Angles α_i between straight-line edges, and angular differences x_i

Any set of control segments satisfying the angle constraints is called a *rotation* at v. There are several ways to define desirable rotations. Since the angles between control segments are fixed. all angular differences of a vertex are dependent. It is easily seen that $x_1 = x_0 + \frac{2\pi}{d_G(v)} - \alpha_0$, and more general $x_i = x_0 + i \cdot \frac{2\pi}{d_G(v)} - \sum_{j=0}^{i-1} \alpha_j$ for $i = 0, \ldots, d_G(v) - 1$. We frequently use $y_i = i \cdot \frac{2\pi}{d_G(v)} - \sum_{j=0}^{i-1} \alpha_j$ to denote the *angular offset* caused by choosing x_0 as the independent variable. A straightforward objective is to minimize the largest absolute angular difference,

$$\max_{i=0,\ldots,d_G(v)-1} |x_i|. \tag{1}$$

We call this a *minimum* rotation.

Theorem 1. *The minimum rotation is unique, and linear-time computable.*

Proof. Since $x_i = x_0 + y_i$, $i = 0, \ldots, d_G(v) - 1$, the maximum absolute angular difference is the maximum of $|x_0 + \min_i y_i|$ and $|x_0 + \max_i y_i|$. This maximum becomes minimum for

$$x_0 = -\frac{\min_i y_i + \max_i y_i}{2},$$

since both absolute values become equal and every other x_0 increases either one.

We next observe that, if the distribution of angles between straight-line edges in a given drawing is well-behaved, the maximum absolute angular difference will be small.

Corollary 1. *In a minimum rotation of a vertex v with degree at least two, the maximum angular difference is* $\max_i |x_i| = \max_{r<s} \frac{1}{2} \cdot \left| (s-r)\frac{2\pi}{d_G(v)} - \sum_{j=r}^{s-1} \alpha_j \right| < \pi.$

Proof. The proof of Theorem 1 showed that $\max_i |x_i| = |x_0 + \min_i y_i| = \frac{1}{2} \cdot |\min_i y_i - \max_i y_i| = \frac{1}{2} \cdot |\max_i y_i - \min_i y_i|$. It follows that there are $r, s \in \{0, \ldots, d_G(v) - 1\}$ such that $\max_i |x_i| = \frac{1}{2}|y_s - y_r|$ and $r < s$, and there is no pair with a larger absolute difference. Since $y_s - y_r = (s-r)\frac{2\pi}{d_G(v)} - \sum_{j=r}^{s-1} \alpha_j$, the equality holds. Clearly, $0 < (s-r)\frac{2\pi}{d} < 2\pi$ and $0 \le \sum_{j=r}^{s-1} \alpha_j \le 2\pi$ for all pairs $r < s \in \{0, \ldots, d_G(v) - 1\}$.

Essentially, the corollary states that the angular differences are half as bad as the given distortion in the straight-line edges. Recall that angular differences less than or equal to $\frac{\pi}{2}$ exclude the high-curvature cases of Fig. 1.

The corollary also indicates that the minimum rotation may be dominated by one pair of straight-line edges. An alternative objective is therefore to minimize the sum of squared angular differences,

$$\sum_{i=0}^{d_G(v)-1} (x_i)^2, \tag{2}$$

weighing contributions more evenly. Such rotations are called *balanced*.

Theorem 2. *The balanced rotation is unique, and linear-time computable.*

Proof. Substituting for x_i reduces objective function (2) to a positive function in x_0. Thus, any x_0 satisfying $\frac{\partial}{\partial x_0} \sum_{i=0}^{d_G(v)-1} (x_0 + y_i)^2 = \sum_{i=0}^{d_G(v)-1} 2 \cdot (x_0 + y_i) = 0$ yields a minimum. Clearly, there is exactly one such x_0, and it is obtained by averaging over all offsets, i.e.

$$x_0 = \frac{-\sum_{i=0}^{d_G(v)-1} y_i}{d_G(v)} = \frac{1}{d_G(v)} \sum_{i=0}^{d_G(v)-1} \left(\sum_{j=0}^{i-1} \alpha_j - i \cdot \frac{2\pi}{d_G(v)} \right).$$

An immediate consequence of the proof explains the name of this rotation.

Corollary 2. *In a balanced rotation,* $\sum\limits_{i=0}^{d_G(v)-1} x_i = 0.$

See Fig. 3 for a comparison of minimum and balanced rotation. For completeness we note that the seemingly less interesting *absolute* rotation, suggested by this corollary, minimizing $\sum_{i=0}^{d_G(v)-1} |x_i|$ is also unique, and that all angular differences for vertices with already optimal local angular resolution equal zero in any of these rotations.

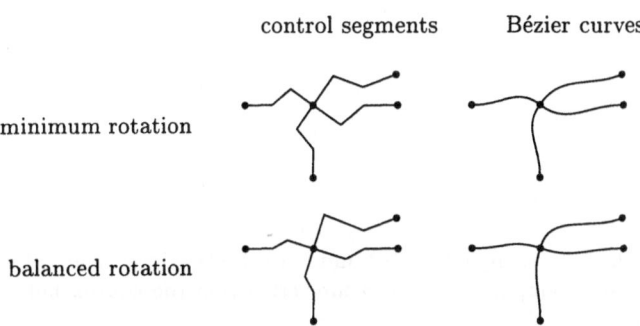

control segments   Bézier curves

minimum rotation

balanced rotation

Fig. 3. Comparison of minimum and balanced rotation

4 Extensions

The following are some of several possible extensions that may be useful when a particular application yields additional requirements on the quality of a drawing.

Arbitrary angle constraints. From Corollary 1 we see that angular differences can grow quite big, when the given straight-line drawing has large angles, which typically results in edges with high curvature. In such situations it may be advantageous not to optimize local angular resolution exactly, but only up to a constant. For example, if there is an angle of at least π, all angles between control segments can be set to $\frac{\pi}{d_G(v)-1}$, except for one which is fixed to π. Note that we can impose any constraint on the angles between control segments, provided they add up to 2π, without affecting Theorems 1 and 2. Figure 4 shows some potentially useful templates for angles between control segments. The middle one is applied to vertices on the convex hull in Fig. 5.

Weighted angular differences. In an application with different types of edges, angular differences of some edges may be more important than those of others. If this importance can be quantified, the objective functions for rotations can be modified to accommodate weights, and analog theorems hold. Weights may be particularly useful when the straight-line edges incident to a vertex are of significantly different length.

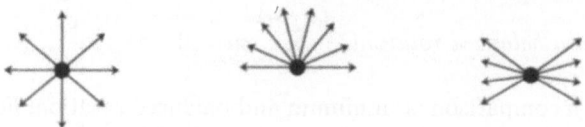

Fig. 4. Example templates for constraints on control segment angles

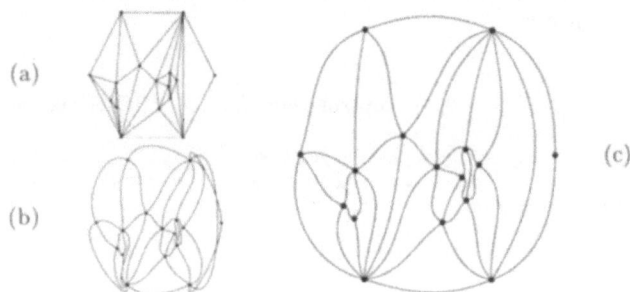

Fig. 5. A straight-line triangulation (a) and curved edges in a balanced rotation (b). Using the half-sided templates of Fig. 4 for vertices on the convex hull significantly improves the drawing (c)

Cyclic ordering. The order of control segments of a vertex need not the same as for the straight-line edges. Figure 6 depicts a situation in which a different ordering would be better. Allowing negative angles α_i, control segments can be put in any order.

Crossings. While there is no control over whether curved edges cross each other when a rotation is determined, crossings already existing in the straight-line drawing can be maintained by replacing them with a dummy vertex. Since angle constraints are arbitrary, we can thus constrain curves to cross, e.g., at a right angle or at the same angle as in the given drawing.

5 Application Examples

5.1 Train Connections

To analyze time table data of a number of European public transport systems (mostly trains), a graph is constructed in the following way [6]: Each station is represented by a vertex, and there is an edge between two vertices if there is at least one non-stop service between them at any time. This graph is analyzed, e.g., with respect to completeness, consistency, or changes between schedules, serving, e.g., quality control, international coordination, and pricing.

Each vertex has a geographic location, thus providing geographical context for visual data exploration. Part of the analysis is an automatic classification into *minimal* and *transitive* edges, corresponding to tracks and high-speed connections passing by minor stations, respectively. By their very nature, many edges run almost in parallel when drawn as straight-lines. In particular, transitive edges overlap each other, and minimal edges (see Fig. 6(a)). Figure 6(b) shows a balanced rotation with the special half-sided template for vertices with an angle of at least π. The unnecessary detour of one minimal edge is due to the unfortunate ordering of straight-line edges. The system we are building will order edges based on their length and classification.

An elaborate algorithm for application-specific curved layout of transitive edges is presented in [6]. Users are very satisfied with the output, but running times are painful (7–45 minutes for the networks compared in Tab. 1). Using the approach described in this paper, we can easily afford to compute a new layout every time a network is displayed and thus support interactive querying. In fact, optimal rotations are computed faster than the editor (LEDA's [19] GraphWin class) renders the network. See also the larger example in Fig. 7.

instance	nodes	edges	minimum	balanced
switzerland	2218	3203	0.36 sec	0.36 sec
italy	2386	4370	0.51 sec	0.51 sec
france	4551	7793	0.81 sec	0.80 sec
germany	7083	9713	1.19 sec	1.18 sec

Table 1. Computation times for rotations (Sun UltraSparc, 440 Mhz, 256 MBytes)

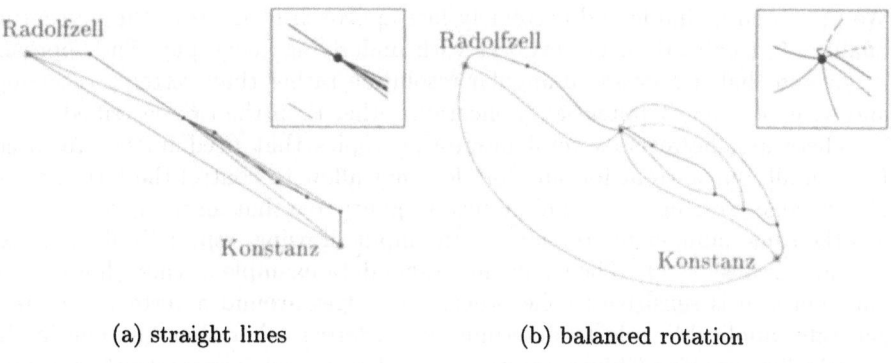

(a) straight lines   (b) balanced rotation

Fig. 6. Small network in southern Germany with one undesirable edge ordering

5.2 Internet Multicast Backbone

Internet connections represent another example of an organically grown, and growing, geographic network. To support their analysis, which is crucial for maintenance and development, advanced systems such as [7, 20] provide effective interactive visualizations of network topologies. The environment of [20] is publicly available,[1] and generates a geometric scene description that can be explored in an interactive browser such as Geomview [2]. We replaced the generation of geometric output to demonstrate how to improve the angular resolution, and thus the perceptual quality, of the resulting visualizations.

The original system visualizes the topology of the MBone, the Internet's multicast backbone, by representing it as a geographic network on the globe, where connections (MBone tunnels) are shown as great circle segments elevated into space (Fig. 8(a)). The angles between great circles correspond to the angles between their projections into the plane tangent to their intersection. Our approach of orienting control segments is therefore easily generalized to deal with great circles on a sphere rather than lines in the plane. Once the inner control points of a curved edge are placed on the sphere, the actual curve is determined as follows. Assume the globe is represented as a sphere with unit radius, and $B(t) : [0,1] \rightarrow \mathbb{R}^3$ is a cubic Bézier curve in space connecting points b_0 and b_3. The curve is projected onto the sphere and elevated into space according to

$$\frac{B(t)}{\|B(t)\|} \cdot (1 + c \cdot \arccos(b_0 \cdot b_3) \cdot \sin(t\pi)).$$

where c is an elevation constant. The resulting curves have essentially the same height as those in [7]. See Fig. 8(b).

6   Conclusions

We presented a simple and extremely fast approach to improve the perceptual quality of visualizations of networks with underlying geography. The approach is novel in that it focuses on angular resolution rather than vertex positioning, and we expect many potential applications other than the two described.

There are, however, several interesting topics that need further research. First of all, our current formulation does not allow to control the introduction of new edge crossings. It would be nice to guarantee that an improved drawing has the same number of crossings as the input drawing, especially if the latter has no crossings at all. The small time table data example already showed that our approach is sensitive to the ordering of edges around a vertex. Is there a generally applicable rule to determine an ordering other than the one in the straight-line drawing? Since rotations are determined independently for each vertex, an optimal rotation may introduce unnecessarily many S-shaped curves, whereas a different rotation may be more pleasing. Can we introduce some form of dependency to make angular differences at both ends of an edge similar? Finally, we would like to improve the heuristic used to determine the length of a control segment. It may be advantageous to have control segments of similar length at a vertex, or at both ends of an edge.

---

[1] `http://oceana.nlanr.net/PlanetMulticast/`

References

1. J. Abello and E. R. Gansner. Short and smooth polygonal paths. *Proc. 3rd Latin American Symp. Theoretical Informatics (LATIN '98)*, LNCS vol. 1380, pp. 151–162. Springer, 1998.
2. N. Amenta, S. Levy, T. Munzner, and M. Phillips. Geomview: A system for geometric visualization. In *Proc. 11th ACM Ann. Symp. Computational Geometry (SoCG '95)*, pp. C12–13. ACM, 1995.
3. R. A. Becker, S. G. Eick, and A. R. Wilks. Visualizing network data. *IEEE Transactions on Visualization and Graphics*, 1(1):16–28, 1995.
4. P. Bézier. *Numerical Control.* John Wiley & Sons, 1972.
5. W. Boehm. On cubics: A survey. *Computer Graphics and Image Processing*, 19:201–226, 1982.
6. U. Brandes and D. Wagner. Using graph layout to visualize train interconnection data. *Proc. 6th Intl. Symp. Graph Drawing (GD '98)*, LNCS vol. 1547, pp. 44–56. Springer, 1998.
7. K. C. Cox, S. G. Eick, and T. He. 3D geographic network displays. *ACM SIGMOD Record*, 24(4):50–54, 1996.
8. D. P. Dobkin, E. R. Gansner, E. Koutsofios, and S. C. North. Implementing a general-purpose edge router. *Proc. 5th Intl. Symp. Graph Drawing (GD '97)*, LNCS vol. 1353, pp. 262–271. Springer, 1997.
9. M. Formann, T. Hagerup, J. Haralambides, M. Kaufmann, F. T. Leighton, A. Symvonis, E. Welzl, and G. Woeginger. Drawing graphs in the plane with high resolution. *SIAM Journal on Computing*, 22(5):1035–1052, 1993.
10. E. R. Gansner and S. C. North. Improved force-directed layouts. *Proc. 6th Intl. Symp. Graph Drawing (GD '98)*, LNCS vol. 1547, pp. 364–373. Springer, 1998.
11. A. Garg. On drawing angle graphs. *Proc. 2nd Intl. Symp. Graph Drawing (GD '94)*, LNCS vol. 894, pp. 84–95. Springer, 1995.
12. A. Garg and R. Tamassia. Planar drawings and angular resolution: Algorithms and bounds. *Proc. 2nd European Symp. Algorithms (ESA '94)*, LNCS vol. 855, pp. 12–23. Springer, 1994.
13. M. T. Goodrich and C. G. Wagner. A framework for drawing planar graphs with curves and polylines. *Proc. 6th Intl. Symp. Graph Drawing (GD '98)*, LNCS vol. 1547, pp. 153–166. Springer, 1998.
14. C. Gutwenger and P. Mutzel. Planar polyline drawings with good angular resolution. *Proc. 6th Intl. Symp. Graph Drawing (GD '98)*, LNCS vol. 1547, pp. 167–182. Springer, 1998.
15. H. Hagen. Bézier-curves with curvature and torsion continuity. *Rocky Mountain Journal of Mathematics*, 16(3):629–638, 1986.
16. G. Kant. Drawing planar graphs using the canonical ordering. *Algorithmica*, 16:4–32, 1996.
17. K. A. Lyons, H. Meijer, and D. Rappaport. Algorithms for cluster busting in anchored graph drawing. *Journal on Graph Algorithms and Applications*, 2(1):1–24, 1998.
18. S. Malitz and A. Papakostas. On the angular resolution of planar graphs. *SIAM Journal on Discrete Mathematics*, 7(2):172–183, 1994.
19. K. Mehlhorn and S. Näher. *The LEDA Platform of Combinatorial and Geometric Computing.* Cambridge University Press, 1999. http://www.mpi-sb.mpg.de/LEDA/.
20. T. Munzner, E. Hoffman, K. Claffy, and B. Fenner. Visualizing the global topology of the MBone. *Proc. 1996 IEEE Symp. Information Visualization*, pp. 85–92, 1996.

Editors' Note: see Appendix, p. 283 for colored figures of this paper

Squarified Treemaps

Mark Bruls, Kees Huizing, and Jarke J. van Wijk

Eindhoven University of Technology
Dept. of Mathematics and Computer Science,
P.O. Box 513,
5600 MB Eindhoven, The Netherlands
email{keesh, vanwijk}@win.tue.nl

Abstract. An extension to the treemap method for the visualization of hierarchical information, such as directory structures and organization structures, is presented. The standard treemap method often gives thin, elongated rectangles. As a result, rectangles are difficult to compare and to select. A new method is presented to generate lay-outs in which the rectangles approximate squares. To strenghten the visualization of the structure, shaded frames are used around groups of related nodes.

1   Introduction

Hierarchical structures of information are everywhere: directory structures, organization structures, family trees, catalogues, computer programs, and so on. Small hierarchical structures are effective to locate information, but the content and organization of large structures is harder to grasp. We present a new visualization method for large hierarchical structures: Squarified Treemaps. The method is based on Treemaps, developed by Shneiderman and Johnson [9, 6]. Treemaps are efficient and compact displays, which are particularly effective to show the size of the final elements in the structure. In a previous paper [10] we introduced Cushion Treemaps, which provide shading as an extra cue to emphasize the hierarchical structure. In this paper we attack another problem of standard treemaps: the emergence of thin, elongated rectangles. We propose a new method to subdivide rectangular areas, such that the resulting subrectangles have a lower aspect ratio. These rectangles use space more efficiently, are easier to point at in interactive environments, and are easier to estimate with respect to size. Because the resulting structures are somewhat harder to grasp, we also introduce an improved method to visualize nested structures. A variant on nested treemaps is presented, where the rectangular enclosures have been replaced by shaded frames. The combination of these two methods leads to displays of hierarchical structures that are efficient and easy to understand.

In section 2 we discuss existing methods to visualize hierarchical structures. The new method for improved subdivision is presented in section 3. The shaded frames are described in section 4. Finally, we discuss the results in section 5.

2   Background

Many methods exist to browse through and to display hierarchical information structures, or, for short, trees. File browsers are the best known example. Usually a listing of the files and directories is used, where the levels in the hierarchy are shown by means of indentation. The number of files and directories that can be shown simultaneously is limited, which is no problem if one knows what to search for. However, if we want to get an overview, or want to answer a more global question, such as: "Why is my disk full?", scrolling, and opening and closing of subdirectories have to be used intensively. During this process it is hard to form a mental image of the overall structure [3].

Many techniques have been proposed to visualize such structures more effectively. An important category are node and link diagrams (fig. 1(a)). Elements are shown as nodes, relations are shown as links from parent to child nodes. Sophisticated techniques have been presented to improve the efficiency and aesthetic qualities of such diagrams, both in 2D and in 3D [7, 5, 1, 2, 8]. Such diagrams are very effective for small trees, but usually fall short when more than a couple of hundred elements have to be visualized simultaneously. The main reason for this limitation is simply that node and link diagrams use the display space inefficiently: Most of the pixels are used as background. Treemaps

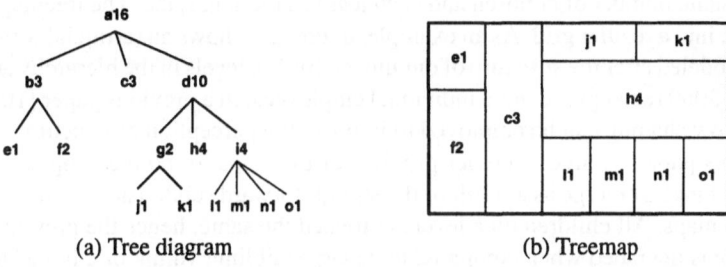

(a) Tree diagram   (b) Treemap

Fig. 1. Tree diagram

[9, 6] were developed to remedy this problem. The full display space is used to visualize the contents of the tree. Here we present an overview of the concept, an in depth treatment is given in the original references. Figure 1(b) shows an example. Each node (as shown in the tree diagram) has a name (a letter) and an associated size (a number). The size of leaves may represent for instance the size of individual files, the size of non-leaf nodes is the sum of the sizes of its children. The treemap is constructed via recursive subdivision of the initial rectangle. The size of each sub-rectangle corresponds to the size of the node. The direction of subdivision alternates per level: first horizontally, next vertically, etcetera. As a result, the initial rectangle is partitioned into smaller rectangles, such that the size of each rectangle reflects the size of the leaf. The structure of the tree is also reflected in the treemap, as a result of its construction. Color and annotation can be used to give extra information about the leaves.

Treemaps are very effective when size is the most important feature to be displayed. Figure 2(a) shows an overview of a file system: 1400 files are shown and one can effortlessly determine which are the largest ones.

However, treemaps have limitations [4]. One problem is that treemaps often fall short to visualize the structure of the tree. The worst case is a balanced tree, where each parent

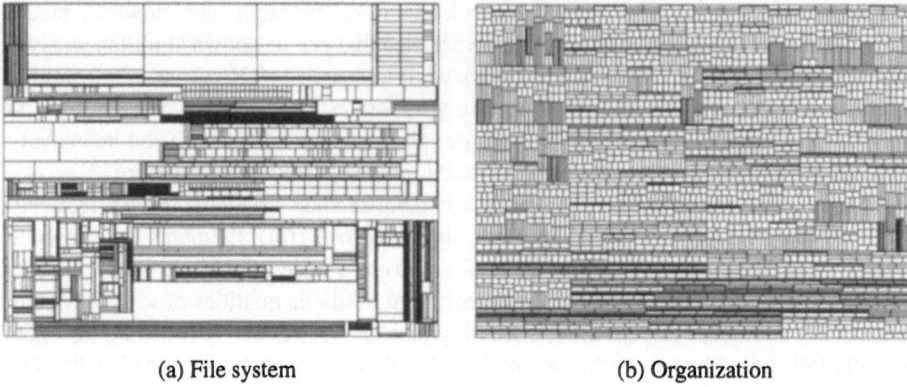

| (a) File system | (b) Organization |

Fig. 2. Treemaps

has the same number of children and each leaf has the same size. The treemap degenerates here into a regular grid. As an example, figure 2(b) shows an (artificial) organization chart, modeled after the structure of our university. Six levels in the hierarchy are shown, the final 3060 rectangles denote individual employees. In a previous paper [10] we have shown how shading can be employed to improve the perception of structure.

In this paper we study another problem of treemaps. In both examples thin, elongated rectangles emerge as a result of the straightforward subdivision technique of standard treemaps. All children on a level are treated the same, hence the presentation of a small file is degraded when compared to its larger siblings (think of a large Unix home directory with a small .cshrc in the top level). The presentation of all nodes and leaves as more square-like rectangles has several advantages:

- display space is used more efficiently. The number of pixels to be used for the border is proportional to its circumference. For rectangles this number is minimal if a square is used;
- square items are easier to detect and point at, thin rectangles clutter up and give rise to aliasing errors;
- comparison of the size of rectangles is easier when their aspect ratios are similar;
- the accuracy of the presentation is improved. A rectangle with a prescribed width of, say, 300 pixels, can only present sizes in coarse steps.

These advantages inspired us to study alternative subdivision techniques, where the aim is to use rectangles that are nearly square for the nodes and leaves. In a late stage of the writing of this paper we found that we are not exceptional in this interest. The concept of squarification has been applied independently of our work for a very clear and effective visualization of a stock market [11].

3   Squarification

How can we tesselate a rectangle recursively into rectangles, such that their aspect-ratios (e.g. $\max(height/width, width/height)$) approach 1 as close as possible? The number of all possible tesselations is very large. This problem falls in the category of NP-hard problems. However, for our application we do not need the optimal solution, a good solution that can be computed in short time is required.

We have experimented with many different algorithms. In this section we present a method that (empirically) turned out to give the best results. The key idea is based on two notions. First, we do not consider the subdivision for all levels simultaneously. This leads to an explosion in computation time. Instead, we strive to produce square-like rectangles for a set of siblings, given the rectangle where they have to fit in, and apply the same method recursively. The startpoint for a next level will then be a square-like rectangle, which gives good opportunities for a good subdivision. Second, we replace the straigthforward subdivision process for a set of siblings of the standard treemap technique (width or height is given, rectangle is subdivided in one direction) by a process that is similar to the hierarchical subdivision process of the standard treemap.

We present our method first with an example, followed by a description of the complete algorithm.

3.1 Example

Suppose we have a rectangle with width 6 and height 4, and furthermore suppose that this rectangle must be subdivided in seven rectangles with areas 6, 6, 4, 3, 2, 2, and 1 (figure 3). The standard treemap algorithm uses a simple approach: The rectangle is subdivided either horizontally or vertically. Thin rectangles emerge, with aspect ratios of 16 and 36, respectively.

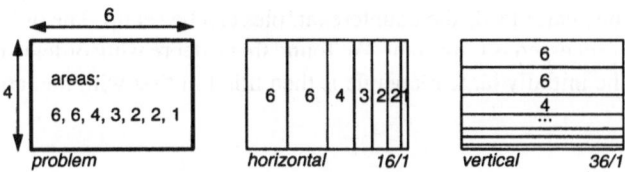

Fig. 3. Subdivision problem

The first step of our algorithm is to split the initial rectangle. We choose for a horizontal subdivision, because the original rectangle is wider than high. We next fill the left half. First we add a single rectangle (figure 4). The aspect ratio of this first rectangle is 8/3. Next we add a second rectangle, above the first. The aspect ratios improve to 3/2. However, if we add the next (area 4) above these original rectangles, the aspect ratio of this rectangle is 4/1. Therefore, we decide that we have reached an optimum for the left half in step two, and start processing the right half.

The initial subdivision we choose here is vertical, because the rectangle is higher than wide. In step 4 we add the rectangle with area 4, followed by the rectangle with area 3 in step 5. The aspect ratio decreases. Addition of the next (area 2) however does not improve the result, so we accept the result of step 5, and start to fill the right top partition.

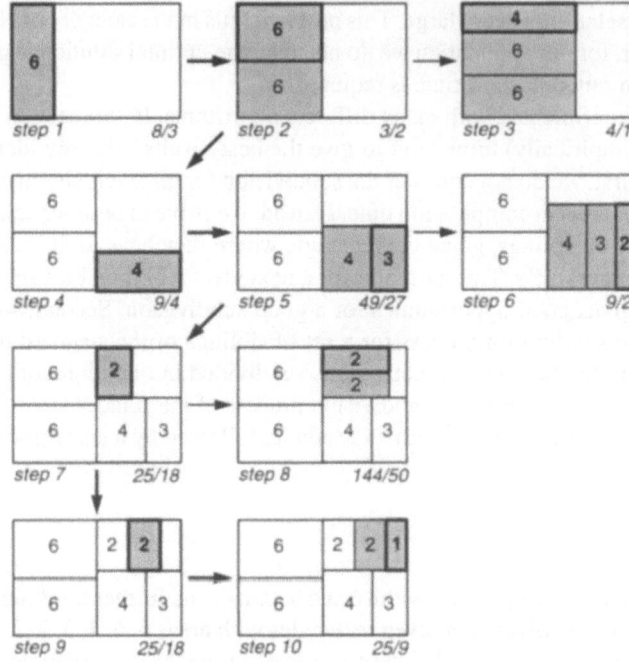

Fig. 4. Subdivision algorithm

These steps are repeated until all rectangles have been processed. Again, an optimal result can not be guaranteed, and counterexamples can be set up. The order in which the rectangles are processed is important. We found that a decreasing order usually gives the best results. The initially large rectangle is then filled in first with the larger subrectangles.

3.2 Algorithm

Following the example, we present our algorithm for the layout of the children in one rectangle as a recursive procedure *squarify*. This procedure lays out the rectangles in horizontal and vertical rows. When a rectangle is processed, a decision is made between two alternatives. Either the rectangle is added to the current row, or the current row is fixed and a new row is started in the remaining subrectangle. This decision depends only on whether adding a rectangle to the row will improve the layout of the current row or not.

We assume a datatype *Rectangle* that contains the layout during the computation and is global to the procedure *squarify*. It supports a function *width()* that gives the length of the shortest side of the remaining subrectangle in which the current row is placed and a function *layoutrow()* that adds a new row of children to the rectangle. To keep the description simple, we use some list notation: ++ is concatenation of lists, [x] is the list containing element x, and [] is the empty list. The input of *squarify()* is basically a list of real numbers, representing the areas of the children to be laid out. The list *row* contains the rectangles that is currently being laid out. The function *worst()* gives the highest aspect ratio of a list of rectangles, given the length of the side along which they are to be laid out. This function is further discussed below.

procedure squarify(**list of real** children, **list of real** row, **real** w)
begin
       real c = head(children);
       if worst(row, w) $\leq$ worst(row++[c], w) **then**
          squarify(tail(children), row++[c], w)
     else
          layoutrow(row);
          squarify(children, [], width());
    fi
end

Let a list of areas R be given and let s be their total sum. Then the function *worst* is defined by:

$$worst(R, w) = \max_{r \in R}(\max(w^2 r / s^2, s^2 / (w^2 r)))$$

Since one term is increasing in r and the other is decreasing, this is equal to

$$\max(w^2 r^+ / (s^2), s^2 / (w^2 r^-))$$

where r^+ and r^- are the maximum and minimum of R. Hence, the current maximum and minimum of the row that is being laid out.

    Applying this algorithm to the data sets of figure 2 results in figure 5. This shows clearly that the algorithm is succesful in the sense that the rectangles are far less elongated and that the black areas with cluttered rectangles have disappeared. However, they also show that the hierarchical structure is far less obvious than with the standard treemap algorithm. The alternating directions scheme aids in providing the viewer with direct cues on the structure, whereas with a less regular scheme these visual cues disappear. When the sizes vary strongly (figure 5(a)), a subtle cue is provided. From the lower-left corner to the upper-right corner the size of the rectangles decreases, which shows more or less which child-nodes have the same parent. However, when the leaves have the same size (figure 5(b)), this cue is not present.

    One way to improve the visualization of the structure is to use cushions (figure 6). However, the global structure is still not obvious, for instance in figure 6(b) it is hard to detect the seven nodes at the highest level. Hence, we have studied additional methods to emphasize the structure.

4   Frames

Nesting was introduced by Shneiderman and Johnson [9, 6] to strengthen the visualization of structure. Each rectangle that represents a non-leaf node is provided with a border to show that its children have the same parent. An example for a binary tree is shown in

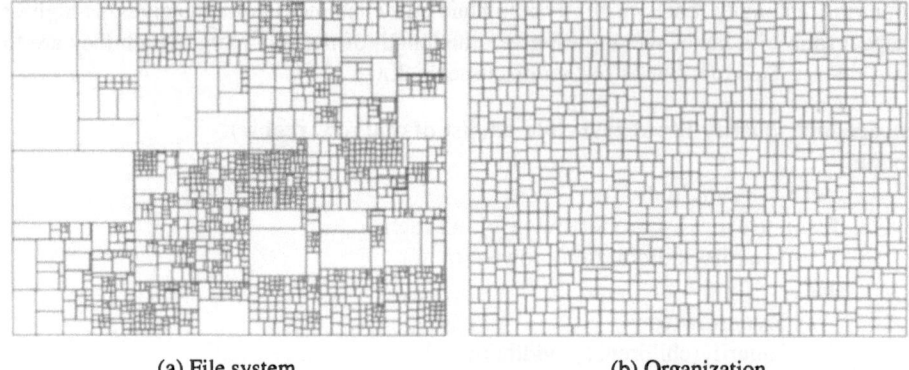

(a) File system   (b) Organization

Fig. 5. Squarified treemaps

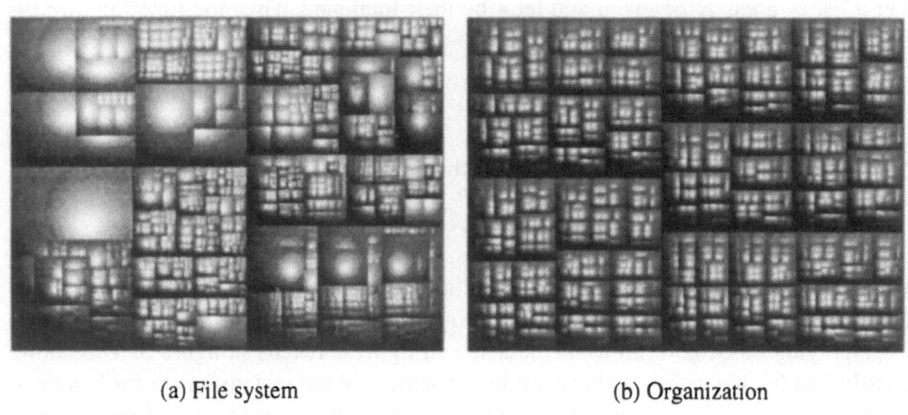

(a) File system   (b) Organization

Fig. 6. Squarified cushion treemaps

figure 7(a). This method has some disadvantages. Extra screen-space is used, and furthermore, it gives rise to maze-like images, which can be puzzling for the viewer.

However, the second disadvantage can be remedied in a similar way as for the visualization of the nodes. We fill in the borders with grey-shades, based on a simple geometric model (figure 8). The width d_l in pixels of a border of level l, with $l = 1, \ldots, n$ is given by:

$$d_l = \lceil w f^{l-1} \rceil,$$

where w is the width of the root level border, and f a factor that can be used to decrease the width for lower level borders. For the profile of the border we use a parabola:

$$z_l(r) = a(r + s_l)^2 + b(r + s_l), \text{ with}$$

$$s_l = \sum_{i=1}^{l-1} d_i,$$

where $z_l(r)$ is the height of the profile for level l, r is the distance from the outside of the border for this level, and a and b are two coefficients that control the shape of the parabola.

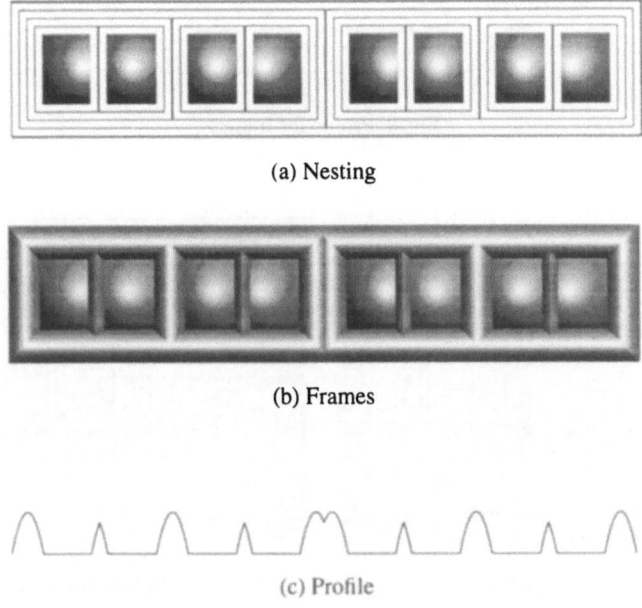

(a) Nesting

(b) Frames

(c) Profile

Fig. 7. Binary tree

Figure 7(b) shows the result for the binary tree. In this (exaggerated) example we used $w = 8$, $f = 1$, $a = -1$, and $b = 16$, which gives a parabola with its top shifted to the interior. Figure 7(c) gives the profile of the frames for this example. Applications of a similar profile are shown in figure 9. This gives the following effects:

- nested borders appear as solid frames;
- the depth of each node in the tree is visualized directly via the height of the frame that surrounds it;
- neighbouring first level boundaries, such as the boundary between node 4 and 5, get an indent, because of the shift of the parabola;
- neighbouring second level boundaries, such as the boundary between node 2 and 3, have a smooth top;
- lower level boundaries have a sharp top.

5   Discussion

We have presented two extensions to the standard treemaps. First, we have shown how rectangles can be forced to be more square. This gives rectangles that are easier to com-

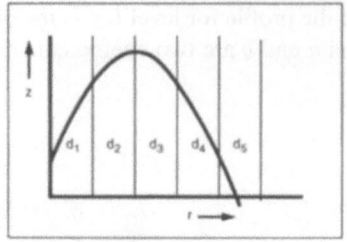

Fig. 8. Profile for frames

(a) File system   (b) Organization

Fig. 9. Framed squarified treemaps

pare, to point at, and that can represent sizes more accurately. Second, we have shown how frames can be used to improve the perception of structure. However, both methods have their limitations. With squarification, the relative ordering of siblings is lost and images tend to be less regular, with less standard patterns, than standard treemaps. When structure of the tree is important and ordering in the data is not, the methods presented here will be very useful.

References

1. A. Bruggemann-Klein and D. Wood. Drawing trees nicely with tex. *Electronic Publishing*, 2(2):101–115, 1989.
2. S.K. Card, G.G. Robertson, and J.D. Mackinlay. The information visualizer, an information workspace. In *Proc. of ACM CHI'91, Conference on Human Factors in Computing Systems*, pages 181–188, 1991.
3. R. Chimera, K. Wolman, and B. Shneiderman. Evaluation of three interfaces for browsing hierarchical tables of contents. Technical Report Technical Report CAR-TR-539, CS-TR-2620, University of Maryland, February 1991.
4. S.G. Eick. Visualization and interaction techniques. In *CHI97 Tutorial notes on Information Visualization*. ACM SIGCHI, March 1997.

5. G.W. Furnas. Generalized fisheye views. In *Proc. of ACM CHI'86, Conference on Human Factors in computing systems*, pages 16–23, 1986.
6. B. Johnson and B. Shneiderman. Treemaps: a space-filling approach to the visualization of hierarchical information structures. In *Proc. of the 2nd International IEEE Visualization Conference*, pages 284–291, October 1991.
7. D.E. Knuth. *Fundamental algorithms, art of computer programming*, volume 1. Addison-Wesley, Reading, MA, 1973.
8. G.G. Robertson, J.D. Mackinlay, and S.K. Card. Cone trees: animated 3d visualizations of hierarchical information. In *Proc. of ACM CHI'91, Conference on Human Factors in Computing Systems*, pages 189–194, 1991.
9. B. Shneiderman. Tree visualization with tree-maps: a 2d space-filling approach. *ACM Transactions on Graphics*, 11(1):92–99, September 1992.
10. J.J. van Wijk and H. van de Wetering. Cushion treemaps - visualization of hierarchical information. In G. Wills and D. Keim, editors, *Proceedings 1999 IEEE Symposium on Information Visualization (InfoVis'99)*, pages 73–78, October 1999.
11. M. Wattenberg. *Map of the Market.* http://smartmoney.com/marketmap/, SmartMoney.com, 1998.

Dynamic Overview Techniques for Image Retrieval *

Pearl Pu and Zoran Pečenović

Ergonomics of Intelligent Systems, DMT/ISR
AudioVisual Communications Lab, LCAV/DSC
Swiss Federal Institute of Technology Lausanne
EPFL, CH-1015 Lausanne, Switzerland
(pearl.pu,zoran.pecenovic)@epfl.ch

Abstract. One difficulty often overlooked in information retrieval systems is that search criteria themselves are often poorly defined. People describe their information needs in many different ways and frequently change their goals depending on the current results of their search. We have investigated the hypothesis that overviews of the space of available solutions are a good way to remedy this situation. Our overview techniques allow users to get a feel for the meaning of categories through randomly chosen examples, find similar images using content search, and to inspect the global distribution of images according to certain criteria. Users thus organize the retrieval task into an iterative browsing process that makes them specify their queries more accurately. As a result they are more satisfied with what the system retrieves.

Keywords: Overview techniques, image retrieval, active query, direct search, exploratory search.

1 Introduction

Users have very diverse and complex ways to express their needs for visual information. Some describe a goal such as "I am looking for a horse for my web page because I am an animal lover." Some use a more pictorial expression such as "I want photographs of bluish landscapes." Yet others look for images of an abstract concept, "I want images to represent intelligence." Existing image retrieval systems assume that users can formulate their needs either in terms of keywords, visual content, or color composition. As pointed out in [7], average users need more exploratory search techniques to browse image collections before a precise query can be formulated. We thus treat image retrieval as an information need clarification process consisting of task description, needs expression both in verbal and visual terms, and exploratory search for targets and their alternatives. In addition, we offer overviews of the entire search space to allow comparison and evaluation of alternatives in combination with content search. This active query approach is in contrary to current systems which are more suited to professional users whose needs for images can be more precisely formulated in their queries.

---

* This work was partly funded by the Swiss National Fund for Scientific Research, No. 21-52439.97. We thank our students: S. Gerlach, V. Tschpp, J. Beck, and others for some of the implemetations.

In this paper, we make use of CIRCUS, a content-based image retrieval system (see Sect. 3.1) using visual features such as color, layout, texture, or any combination of them. Furthermore, to support all three stages of active query, we have added a set of dynamic overview techniques that allow users to become familiar with the collection of the images before specifying the query, either using keywords, exploratory, or direct search methods.

2   Current Methods for Image Retrieval

To understand how non-professionals look for images, we conducted a user study on a set of 40 computer literate subjects. We asked them several questions among which to tell us the type of images they are looking for, describe how they usually do the searching. The following give an idea of the different answers recorded:

- I look for images for presentations. I have a Corel CD of images. I browse the book that comes with the CD until I find one that I like.
- I looked for images to represent intelligence as a logo for my research group on intelligent systems. I used the WWW based search engines. I finally discovered Corbis and was able to type the word intelligence.
- I was looking for images of paintings of nature.
- I like to find images of nature with lakes in a bluish background, definitely not dark reddish background such as those in a sunset.
- I'd like to find red images because I like the color red.

To analyze how some of the existing systems accommodate users' needs for images, we searched for images according to the users' task descriptions at two web sites where images can be retrieved by keywords, by browsing a category taxonomy, or by content.

2.1   Picture experience at Corbis

This site (http://www.corbis.com) offers 1,000,000 images online and supports both keyword based search and category browsing. Users can choose to search images in any of the categories: all, digital images, e-cards, and prints. Keywords search supports single word, or composition of words using connectors and, or, and not. However, keyword based search cannot be used in addition to category browsing, or vise versa.

Typing the word "intelligence," we obtained 690 images. Some of the examples are shown in Fig. 1. Our user was quite happy with the results. Typing the phrase "paintings of nature" resulted in only one image. Not satisfied with the image, we typed in the words "painting and nature" (word order is not important), 5 images were found. Still unsatisfied with these few possibilities, we then typed in "painting," and obtained 720 images. But surprisingly, a large number of the initial results are images about painting, and not of paintings as shown in Fig. 2.

Realizing that recently a category search method has been added to the site, we drilled down following Art and Architecture → Fine Art → Paintings and found 1748 images of paintings. Browsing this collection in a linear fashion, we found several images of paintings of nature. Unfortunately, we were unable to find more images of the same kind using content search.

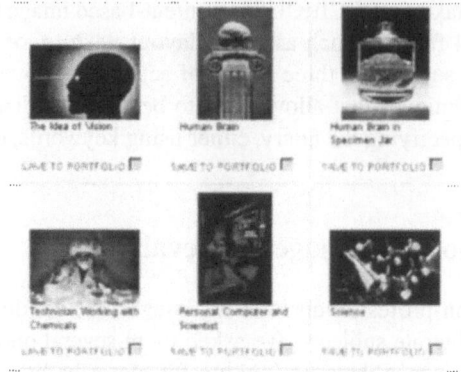

Fig. 1. Images representing "intelligence"

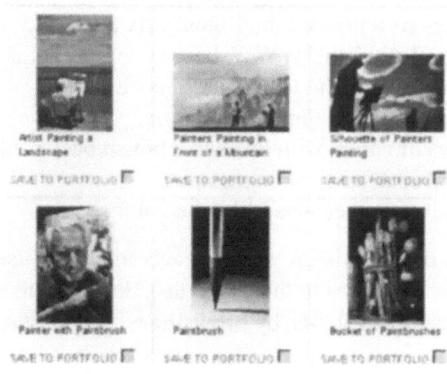

Fig. 2. Images about painting

Typing in "red," we obtained 720 out of 13,000 images, of which many images are indeed red in color, but few of the initial sets contain images of the Red Square Fig. 3). The results of red images improved when we selected the digital picture category.

Fig. 3. Some results for "red" query from Corbis (only the second actually red).

2.2   QBIC system

QBIC [5] is a content-based image retrieval system. The online demo[1] offers three collections: U.S. stamps, stock photography and trademarks. Users can search an image collection by randomly sampling through it or by typing keywords. Content search can be used to find more images of similar visual qualities of a sample image by clicking on color, layout, or texture. It is not possible, however, to use combinations of these visual features.

Figure 4 shows the images that we found by typing lake and then clicking on one image that was our near target, shown as the first one on the upper-left corner. This is a significant improvement from the search method that uses keywords only. Here users have the option to either describe needs in words, in visual aspects, or a combination of the two. Unfortunately QBIC's keyword search is not very powerful since no results

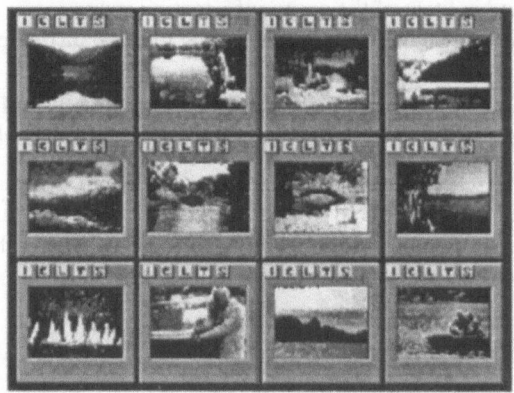

Fig. 4. Images similar to the top-left lake with a bluish background.

were returned for "intelligence," only four images returned for "red," and three returned for "painting".

When users describe their image needs abstractly, and once they have a good idea of the database contents, the keyword based method is natural and effective. The images retrieved corresponded to the subject's previous notion of intelligence: brains, eyes, images of scientific experiments. Furthermore, serendipitous findings are thought provoking and educational. For example, the user has never thought of using an image of Einstein to represent intelligence. However, quite often users got false hits because the image collection's taxonomy is not externalized into visual structures to allow accurate formulation of the keyword query.

Furthermore, keyword and category based methods supported poorly query formulation when visual aspects of images are important to a user's need. As indicated in our study, both paintings of nature and red images took a while to find satisfactory results.

---

[1] http://wwwqbic.almaden.ibm.com/stage

3   Overview Techniques For Image Retrieval

Overview techniques have been used in some document retrieval systems (Starfield [1] and FilmFinder [2]), text retrieval (Tilebar [6]), software visualization (SeeSoft [9]), visualization of web navigation (WebBook and WebForager [4]), and visualization of a collection of photographs (DynamicTimeline [8]). These techniques have shown considerable successes to help users navigate in a complex information space, orient themselves, zoom in on interesting details, and select the targeted items. Furthermore, overview techniques allow users to easily locate, compare and evaluate alternative data items. This feature is especially useful when users have formulated a query that results in a futile search.

The key in designing powerful overview techniques lies in providing users with the following set of characteristics:

1. A compact and dynamic display technique that allows users to see the structure, organization, and content of the entire information space;
2. An easy and real-time navigation method to get to any data of the information space to examine details;
3. A multi-attribute scatter plot to dynamically organize the information space to tradeoff space so that users are able to compare and evaluate alternatives in order to select optimal solutions.

In this paper we will present three overview techniques for combining exploratory and direct search methods into one system for image retrieval. We first briefly describe the underlying content-based image retrieval system, CIRCUS.

3.1   The CIRCUS IR System

The retrieval model used by CIRCUS[2] is an adaptation to image retrieval of the Latent Semantic Indexing [3] method. The core idea is to extract a compact, useful representation of the relations between terms and documents. This is achieved through a lower-rank approximation of the term-document occurrence matrix. LSI shows a slightly higher semantic level and better performance than using term weighted vector space models (*idf-tdf* and extensions). However its translation to image retrieval does raise the fundamental issue, namely the identification of the "terms" images are made of. We opted for a low-level, visual, description of the image including vector quantized color histograms, wavelet based texture and layout descriptors. Since LSI permits the use of any countable descriptor, we include as well higher-level textual annotations. The visual features are extracted both globally and locally.

CIRCUS offers a query construction interface allowing query: by example (Fig. 5a), by color (Fig. 5b) and by sketch (not complete yet). It also represents query results in simple lists (Fig. 5a) and in a 2-D "orbit diagram" (Fig. 6). This 2-D tradeoff space represents resulting images according to their similarity along several user selectable axes (color,texture,layout,keywords). This view gives the user a chance to understand how the system perceives the above mentioned notions of similarity and helps her/him to come up with more accurate and meaningful queries.

---

[2] A simple demo and further information is available at http://lcavwww.epfl.ch

a. Query by example and results.　　　　b. Query by color and results.

Fig. 5. Query specification tools and sample results.

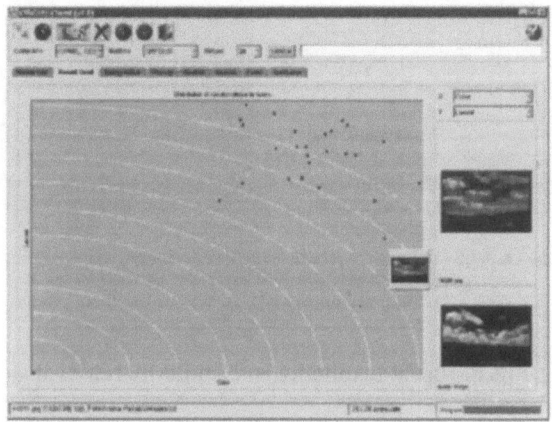

Fig. 6. Orbit: color on X; layout on Y axis.

3.2 Visual taxonomy

The first approach to creating an overview of an image collection summarizes the collection into a visual taxonomy using a single screen space. In one collection, we visualize 9 categories with a total of 650 starting images. An example image from each category is displayed along with the name and the number of images contained in each category. Users can browse through the collection by using the random sampling button so that the surface images change accordingly (Figure 7). If an image somewhat satisfies the user' needs (the near-target image), s/he can use content search to get similar ones in the same or in any combination of categories. Figure 8a shows the first 25 matches after we selected "painting" and "nature". The results are ordered in concentric circles with the example image either in the upper-left corner or more adequately in the center of the display, and the closest ones immediately around it. In the nature category, we are able to single out landscape photos with bluish/gray background (Fig. 8b).

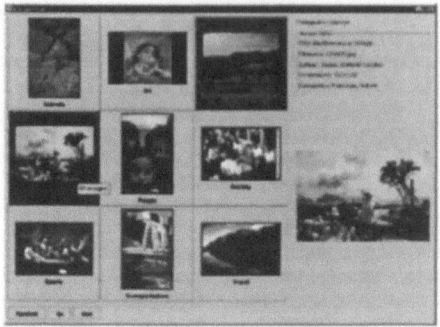

Fig. 7. Browsing the visual taxonomy

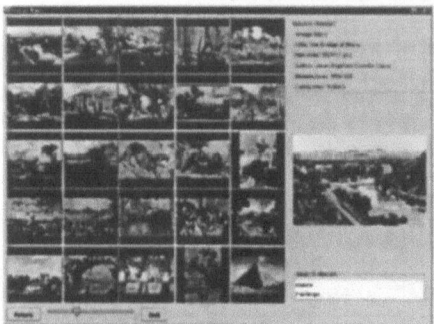

a)paintings or photos of nature   b)images of lake scenery in a bluish background

Fig. 8. Visual taxonomy and similarity retrieval

3.3   Real-time display and Galaxy overview

The second approach displays an overview of an image collection as a galaxy of thumb-nails (shown in Fig. 9). It is a real-time display system offering zoom and pan capabilities. The axes meanings can be defined by the user among a choice of visual and alpha-numeric attributes. The characteristics of an image that map to numerical values (such as hues, lightness or "texturedness") are displayed with a continuous axis. Other dis-crete characteristics like annotation, class or color correspondences to a fixed color-map are displayed on a discrete axis with additional either random jitter or jitter according to a third attribute (usually name of image) for spacing. For instance when the average hue (vertical) and category (horizontal) axes are chosen, users can easily spot images of particular color properties in a given category. The horizontal displacement of the stars within the same category is in this case given by the image name.

For example, natural, bluish photographs of landscapes are located in the lower-left corner. Users can navigate by zooming in on a region of interest and/or panning through the constellations, either by direct manipulation or by clicking on a thumbnail. Once the images become large enough, some details are displayed next to them (author, title, etc). The image at the center of the view becomes the current target and is automatically

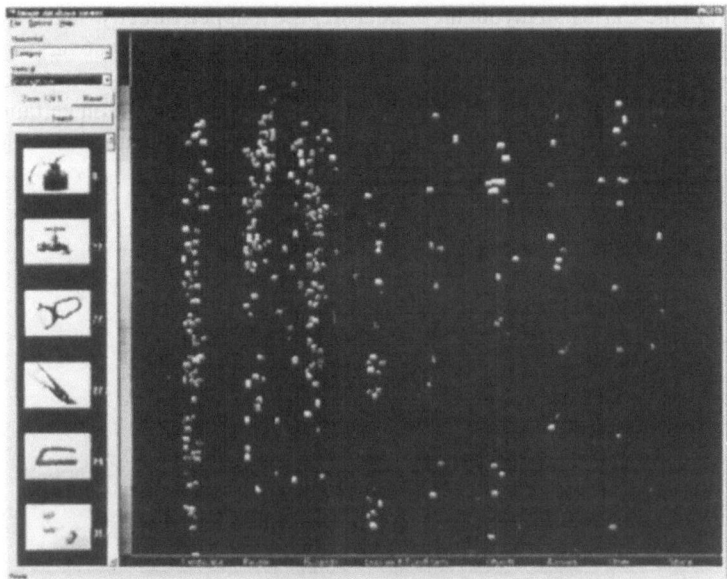

Fig. 9. A galaxy of images spread by Category and Name on X axis, and by Hue on Y axis. (Reproduced in color at the end of the volume.)

centered into the viewing area. The user can then execute a similarity based search to get more similar images.

This overview also combines the navigation and tradeoff space into a single display. That is, for users who had vague search criteria, the tradeoff visualization helps them decide what is the optimal space to explore. For example, consider a user who is looking for images of tools on uniform backgrounds. Browsing in the galaxy where the horizontal axis is category and the vertical axis is hue, s/he will soon realize that there are more photographs of tools on bright then on dark backgrounds and will narrow down the field of view to the appropriate region. S/he will find a number of near-optimal images, then it is sufficient to track on a good candidate and CIRCUS will display a list of similar ones in the side window.

3.4 Overview 3: Combining all

While many systems can support different search methods, this technique allows the simultaneous use of keyword, category, and color composition based search methods in any combination. Selecting the category painting and using the sliders to define red color, the system returns red paintings in a linear nearest-first list, thus providing an overview of the red paintings. Once a suitable image is identified a similarity search can be executed and the cluster of red paintings can be visualized in 2D space using radial coordinates Fig. 10) or Cartesian coordinates (Fig. 11). The position of each thumbnail is determined by the level of similarity according to the user defined axes (color, layout, texture), additionally the size of the thumbnail can be assigned a third dimension of similarity. In the Cartesian layout the similarities are mapped directly on the axes; in the radial layout, the choice of the mappings is constrained by the largest varying attribute being mapped to the angle coordinate. These representations allow easy browsing of alternatives. This approach also makes the machine's measurement on certain qualities apparent for users and thus facilitates subsequent query specifications. A small user

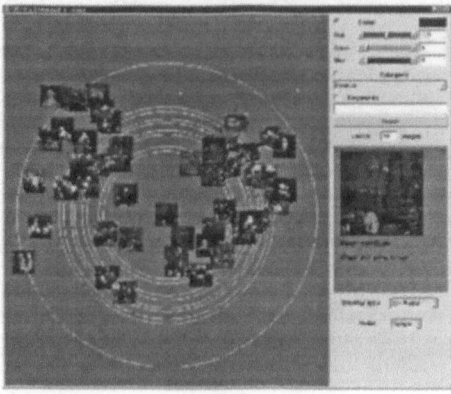

Fig. 10. Results satisfying category Painting, color red and similarity to a given image (center) in 2D radial coordinates.

Fig. 11. Blue "sports and entertainment" images in Cartesian coordinates. (Reproduced in color at the end of the volume.)

study showed that satisfaction with the results increased with the use of overviews and comparative result representations. The other major benefits of such approaches was the ability they confer to the user to understand the systems functioning allowing them to improve future query constructions.

4 Conclusion

We have argued that most available image retrieval systems lack some important functionalities, namely the integration of visual & semantic query capabilities, the ability to present an overview of the available data to the user, and finally a way of communicating to the user the systems conception of match and similarity. We have then shown several overview techniques combining visual taxonomy, real-time display, navigation tools, and multi-attribute scatter plot methods into an exploratory and direct search based image retrieval system. This approach models users' information needs as a process, thus offering an open taxonomy to reduce false hits, a dynamic display of an image collection in Cartesian space to show alternatives and tradeoff analysis, and a hybrid method to match and optimally exploit humans' verbal and visual abilities.

References

[1] C. Ahlberg and B. Shneiderman. Visual information seeking: Tight coupling of dynamic query filters with starfield displays. In *Proceedings of ACM CHI'94, Information Visualization*, volume 1, pages 313–317, 1994.

[2] C. Ahlberg and B. Shneiderman. Visual information seeking using the filmfinder. In *Proceedings of ACM CHI'94, VIDEOS: Part I: Browsing Navigation*, volume 2, page 433, 1994.

[3] M. W. Berry, S. Dumais, and G. O'Brian. Using linear algebrà for intelligent information retrieval. *SIAM Review*, 37:573–595, Dec. 1995.

[4] S. K. Card, G. G. Robertson, and W. York. The webbook and the web forager: Video use scenarios for a world-wide web information workspace. In *Proceedings of ACM CHI 96, VIDEOS: World Wide Web*, volume 2, pages 416–417, 1996.

[5] C. Faloutsos et al. Efficient and effective querying by image content. *Journal of Intelligent Information Systems*, 3:231–262, 1994.

[6] M. A. Hearst and J. O. Pedersen. Visualizing information retrieval results: A demonstration of the tilebar interface. In *in Proceedings of ACM CHI'96, VIDEOS: Visualization*, volume 2, pages 394–395, 1996.

[7] A. Kuchinsky, C. Pering, M. Greech, D. Freeze, B. Serra, and J. Gwizdka. Fotofile: A consumer multimedia organization and retrieval system. In *Proceedings of ACM CHI'99*, 1999.

[8] R. Kullberg. Dynamic timelines: Visualizing the history of photograph. In *Proceedings of ACM CHI'96, VIDEOS: Visualization*, volume 2, pages 386–387, 1996.

[9] J. L. Stephen and S. G. Eick. High interaction data visualization using seesoft to visualize program change history. In *Proceedings of ACM INTERCHI'93, Video: Visualisation*, page 517, 1993.

Editors' Note: see Appendix, p. 284 for colored figures of this paper

Drawing Relational Schemas *

Giuseppe Di Battista, Walter Didimo,
Maurizio Patrignani, and Maurizio Pizzonia
{gdb,didimo,patrigna,pizzonia}@dia.uniroma3.it

Dipartimento di Informatica e Automazione, Università di Roma Tre,
Via della Vasca Navale 79, 00146 Roma, Italy.

Abstract. A wide number of practical applications would benefit from automatically generated graphical representations of relational schemas, in which tables are represented by boxes, and table attributes correspond to distinct stripes inside each table. Links, connecting two attributes of two different tables, represent relational constraits or join paths, and may attach arbitrarily to the left or to the right side of the stripes representing the attributes. To our knowledge no drawing technique is available to automatically produce diagrams in such strongly constrained drawing convention. In this paper we provide a polynomial time algorithm solving this problem and test its efficiency and effectiveness against a large test suite.

1   Introduction

The tasks of designing, maintaining, updating, and querying databases require users and administrators to cope with the complexity of the relational schemas describing the structure of the data. A graphical representation of such schemas greatly improves the friendliness of database applications and it is essential for producing high-quality understandable documentation. For this reason many commercial tools have some diagramming facilities (see Figure 1 for an example) that rely on the user to nicely place tables and their relationships on the screen. However, drawing diagrams manually is time consuming and the aesthetic results are often unsatisfactory.

Unfortunately, to our knowledge, no drawing technique is available to automatically produce high quality diagrams of this kind. In fact, such diagrams are strongly constrained: each table of the relational schema is usually represented by a box composed by a vertically ordered sequence of attributes, topped by the table name. Edges represent constraints or join paths between tables. An edge linking an attribute of one table to an attribute of a different table may attach arbitrarily to the left side or to the right side of the boxes, and should incide the box at the level of the attribute name.

---

* Work partially supported by: "Progetto Algoritmi per Grandi Insiemi di Dati: Scienza e Ingegneria", MURST Programmi di Ricerca di Rilevante Interesse Nazionale.

54

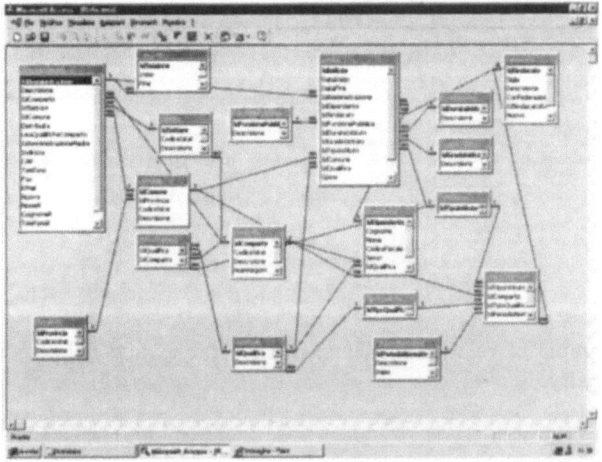

Fig. 1. A screen snapshot of Microsoft Access©. The example is taken from a real life application. Lines represent referential integrity constraints (see Section 2).

Actually, even if the link between the database research area and the graph drawing one is strong, the interest has been so far mainly focused on the visualization of Entity-Relationship diagrams and Data-Flow diagrams, that are relatively simpler to draw automatically than the relational schema diagrams (see, e.g. [5, 3, 11]).

The results presented in this paper can be summarized as follows: (i) We formulate the problem of automatically generating relational schema diagrams as a constrained orthogonal graph drawing problem, and we address it within the "classical" topology-shape-metric approach [17, 18, 11], showing how this approach can be tailored to take into account the complex constraints originated by this type of diagrams (Section 3). (ii) We give a polynomial time algorithm for constructing relational schema diagrams. The algorithm relies on several variations of existing graph drawing techniques, giving new highlights on their practical applicability (Section 3). (iii) We present an implementation of the algorithm and show its efficiency and effectiveness performing an experimental test with a random generated test suite (Section 4). Basic definitions and background are given in Section 2 and open problems are outlined in Section 5.

2   Background

Since our target is to pictorially show only few, well determined, features of database schemas, we give a simplified model of them. We call *table* an ordered set of named *attributes*. We call *relational schema* a set of named tables and a set of pairs of attributes, called *links*. In our model tables and attributes are in one-to-one correspondence with the homonymous concepts of the database research field while the concept of link is new.

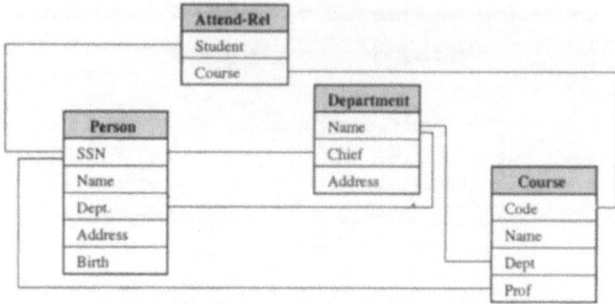

Fig. 2. An example of relational schema with 4 tables and 6 links. Links represent referential integrity.

A link is an abstract placeholder. Its purpose it to represent *join paths* and/or *referential integrity constraints*. A join path is a frequently used join operation between two tables, based on the equality of the two attributes (represented in our model as the extremes of the link). A referential integrity constraint states that the legal values for a given attribute a are the values that appear in the key k of specified table (represented in our model as a link between a and k). Further details may be found in [2]. From the point of view of our algorithm it is not relevant to distinguish the two cases.

We consider pictorial representations of a relational schema (see Figure 2) with the following properties: (a) Each table is represented as a box and its attributes are sequentially listed in the box. Each attribute corresponds to a horizontal stripe of the box. The top stripe of each table is reserved for its name. All stripes have the same height. (b) Each link (a, b) is represented as a polygonal line between the boxes of the two tables containing attributes a and b, respectively. The polygonal line incides the extremal boxes at the heights of the stripes associated with a and b, and all its segments are either horizontal or vertical (*orthogonal standard*).

We call RO-drawing (*Relational-schema Orthogonal drawing*) a layout of a relational schema that respects the above properties.

We now recall basic graph drawing definitions. We assume some familiarity with graph theory and connectivity [13].

A *plane drawing* Γ of a graph G maps each vertex of G into a point of the plane, and each edge of G into a Jordan curve between the two points associated with the end-vertices of the edge. A drawing Γ of G is *planar* if it does not contain crossings between edges. A graph is *planar* if it admits a planar drawing. A planar drawing Γ of G induces for each vertex v of G a circular clockwise ordering of the edges incident on v. Two planar drawings of G are said to be *equivalent* if for each vertex v of G they induce the same ordering of the edges around v. An *embedding* ϕ of G is a class of equivalent planar drawings of G. In other words, we can regard an embedding of G as the choice of a clockwise ordering of the edges around every vertex. An *embedded graph* G_ϕ is a planar graph G with a given embedding ϕ.

An *orthogonal drawing* of G is a plane drawing of G such that all edges are represented as chains of horizontal and vertical segments. An *orthogonal representation* (or *shape*) of G is an equivalence class of planar orthogonal drawings such that all the drawings of the class: (i) have the same sequence of left and right turns (*bends*) along the edges, and (ii) two edges incident at a common vertex determine the same angle. Roughly speaking, an orthogonal representation defines a class of orthogonal drawings that may differ only for the length of the segments of the edges.

One of the most popular technique for computing orthogonal drawings of a graph G is the so called *topology-shape-metrics* approach [4, 17, 11]. It consists of three consecutive steps:

Topology: In this step a topology for G is computed. Namely, if G is planar an embedding ϕ of G is determined in linear time, by applying a well-known planarity testing algorithm [14, 10]. If G is not planar, an embedding can be computed for it by adding a minimal number of dummy vertices to replace crossings. Such operation is usually called *planarization*. The number of crossings depends on the planarization technique, and it may be $\Omega(n^4)$. However, in practice this number is usually much smaller. For a survey on planarization techniques see [11].

Shape: During this step, an orthogonal representation H of G_ϕ is computed within the embedding ϕ. A famous algorithm for constructing an orthogonal representation of an embedded graph with vertices having at most four incident edges is presented in a work by Tamassia [17]. Such algorithm computes an orthogonal drawing that has the minimum number of bends within the given embedding. Extensions of Tamassia's algorithm to general embedded graphs are provided in [18, 12].

Metrics: In this step a final geometry for H is determined. Namely, a *compaction* algorithm assigns coordinates to vertices and bends of H with the purpose of reducing as much as possible the area (or the total edge length) of the final drawing, while preserving its planarity [11].

The Topology-Shape-Metrics approach allows us to deal with topology, shape, and geometry of the drawing separately, so simplifying the whole drawing problem. However, decisions taken in early steps cannot be changed, thus overall optimization is not achieved in general. For instance, introducing cross vertices forces crossings to appear on specific edge pairs, thus the total number of bends may be not optimal.

3 RO-Algorithm

In this section we describe a polynomial time algorithm for computing drawings of relational schemas within the RO-drawing convention described in Section 2. The algorithm is based on the topology-shape-metrics approach and exploits and modifies several existing graph drawing techniques. In particular, it makes

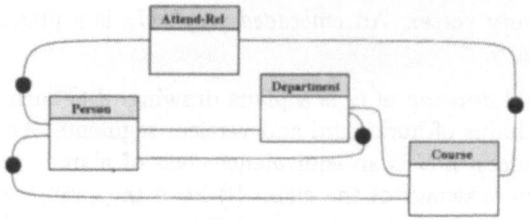

Fig. 3. Four u-turns in a partially computed RO-drawing, the round vertices are u-vertices.

careful use of a constrained planarization technique and of a variation of the algorithm in [6] for computing an orthogonal drawing of the relational schema.

Let S be a relational schema. The *underlying* graph of S is the graph G_S whose vertices are the tables of S and whose edges are the links of S. We say that S is *connected* when G_S is connected. We assume that S is always connected. If S is not connected we can apply the algorithm we describe hereunder to every connected component, and then arrange all obtained drawings on the plane by using any packing heuristics [9, 15].

The RO-Algorithm consists of three main steps:

Constrained Planarization A planarization is performed on the underlying graph G_S of S. The purpose of this step is to obtain a planar embedding of G_S such that the order of the edges around each vertex v_T, representing a table T, is compatible with the drawing standard described in Section 2 and the specific sequence of attributes of T. The output of this step is an embedded graph G'_S where dummy vertices of degree four are introduced to replace crossings (*cross-vertices*). Each link of S is represented in G'_S as an alternating chain of edges and cross-vertices.

U-Turns Assignment This step deals with the left-to-right development of the drawing. From this perspective the edges can be classified into two types: Edges that monotonically follow the left-to-right direction and edges that have to perform one or more "u-turn". In this step a (possibly empty) sequence of u-turns is associated with each edge trying to minimize their total number. U-turns are represented in G'_S with a particular kind of dummy vertices (*u-vertices*) of degree two that split the edges (see Figure 3).

Orthogonalization For each vertex of G'_S a pattern, among the ones depicted in Figure 4, is applied according to the type of the vertex. Once all vertices of G'_S have been considered, an appropriate sequence of 90 degrees bends (left or right) is associated with each edge, so describing an orthogonal representation H. To obtain the final RO-drawing from H the length of the edges and the size of the vertices are computed, heuristically "minimizing" the total edge length, and avoiding overlaps. Finally, cross-vertices and u-vertices are removed so that each link is again represented by exactly one edge.

3.1 Constrained Planarization

In the Constrained Planarization step, a planarization is performed on the underlying graph G_S of S. Let v_T be a vertex representing table T in G_S and let $a_1, \ldots, a_k$ be the attributes of T. Our algorithm partitions the edges incident on v_T into $2k$ possibly empty sets $l_1, \ldots, l_k, r_1, \ldots, r_k$, where the edges of $l_i \cup r_i$ represent the links incident on attribute a_i. Edges of l_i (r_i) will enter v_T from the left (right) in the final drawing. An edge of l_i (r_i) is a *left edge* (*right edge*) for v_T. We aim at computing a planarization of G_S with the following constraints for each v_T: (i) The edges of the same set should appear contiguously in the circular order around v_T; (ii) Sets $l_1, \ldots, l_k, r_k, \ldots, r_1$ should appear in this counter-clockwise order around v_T.

We solve the above problem carefully exploiting a constrained planarization technique that allows to specify a set of uncrossable edges. Primitives of this kind are available, for instance, within the GDToolkit library [1].

Namely, Graph G_S is mapped into a new graph P_S in which each vertex v_T, associated with a table T with k attributes, is represented by a $(k+2)$-vertex path whose vertices and edges are called τ-*vertices* and τ-*edges*, respectively. The vertices of the path are $\{v_{\text{north}}, v_1, \ldots, v_k, v_{\text{south}}\}$, where v_i is associated with attribute a_i ($i = 1, \ldots, k$). The edges of the path are $(v_{\text{north}}, v_1), (v_1, v_2), \ldots,$ (v_k, v_{south}). The edges representing links incident on attribute a_i are made incident on v_i. Intuitively, τ-vertices and τ-edges represent the structure of a table, and vertices v_{north} and v_{south} represent the upper and bottom part of the table, respectively.

Now, we run a planarization on P_S with the constraint that every τ-edge in P_S is uncrossable. Intuitively, this is done to have no edges that intersect tables in the final drawing. After the planarization, a contraction operation is applied to all the τ-vertices and τ-edges associated with the same table. The result of this phase is an embedded graph G'_S whose vertices may either represent a table or a cross. It is possible to show that the constraints described at the beginning of the subsection are enforced.

3.2 U-Turns Assignment

This step associates a (possibly empty) sequence of u-vertices with each edge of G'_S. A two phases procedure is adopted.

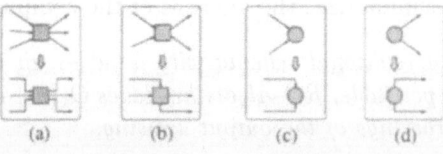

Fig. 4. Patterns for translating the description of a graph with cross-vertices and u-vertices into an orthogonal representation. (a) Vertex representing a table; (b) Cross-vertex; (c,d) u-vertices.

First, we assign an orientation to the edges. Such orientation describes the left-to-right development of the drawing. Consider edge $e = (v_{T'}, v_{T''})$, where none of $v_{T'}$, $v_{T''}$ is a cross vertex. Four cases are possible. If e is a right edge for $v_{T'}$ and a left edge for $v_{T''}$, then e is oriented from $v_{T'}$ to $v_{T''}$. If e is a left edge for $v_{T'}$ and a right edge for $v_{T''}$, then e is oriented from $v_{T''}$ to $v_{T'}$. If e is a right edge for $v_{T'}$ and a right edge for $v_{T''}$, then e is split into two edges both oriented outgoing from $v_{T'}$ and $v_{T''}$. If e is a left edge for $v_{T'}$ and a left edge for $v_{T''}$, then e is split into two edges both oriented incoming in $v_{T'}$ and $v_{T''}$.

A special technique is used for orienting the edges around cross-vertices imposing two of the incident edges to be ingoing and the other two outgoing. Orientation is then propagated from the vertices representing tables through the cross-vertices (possibly inserting u-vertices to avoid conflicts) until all edges are oriented.

Second, consider the obtained orientation and the inserted u-vertices. Two cases are possible. Either the embedded directed graph can be drawn left-to-right within the given embedding and with edges that monotonically follow the left-to-right orientation or not. In the first case we just go to the Orthogonalization step. In the second case we insert into G'_S the minimum number of u-turns that are needed to do that. Such a problem has been studied in [8] and can be solved in polynomial time using the flow techniques described in that paper.

3.3   Orthogonalization

The output of the U-Turns Assignment Step is an embedded directed graph with some dummy vertices called cross-vertices and u-vertices. Further, such a directed graph is drawable monotonically in the left-to-right direction preserving the embedding. To compute an orthogonal representation of G'_S we first draw it monotonically with the technique shown in [8] and then apply the patterns depicted in Figure 4, obtaining an orthogonal representation.

Once an orthogonal representation is constructed we shrink it in a limited area by using a variation of the compaction technique shown in [6]. Such technique allows to assign to each vertex exactly the required size and to arrange the edges to incide at the right height.

3.4   Time Complexity

The following result summarizes the analysis of the computational complexity:

Theorem 1 *Given a relational schema with n tables, m links, and a bounded number of attributes per table, RO-Algorithm takes $O((n+c)^2 \log(n+c))$, where c is the number of crossings of the output drawing.*

Proof. (sketch) The Constrained Planarization step takes $O(m(n+c))$ time, because it executes $O(m)$ times a breadth first search algorithm for computing shortest paths, as explained in [11]. After this step, the number of vertices of the graph is $N = n + c$. The U-Turn Assignment step takes $O(N)$ time to make

G'_S oriented. The application of the flow technique described in [8] for inserting the minimum number of u-turns takes $O(N^2 \log N)$ time. The Orthogonalization step takes $O(N)$ time to produce the orthogonal representation of the planarized G'_S and $O(N^2 \log N)$ time to compact the drawing by using flow techniques, as described in [11, 6]. Hence the statement follows. □

Note that even if c may be $\Omega(n^4)$, relational schemas observed in practice are quite sparse ($m = O(n)$), and "almost planar", so c is much smaller.

4   Implementation and Experiments

We implemented the algorithm for computing RO-drawings, described in Section 3. The implementation is written in C++ and uses the GDToolkit graph drawing library (*http://www.dia.uniroma3.it/~gdt*) which is based on LEDA [16].

In order to evaluate the effectiveness of the algorithm, we tested it over a set of 900 randomly generated relational schemas, with up to 90 tables. Namely, for each fixed number n of tables in the range 10–90, we considered 10 different relational schemas; each one of these schemas has been generated as follows: We denote the tables of the schema by $T_1, \ldots, T_n$. For each table T_i $(i = 1, \ldots, n)$, we randomly chose the number k_i of attributes of T_i, and sequentially enumerate them. The choice of k_i is done with a uniform probability distribution in the range $1 - 10$. We denote by $A_i = \{a_{i1}, \ldots, a_{ik_i}\}$ the set of the enumerated attributes of T_i. At the general step of the algorithm we randomly select two distinct tables T_i and T_j $(i, j \in \{1, .., n\})$, and two their attributes a_{ir}, a_{js}, where $r \in \{1, \ldots, k_i\}$ and $s \in \{1, \ldots, k_j\}$. We add the link (a_{ir}, a_{js}) to the relational schema. We add a total number of links that is randomly chosen in the range n–$2n$. When all links have been added, we check if the relational schema is connected. If it is not connected we discard the schema and restart the generation all over again, and so until a connected schema is obtained.

From the experiments, the space occupied by the drawings appears to increase quadratically with the number of tables, which is coherent with the results of other experiments in previous works [6]. In Figure 5 (a) the chart of the total area occupied by the drawings of the schemas of the test suite is shown. In Figure 5 (b) a drawing of a relational schema of the test suite is put in evidence. In Figure 6 it is shown an example of RO-drawing of a relational schema with 18 vertices taken from real life.

5   Conclusions and Open Problems

We have presented an algorithm for automatically drawing diagrams representing relational database schemas. We have also implemented and experimented such algorithm on a test suite of randomly generated relational schemas. Several problems remain open: (a) We are currently integrating the implementation within the Microsoft OLE platform. We would like to set up a web service that allows any user to exploit ability of our algorithm in drawing relational schemas.

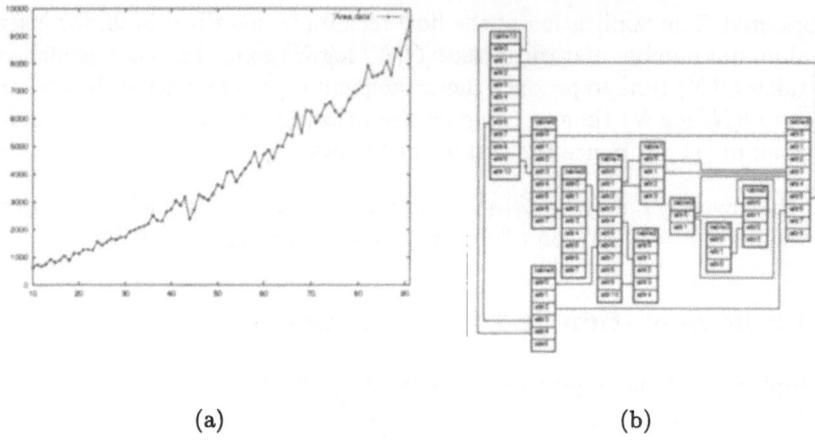

(a)   (b)

Fig. 5. (a) Chart of the area occupied by the drawings of the schemas of the test suite with respect to the number of tables (x-asis). (b) A drawing of a relational schema of the test suite.

The user will be free to adopt several types of interchange formats. (b) Some more fine tuning could be performed on the algorithm. We plan to do it by further enriching the test suite. (c) It would be interesting to understand how to modify the algorithm if the attributes of the tables can be arbitrarily permuted in such a way to improve the aesthetic quality of the final drawing. (d) We plan to modify the technique described in this paper in order to obtain drawings of similar widely employed diagram standards (as, for example, UML diagrams) .

References

1. Gdtoolkit: An object-oriented library for handling and drawing graphs, 1999. Third University of Rome, http://www.dia.uniroma3.it/~gdt.
2. P. Atzeni, S. Ceri, S. Paraboschi, and R. Torlone. *Database Systems: Concepts, Languages and Architetures.* McGraw Hill, London, United Kingdom, 1999.
3. C. Batini, E. Nardelli, M. Talamo, and R. Tamassia. GINCOD: a graphical tool for conceptual design of data base applications. In A. Albano, V. D. Antonellis, and A. D. Leva, editors, *Computer Aided Data Base Design*, pages 33–51. North-Holland, New York, NY, 1985.
4. C. Batini, E. Nardelli, and R. Tamassia. A layout algorithm for data flow diagrams. *IEEE Trans. Softw. Eng.*, SE-12(4):538–546, 1986.
5. C. Batini, M. Talamo, and R. Tamassia. Computer aided layout of entity-relationship diagrams. *Journal of Systems and Software*, 4:163–173, 1984.
6. G. D. Battista, W. Didimo, M. Patrignani, and M. Pizzonia. Orthogonal and quasi-upward drawings with vertices of arbitrary size. In J. Kratochvil, editor, *Graph Drawing (Proc. GD '99)*, Lecture Notes Comput. Sci. Springer-Verlag, 1999. to appear.

62

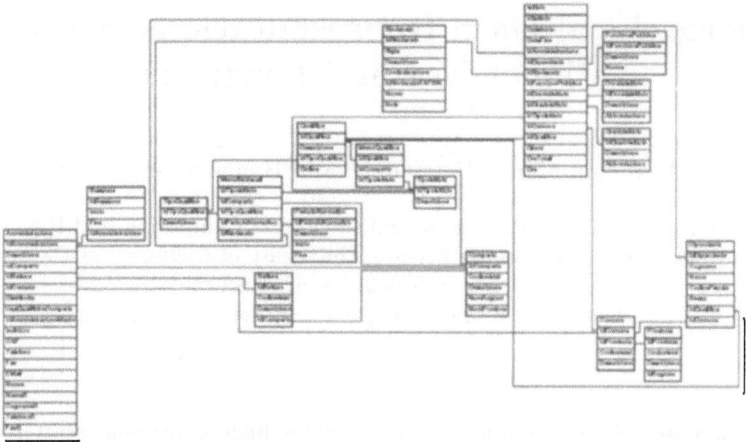

Fig. 6. A drawing of a relational schema, from real life, computed by our implementation of `RO-Algorithm`.

7. G. D. Battista, S. Diglio, M. Lenti, and M. Simoncelli. Queryviewer: A java system for drawing the result of a query, 1998. Third University of Rome, http://www.dia.uniroma3.it/~lenti/QueryViewer/.

8. P. Bertolazzi, G. Di Battista, and W. Didimo. Quasi-upward planarity. In S. H. Whitesides, editor, *Graph Drawing (Proc. GD '98)*, volume 1547 of *Lecture Notes Comput. Sci.*, pages 15–29. Springer-Verlag, 1998.

9. B. Chazelle. The bottom-left bin-packing heuristic: an efficient implementation. *IEEE Trans. Comput.*, C-32:697–707, 1983.

10. N. Chiba, T. Nishizeki, S. Abe, and T. Ozawa. A linear algorithm for embedding planar graphs using PQ-trees. *J. Comput. Syst. Sci.*, 30(1):54–76, 1985.

11. G. Di Battista, P. Eades, R. Tamassia, and I. G. Tollis. *Graph Drawing*. Prentice Hall, Upper Saddle River, NJ, 1999.

12. U. Fößmeier and M. Kaufmann. Drawing high degree graphs with low bend numbers. In F. J. Brandenburg, editor, *Graph Drawing (Proc. GD '95)*, volume 1027 of *Lecture Notes Comput. Sci.*, pages 254–266. Springer-Verlag, 1996.

13. F. Harary. *Graph Theory*. Addison-Wesley, Reading, MA, 1972.

14. J. Hopcroft and R. E. Tarjan. Efficient planarity testing. *J. ACM*, 21(4):549–568, 1974.

15. K. Jansen. An approximation scheme for bin packing with conflicts. In *Proc. 6th Scand. Workshop Algorithm Theory*, volume 1432 of *Lecture Notes Comput. Sci.*, pages 35–46. Springer-Verlag, 1998.

16. K. Mehlhorn and S. Näher. LEDA: a platform for combinatorial and geometric computing. *Commun. ACM*, 38(1):96–102, 1995.

17. R. Tamassia. On embedding a graph in the grid with the minimum number of bends. *SIAM J. Comput.*, 16(3):421–444, 1987.

18. R. Tamassia, G. Di Battista, and C. Batini. Automatic graph drawing and readability of diagrams. *IEEE Trans. Syst. Man Cybern.*, SMC-18(1):61–79, 1988.

Hierarchical Data Representations Based on Planar Voronoi Diagrams

Shirley Schussman, Martin Bertram, Bernd Hamann, and Kenneth I. Joy

Center for Image Processing and Integrated Computing (CIPIC),
Department of Computer Science, University of California at Davis
Davis, CA 95616-8562, USA
{schussms, bertram, joy, hamann}@cs.ucdavis.edu

Abstract. Multiresolution representation of high-dimensional scattered data is a fundamental problem in scientific visualization. This paper introduces a data hierarchy of Voronoi diagrams as a versatile solution. Given an arbitrary set of points in the plane, our goal is the construction of an approximation hierarchy using the Voronoi diagram as the essential building block. We have implemented two Voronoi diagram-based algorithms to demonstrate their usefulness for hierarchical scattered data approximation. The first algorithm uses a constant function to approximate the data within each Voronoi cell, and the second algorithm uses the Sibson interpolant [14].

1 Introduction

This paper presents a new solution for constructing multiresolution data representations: data hierarchies based on Voronoi diagrams. This approach is motivated by the need to interactively explore very large data sets that consist of scattered or arbitrarily gridded data. A hierarchy of Voronoi diagrams is a natural solution for a number of reasons. First, Voronoi diagrams define a "natural mesh" for scattered data, data without explicit point connectivity. Second, point insertion and deletion operations for Voronoi diagrams are expected constant-time operations [10, 3]. In addition, Voronoi cells can be sorted in depth in linear time, which is important for volume visualization, and Voronoi diagrams can be extended to n dimensions. Although the Voronoi diagram's dual–the Delaunay triangulation–has the same properties, the Voronoi diagram provides a more intuitive tessellation.

2 Related Work

A number of approaches have been developed during the past two decades to visualize scientific data that is scattered [5] or defined on very large and often highly irregular grids. The most common methods for hierarchical data representations are based on mesh reduction. These techniques associate a mesh with the data sites, apply various reduction techniques to the mesh, and use reduced meshes as basis for visualization.

Several data decimation and hierarchical schemes have been developed over the past few years by the computer graphics and visualization communities. Schroeder et al. [13] and Renze and Oliver [12] have developed algorithms that simplify a mesh by removing vertices. Removing a vertex creates a hole in the mesh that must be re-triangulated, and several strategies may be used.

Hoppe [7,8] and Hoppe and Popović [11] describe a progressive-mesh representation of a triangle mesh. This is a continuous-resolution representation based on an edge-collapse operation. The data reduction problem is formulated in terms of a global mesh optimization problem ordering the edges according to an energy function to be minimized. As edges are collapsed, and the priorities of the edges in the neighborhood of the transformation are recomputed. The result is an initial coarse representation of a mesh, and a linear list of edge-collapse operations. Garland and Heckbert [6] utilize a different strategy, based on quadratic error metrics for efficient calculation of a hierarchy. Hoppe [9] has extended this method to multidimensional meshes with appearance attributes.

Trotts et al. [17,16] and Staadt and Gross [15] have extended the edge collapse paradigm to tetrahedral meshes. Cignoni et al.[1] also treat the tetrahedral mesh problem. They use a top-down Delaunay-based procedure to define a tetrahedral mesh that represents a three-dimensional set of points. The mesh is refined by selecting a data point whose associated function value is poorly approximated by an existing mesh and inserting this point into the mesh. The mesh is modified locally to preserve the Delaunay property.

This paper presents a new technique that produces a hierarchy of Voronoi diagrams. These diagrams can be used to approximate massive data sets by utilizing functional approximations over the Voronoi cells. We generate a hierarchy by inserting points into the Voronoi diagram that represent the largest error in individual cells. We construct two interpolation methods, one based on constant functions and the other one based on the Sibson interpolant [?,4]. We discuss their advantages and disadvantages.

Our algorithm is a top-down approach that produces a hierarchy of Voronoi diagrams, where each diagram has an associated approximation that is within a certain threshold of the original data. The error calculations are local, which makes the algorithm fairly efficient. Our algorithm utilizes only the original data points, which allows for a very compact representation.

3 Voronoi Hierarchies

A Voronoi hierarchy consists of a set of Voronoi diagrams and interpolating functions defined on the Voronoi diagrams that approximate a given data set at different resolutions and qualities of approximation. Any implementation requires methods for selecting points from a given finite data set, choosing an interpolant for each Voronoi cell, and choosing an error metric to determine the overall accuracy of each level in the Voronoi hierarchy.

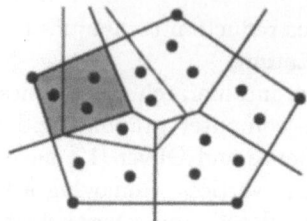

Fig. 1. Example of vertex insertion. Black lines: current Voronoi diagram; black polygon: convex hull of point set; grey region: Voronoi cell with maximal error; grey point: vertex p_{c_j} with maximal error ε_{max}; grey lines: Voronoi cell to be inserted.

3.1 Refinement of a Voronoi Diagram

Given a data set $D = \{p_1, \ldots, p_n\}$ and n associated functions values $f_1, \ldots, f_n$. We define a sequence of Voronoi diagrams $V^{n_0}, V^{n_1}, \ldots, V^{n_h}$, where each V^i is defined by i points selected from D. The initial Voronoi diagram V^{n_0} is defined by the n_0 points of D that lie on the boundary of the minimal point set defining the closed boundary polygon of the convex hull of D.

A Voronoi diagram, V^{n_k}, is refined by inserting additional data points of D into the Voronoi diagram in cells of high error as is shown in Figure 1. More specifically, a set of Voronoi cells with high error is identified, and a point is inserted into each cell c_j in that set. The error ε_{c_j} is calculated using the L^1 norm, an average of the error over all n_{c_j} data points of D lying in cell c_j. We define the error of cell c_j as

$$\varepsilon_{c_j} = \frac{1}{n_{c_j}} \sum_{p_i \in c_j} \|f(c_j, p_i) - f_i\|. \tag{1}$$

where $f(c, p)$ is used to represent an interpolant over cell c_j containing p_i and where f_i denotes the associated function value at p_i. Once ε_{t_j} is calculated, it is compared with some threshold value to determine whether or not it belongs to the set of cells to be refined. It is convenient to determine the threshold as a percentage of the average global error ε_{avg}, which we define as

$$\varepsilon_{avg} = \frac{1}{n} \sum_{j=1}^{n_k} \sum_{p_i \in c_j} \|f(c_j, p_i) - f_i\|. \tag{2}$$

Once the set of cells to be refined is identified, a point $p_{c_j} \in D$ is inserted into each cell c_j. Ideally, p_{c_j} would define a cell that would eliminate or at least minimize the error in the resulting local cell configuration. Rather than searching exhaustively for the ideal point, we simply choose the point p_{c_j} in c_j with the highest error ε_{max}, where

$$\varepsilon_{max} = \max_{p_i \in c_j} \|f(c_j, p_i) - f_i\|. \tag{3}$$

3.2 Constructing the Hierarchy

The following pseudocode describes the basic algorithm for generating the hierarchy of Voronoi diagrams.

Algorithm: Voronoi Hierarchy Construction:

Input:

- Set of n points $p_i = (x_i, y_i)$ and associated scalar or vector function values f_i, $i = 1, \ldots, n$
- Number of levels, h, to be calculated and error tolerances for all levels, called ε_k, $k = 1, \ldots, h$

Output:

- Set of h Voronoi diagrams, where the global error associated with each Voronoi diagram V^{n_k} is smaller than the level-specific global error tolerance ε_k

Steps of the Algorithm:

- Determine minimal point set defining boundary polygon of convex hull of given points.
- Create initial Voronoi diagram for this minimal point set.
- Compute global approximation error for initial Voronoi diagram, called V^{n_0}.
- Assuming that the global approximation error of V^{n_0} is larger than ε_1, determine a set of cells in V^{n_0} with high error.
- For each cell in this set, choose an appropriate data point in D that lies in it and update the diagram accordingly.
- Check whether global approximation error of refined Voronoi diagram still exceeds ε_1.
- Continue process of point selection and insertion until diagram's global error approximation is smaller than ε_1; call this Voronoi diagram V^{n_1}.
- Construct Voronoi diagrams $V^{n_2}, \ldots, V^{n_h}$ in the same manner.

3.3 Point Insertion and Selection

One reason Voronoi hierarchies are a general and suitable form for multiresolution representations is that there are multiple ways to choose points for refinement. Rather than exploring all possibilities, which is a research topic in its own right, we developed the method described in Section 3.1. This method is designed to be fast, to refine Voronoi diagrams adaptively, and to quickly capture patterns in the underlying data. It can also be extended to higher dimensional domains, as can be seen by all equations.

The method of point insertion described in Section 3.1 refines a Voronoi diagram in a way that captures high-gradient regions and discontinuities very early in the refinement process. The point insertion strategy detects "extreme

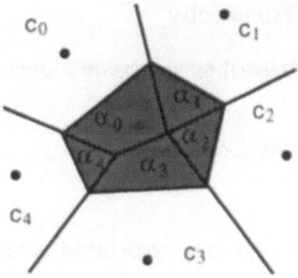

Fig. 2. Example of Sibson interpolant. Black lines: current Voronoi diagram; c_i: cells in current Voronoi diagram; grey point: simulated insertion point; grey region: simulated cell resulting from point insertion; α_i: regions where simulated cell overlaps c_i

values" first, since ε_{max} is always the maximum or minimum function value in a cell c_j. If the point p_j defining c_j was defined by a maximum value in a previous iteration, then p_{c_j} is a minimum value and vice versa.

A problem arises when there are multiple points in cell c_j with the same maximal value ε_{max}. In this case, a random point is selected from the set of candidates for insertion. Random selection is crucial in this context, as the order of indices of the points in D should not bias point insertion.

Selecting an appropriate threshold value to determine which cells should be refined is another issue. As stated in Section 3.1, if a cell error ε_{c_j} is greater than a threshold value, the cell should be refined. In the context of data approximation, high threshold values are good since only areas of high error will be refined. Unfortunately, it is difficult to determine a good threshold value for an arbitrary data set. Another approach is to only insert a point in a cell with maximal error.

3.4 Interpolation Functions

In principle, any function that interpolates values at the cell centers can be used. We implemented two interpolants for comparison, a piecewise constant and the Sibson interpolant The strengths and weaknesses of both interpolants are compared.

Using a constant value per cell, where the value is that of the defining point for the cell, is a simple and efficient interpolant. The constant function can be rendered quickly, which means that more data points can be rendered in the same amount of time. A piecewise constant interpolant representation permits the representation of discontinuities, but cannot represent smoothly varying data well.

The Sibson interpolant is a smoothly varying function that smoothly represents the underlying data. The Sibson interpolant is based on blending the function values f_j associated with the points defining a Voronoi diagram. The resulting interpolation defines a smooth function that is C^1-continuous everywhere except at the points themselves. The interpolating function $f(p)$ is evaluated at a

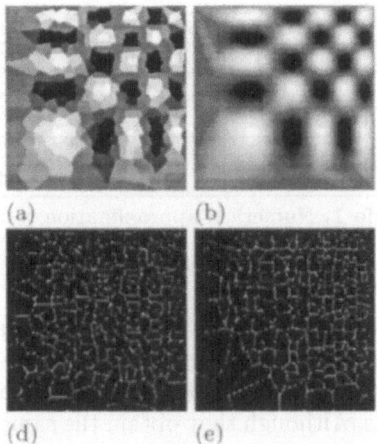

Fig. 3. (a) Piecewise constant function used to evaluate analytical function with 270 points; (b) Sibson's interpolant used to estimate same function with 270 points; (c) and (d) show corresponding Voronoi meshes for (a) and (b), respectively

point p by "simulating its insertion" into the Voronoi diagram, without actually changing the Voronoi diagram, and by estimating the areas a_j cut away from Voronoi cells c_j in a local neighborhood. The value of the Sibson interpolant at p is defined as

$$f(p) = \frac{\sum_j a_j f_j}{\sum_j a_j},$$

which is illustrated in the Figure 2.

4 Results

We present the results of the constant function and Sibson interpolant based algorithms for three data sets shown in Figures 3, 5 and 6, along with their numerical performance data in Table 1. We also show a hierarchy using the Sibson interpolant in Figure 4. All of the input data sets are defined on a 250x250 uniformly spaced rectilinear grid representing color and grey scale images.

The first greyscale data set, see Figure 3, was generated by evaluating the function

$$\omega(x, y) = sin(x^2) \, sin(y^2), \quad x, \, y \, \epsilon \, [0, \, 4].$$

Although both algorithms pick up the pattern quickly, the Sibson interpolant obtains better results because it can represent smooth functions well.

The remaining data sets are color images produced by the Hubble Space Telescope, courtesy of NASA. Figure 4 shows a Voronoi hierarchy with four levels. The basic pattern is represented well with only 100 points, as is shown is Figure 4(b). Successive levels refine the center and represent additional stars.

Dataset	No. Tiles	Piecewise Constant		Sibson Interpolant	
		L^1 Error [%]	L^2 Error [%]	L^1 Error [%]	L^2 Error [%]
ω	270	7.1	21	2.9	3.4
"Cat's Eye"	580	2.3	8.3	2.3	2.9
"Cygnus Loop"	620	8.2	19	5.3	6.4
	1740	5.2	11	4.4	5.3

Table 1. Numerical approximation results.

Figure 4(e) shows a good approximation of the original data set with 2000 points, which is only three percent of the original data points.

Figure 5 shows the effectiveness of both algorithms on the Cat's Eye Nebula data set using 580 points. Although they obtain the same numerical performance for the L^1 norm, the constant function algorithm detects more features. Namely, it detects the second elliptical path in the center, the one whose primary axis has a negative slope. As is seen from the Voronoi cells in Figure 5(e), the Sibson interpolant algorithm has fewer cells in that region. The Sibson interpolant, a smooth interpolant, places its points around discontinuous regions, which is the only way the Sibson interpolant can represent discontinuities.

Another data set is an image of the Cygnus Loop Nebula, see Figure 6. Although the Sibson interpolant is better visually and numerically, it depicts fewer stars than the constant function algorithm. The lack of stars results from the point insertion technique. Instead of using a threshold value for the entire data set, like the constant function algorithm, it only inserts a point into the cell with the worst error before it updates the Voronoi diagram. Since the cells with missing stars never counted at the worst cell, they were never refined.

5   Conclusions and Future Work

We have introduced a method for the hierarchical, gridless representation of planar scattered data. The method is straightforward and can be generalized to higher dimensions. We believe that Voronoi diagram-based approaches provide an appropriate framework for constructing hierarchical approximations for gridless, scattered data. Voronoi diagrams provide flexibility and enable adaptive and localized refinement. Voronoi diagram hierarchies require rather involved underlying data structures for their efficient manipulation, but we are convinced that this is acceptable due to the gain in flexibility. We plan to extend our implementations to volumetric data, and eventually to time-varying data. We will develop efficient ray-casting and isosurface extraction methods for Voronoi diagram hierarchies of volumetric data sets.

6   Acknowledgments

This work was supported by the National Science Foundation under contracts ACI 9624034 and ACI 9983641 (CAREER Awards), through the Large Scien-

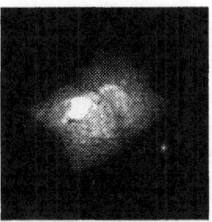

(a) Original image of a dying sun

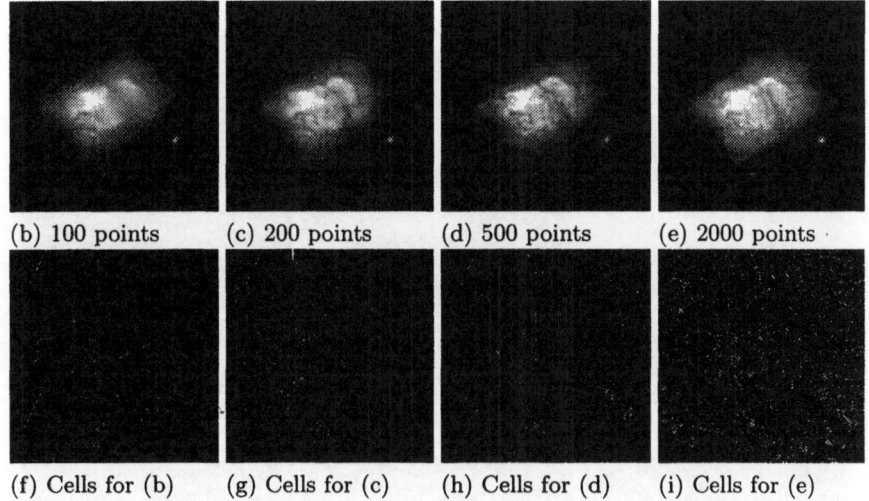

(b) 100 points   (c) 200 points   (d) 500 points   (e) 2000 points

(f) Cells for (b)   (g) Cells for (c)   (h) Cells for (d)   (i) Cells for (e)

Fig. 4. Approximations of the dying sun image with 100, 200, 500 and 2000 points

tific and Software Data Set Visualization (LSSDSV) program under contract ACI 9982251, and through the National Partnership for Advanced Computational Infrastructure (NPACI); the Office of Naval Research under contract N00014-97-1-0222; the Army Research Office under contract ARO 36598-MA-RIP; the NASA Ames Research Center through an NRA award under contract NAG2-1216; the Lawrence Livermore National Laboratory under ASCI ASAP Level-2 Memorandum Agreement B347878 and under Memorandum Agreement B503159; and the North Atlantic Treaty Organization (NATO) under contract CRG.971628 awarded to the University of California, Davis. We also acknowledge the support of ALSTOM Schilling Robotics, Chevron, Silicon Graphics, Inc. and ST Microelectronics, Inc. We thank the members of the Visualization Group at the Center for Image Processing and Integrated Computing (CIPIC) at the University of California, Davis.

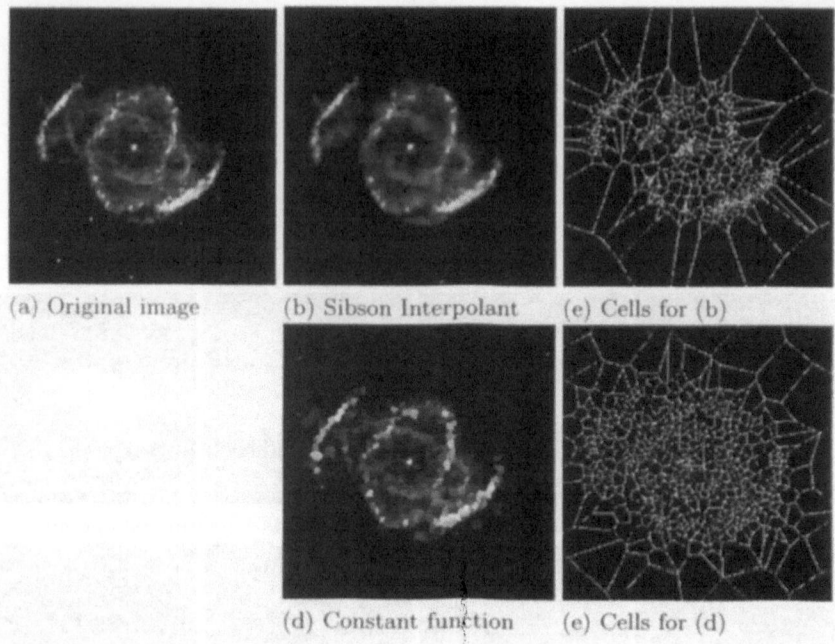

(a) Original image   (b) Sibson Interpolant   (e) Cells for (b)

(d) Constant function   (e) Cells for (d)

Fig. 5. Approximations of the Cat's Eye Nebula made with 580 points

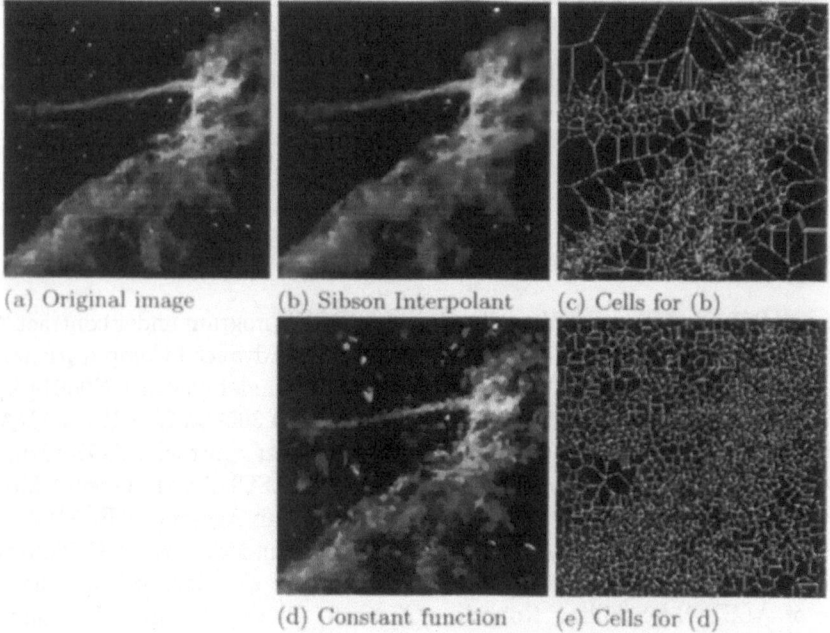

(a) Original image   (b) Sibson Interpolant   (c) Cells for (b)

(d) Constant function   (e) Cells for (d)

Fig. 6. Approximations of the Cygnus Loop Nebula made with 1740 points

References

1. P. Cignoni, L. De Floriani, C. Montoni, E. Puppo, and R. Scopigno. Multiresolution modeling and visualization of volume data based on simplicial complexes. In Arie Kaufman and Wolfgang Krueger, editors, *1994 Symposium on Volume Visualization*, pages 19–26. ACM SIGGRAPH, October 1994.
2. L. De Floriani, P. Marzano, and E. Puppo. Hierarchical terrain models: Survey and formalization. In *Proc. IEEE Sympos. Applied Comput.*, pages 323–327, 1994.
3. O. Devillers. Improved incremental randomized Delaunay triangulation. In *Proc. 14th Annu. ACM Sympos. Comput. Geom.*, pages 106–115, 1998.
4. Gerald Farin. Surfaces over dirichlet tessellations. *Computer Aided Geometric Design*, 7(1-4):281–292, June 1990.
5. R. Franke and G. M. Nielson. Scattered data interpolation and applications: A tutorial and survey. In H. Hagen and D. Roller, editors, *Geometric Modeling*. Springer-Verlag, 1991.
6. Michael Garland and Paul S. Heckbert. Surface simplification using quadric error metrics. In Turner Whitted, editor, *SIGGRAPH 97 Conference Proceedings*, Annual Conference Series, pages 209–216. ACM SIGGRAPH, Addison Wesley, August 1997.
7. Hugues Hoppe. Progressive meshes. In Holly Rushmeier, editor, *SIGGRAPH 96 Conference Proceedings*, Annual Conference Series, pages 99–108. ACM SIGGRAPH, Addison Wesley, August 1996.
8. Hugues Hoppe. View-dependent refinement of progressive meshes. In Turner Whitted, editor, *SIGGRAPH 97 Conference Proceedings*, Annual Conference Series, pages 189–198. ACM SIGGRAPH, Addison Wesley, August 1997.
9. Hugues Hoppe. New quadric metric for simplifying meshes with appearance attributes. In David Ebert, Markus Gross, and Bernd Hamann, editors, *IEEE Visualization 99*, pages 59–67. IEEE, November 1999.
10. Arne Maus. Delaunay triangulation and the convex hull of n points in expected linear time. *BIT*, 24(2):151–163, 1984.
11. Jovan Popović and Hugues Hoppe. Progressive simplicial complexes. In Turner Whitted, editor, *SIGGRAPH 97 Conference Proceedings*, Annual Conference Series, pages 217–224. ACM SIGGRAPH, Addison Wesley, August 1997.
12. Kevin J. Renze and James H. Oliver. Generalized unstructured decimation. *IEEE Computer Graphics & Applications*, 16(6):24–32, November 1996.
13. William J. Schroeder, Jonathan A. Zarge, and William E. Lorensen. Decimation of triangle meshes. *Computer Graphics*, 26(2):65–70, July 1992.
14. R. Sibson. Locally equiangular triangulation. *The Computer Journal*, 21:243–245, 1978.
15. Oliver G. Staadt and Markus H. Gross. Progressive tetrahedralizations. In David Ebert, Hans Hagen, and Holly Rushmeier, editors, *Proceedings of Visualization 98*, pages 397–402. IEEE Computer Society Press, Los Alamitos, California, October 1998.
16. Issac J. Trotts, Bernd Hamann, and Kenneth I. Joy. Simplification of tetrahedral meshes. *IEEE Transactions on Visualization and Computer Graphics*, 5(3):224–237, 1999.
17. Issac J. Trotts, Bernd Hamann, Kenneth I. Joy, and David F. Wiley. Simplification of tetrahedral meshes. In David Ebert, Hans Hagen, and Holly Rushmeier, editors, *Proceedings of Visualization 98*, pages 287–296. IEEE Computer Society Press, Los Alamitos, California, October 1998.

Skeleton Graph Generation for Feature Shape Description

Freek Reinders, Melvin E.D. Jacobson, and Frits H. Post

Delft University of Technology
email: {k.f.j.reinders, m.e.d.jacobson, f.h.post}@cs.tudelft.nl

Abstract. An essential step in feature extraction is the calculation of attribute sets describing the characteristics of a feature. Often, attribute sets include the position, size, and orientation of the feature. These attributes are very important, but they do not provide a good approximation of the shape of a feature. For better shape description, a more sophisticated method is needed.

This paper describes a method that extracts a binary skeleton of a feature, and transforms it into a graphical representation: the skeleton-graph. This graph represents the original skeleton with controlled precision, and contains the essential topology and geometry of the skeleton. In addition, distance information is used to generate a simplified reconstruction of the original 3D feature shape, which can also be used as an iconic object for visualization.

Keywords: Feature Extraction, Shape Description, Skeleton Attributes, Graph Simplification.

1   Introduction

Feature extraction is an approach to visualization aiming at automatic recognition of important features (structures, objects, or regions) in scientific data sets. Rather than leaving the recognition of the interesting features entirely to the visual inspection by the user, this task is performed automatically. The extracted features are characterized by quantitative descriptions or attributes. The features are directly related to physical entities and phenomena studied, and thus dependent on the application. In the field of computational fluid dynamics, common examples of features are vortices, shock waves, and recirculation zones.

Most feature extraction techniques are based on a classification or segmentation of the data, to identify the parts of the data sets that belong to the features. If the data are defined on the nodes of a grid, a filter can be used that selects data items which satisfy a certain selection criterion. Other techniques for segmentation include region growing, and edge detection. The result of a segmentation of a grid data set is a binary grid in which feature nodes are marked. Adjacent marked grid nodes can then be clustered into coherent regions, and for each region certain attributes are calculated [8].

The quantification by calculating attributes is essential to the process of feature extraction. The attribute sets give a quantitative description of a feature, describing its most important characteristics, such as its position, and size. The attribute sets can be used to evaluate a feature or to compare it with other features, for instance in order to track features in time-dependent data sets [7]. A frequently used attribute set is the ellipsoid fit around the feature. The ellipsoid provides a good indication of position, orientation, and size, but is a crude (first order) approximation of shape. Sometimes a more accurate description for shape is needed.

Skeletonization, or Medial Axis Transformation (MAT) provides a more sophisticated method to characterize the shape of a feature. The skeleton of an object can be defined as the locus of points that lie at the center relative to the object's boundary. Thus, the skeleton is a thinner version of a 3D object, which still preserves the topology and geometry of the object. Therefore, it is an efficient and compact shape descriptor of features.

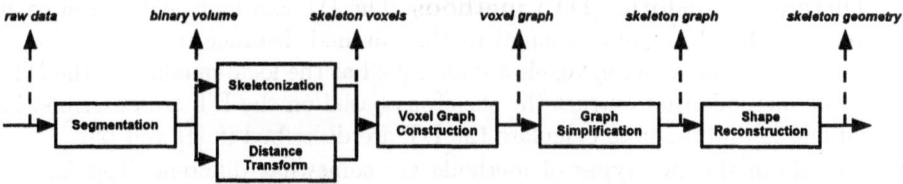

Fig. 1. Pipeline for skeleton graph generation.

This paper presents a method for skeleton attribute calculation of extracted features. Figure 1 illustrates the process of determining for these shape attributes. The input is a regular segmented data volume (a voxel grid), in which binary data represent the voxels belonging to a number of feature objects. The skeleton voxels of the objects are determined by an existing skeletonization method. The resulting voxels are then connected into a voxel graph of skeleton-nodes and edges. The graph initially contains all skeleton voxels, but it can be simplified. First, the topology is extracted by identifying special nodes such as end-nodes, junction-nodes, and loop-nodes. Next, geometry is approximated to a given tolerance by inserting extra curve-nodes and profile-nodes. Third, a reconstruction is made of the 3D shape of the original object.

The combination of skeleton and distance information provides an excellent way to approximately reconstruct the shape of the original object. The skeleton graph is "fleshed out" by wrapping spherical and conical volumes around the edges and nodes of the graph. The size of these volumes is determined by the distance to the surface, which is stored at every skeleton node. The result is a simplified geometric object, which may be used as an iconic representation describing the feature shape and which is superior to the crude approximation by a fitted ellipsoid.

The paper is organized as follows. Sections 2 through 5 describe each step in the process of skeleton graph generation. Section 6 shows a number of applications for this method. Finally, some conclusions and topics for future research are given in section 7.

2 Skeletonization

The first step is the skeletonization of the segmented binary volume. The result is a set of skeleton voxels at the center of the object. Many skeletonization algorithms have been published, especially in the image processing literature. Most algorithms are only for 2D data, but some can be extended to 3D or higher dimensions.

The skeletonization algorithms found in the literature can be classified in two categories:

- **Topological thinning methods.** These methods are based on removing voxels from the surface of the object, by identifying the so-called simple points, i.e. points that will not change the topology of the object when they are removed. Methods (2D and 3D) based on this concept are described in [4] and [6].
- **Distance transform (DT) methods.** The DT can be calculated in each voxel of the object and is equal to the minimal distance to the surface of the object. The skeleton voxels are identified as the local maxima of the DT. Some methods that extract the skeletons based on the DT are described in [5] and [9]. Methods to calculate the DT are described in [1] and [2].

The results of the two types of methods are somewhat different. Topological thinning guarantees the connectivity of the skeleton voxels, while DT-methods in general do not. For our purposes we want a method that guarantees the connectivity.

The algorithm we use is a topological thinning algorithm based on a hit-or-miss evaluation using sets of masks [3]. The mask sets can be used to manipulate a binary object in several ways, of which skeletonization is only one. A mask set consists of a number of 3x3x3 masks with zeroes, ones and "don't cares". Each mask indicates a configuration that must remain in the object. If the 3x3x3 neighborhood of a voxel matches one of the masks, it is a skeleton voxel and should remain in the object, otherwise it is removed. Thus, the object surface is peeled off iteratively until only the skeleton voxels remain.

The characteristics of a skeleton obtained in this way depends on the mask set used. We have three different mask sets to produce different types of skeletons: surface skeletons, line skeletons and point skeletons (see Figure 2). Here, we will only concentrate on the line skeletons. At each skeleton voxel also the distance information is stored, indicating the distance from the voxel to the object surface. We have used a chamfer distance transformation [1, 2], which is calculated together with the skeletonization (see Figure 1).

3 Voxel Graph Construction

After the skeleton voxels have been determined, a *voxel graph* can be constructed by connecting neighboring voxels. In the voxel graph, all skeleton voxels are nodes, and adjacent voxels are connected by edges. The basic approach for constructing the voxel graph is to traverse all voxels of the skeleton and to connect

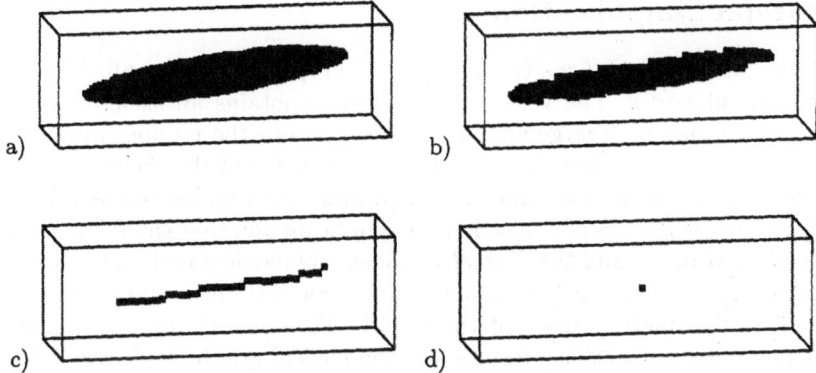

Fig. 2. Skeletonization of a binary volume, resulting in three types of skeletons, depending on the masks used: a) binary volume, b) surface skeleton, c) line skeleton, and d) point skeleton.

neighboring voxels. The voxel graph is a structure that is easy to manipulate and to analyze. Using the voxel graph, the number of nodes and edges can be reduced, while preserving the basic structure of the skeleton.

The connection of neighboring voxels is not a straightforward process. Simply connecting the adjacent skeleton voxels causes problems as shown in Figure 3a: a "zero-loop" occurs at the junction. All three voxels at the junction have more than two edges and may be classified as a junction-node (see section 4.2), while there is in fact only one junction. Lee, Kashyap and Chu [4], solved this problem by directly classifying the junction node as it is encountered, and classifying all its neighboring nodes as regular. However, the choice of the junction node will depend on the starting point of the traversal, which will result in slightly different graphs, as illustrated in Figure 3b, c, and d.

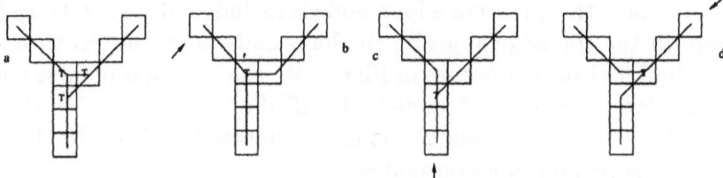

Fig. 3. Problems when constructing the voxel graph: a) zero-loop, b, c, and d) the junction node depends on the traversal starting point.

We solved the problem of zero-loops by assigning a priority ordering to the connections. In a 3D skeleton, each voxel has 26 neighbors: 6 face-connected, 12 edge-connected, and 8 vertex-connected. We give priority to the connection with the nearest neighbor, so face-connected neighbors are connected before edge-connected neighbors and these are connected before the vertex-connected neighbors. In case of a junction, this connection priority will always lead to the graph shown in Figure 3b, which we believe is the best solution.

4   Graph Simplification

After the construction of the voxel graph, the graph can be simplified by removing redundant nodes. The voxel graph initially contains all skeleton voxels as nodes, which may be a large amount. We can reduce the number of nodes and edges while still preserving the topology and geometry of the skeleton. The task is to determine which nodes should be kept and which nodes can be removed.

There are two classes of nodes that are significant and that should be retained: the *topological* nodes and the *geometric* nodes. Topological nodes are nodes that are necessary to preserve the topology, and geometric nodes are necessary to preserve the geometric shape of the object. The identification of topological nodes and their connectivity results in a topological graph, while the detection of geometric nodes results in the geometric graph. The final skeleton graph is a combination of the two graphs.

4.1   Topological Graph

The topological graph holds the basic structure of the object and can be determined by finding the topological nodes. The topological nodes are identified by counting the number of edges connected to a node in the voxel graph:

- *End node*: one edge.
- *Regular node*: two edges.
- *Junction node*: three or more edges.

The end nodes and junction nodes are the nodes that determine the topology of the object. Thus, the topological graph is created by simply removing the regular nodes, and connecting the remaining nodes by edges. However, there is a problem in case of loops.

An additional type of topological node, the *loop node*, is needed to describe loops in the graph. Figure 4 shows a number of situations with loops in the graph. In the case of Figure 4a, the removal of all regular nodes will remove all nodes. To overcome this problem a loop node is included at an arbitrary location on the loop. In the topological graph the loop node is connected to itself with an edge. A similar situation occurs in Figure 4b; the junction node is connected to itself. Therefore, this junction node is classified as a loop node. In the Figures 4c and 4d, also a loop exists, but no node is connected with itself, therefore all nodes in the loop remain junction nodes.

The number of loops is an important variable in the topological graph, because it may provide a cue for comparison of two skeletons. It can be calculated with the *Euler formula*:

$$V - E + F = C - H \tag{1}$$

with V the number of nodes, E the number of edges and F the number of faces. $C - H$ form the so-called *Euler number*, with H the number of holes in an object and C the number of connected objects in a scene. Because we only work with lines F can be set to 0, and because the graph is always a single object, C can be set to 1. Substitution gives the number of loops (holes) in the graph:

$$H = 1 - V + E \tag{2}$$

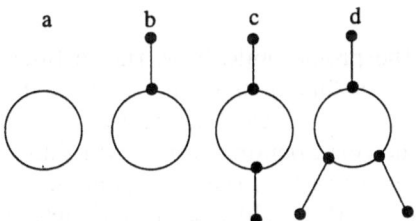

Fig. 4. Loops in the topology of the skeleton.

4.2 Geometric Graph

The geometric graph refines the edges in the topological graph by determining certain geometric nodes. The topological graph is not a sufficient approximation of the shape, especially if the shape is strongly curved. Geometric nodes are key points for describing the shape of the skeleton. The geometry can vary in two ways: the skeleton line is curved, or the profile of the object surface is curved. Hence, we distinguish the following two types of geometric nodes:

- *Curve nodes*: where the skeleton line bends.
- *Profile nodes*: where the surface profile changes.

The geometric nodes are inserted on the edges of the topological graph. For each edge the regular nodes are traversed, and curve or profile nodes are inserted. Insertion of geometric nodes depends on geometric tests, which will be described below. If a new node is inserted, the two new edges are handled recursively in the same way until all geometric tests are satisfied.

Curve nodes

The test for finding the curve nodes uses the maximum distance d_{max} between the voxel nodes and the edge, and compares d_{max} to a distance threshold T_{dist} (see Figure 5). A curve node is added at the location of the maximum when $d_{max} > T_{dist}$. Two new edges are created and both are tested recursively. The process terminates when all intermediate nodes fall within T_{dist} of the corresponding edges.

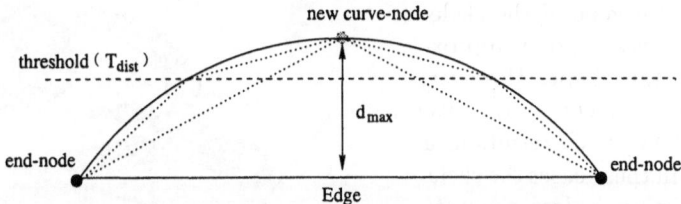

Fig. 5. Finding the curve nodes recursively.

The distance threshold provides a measure of precision for the approximation of the original voxel graph. A zero threshold results in a geometric graph that is equal to the original voxel graph. When the threshold is very large, no curve nodes are inserted and the topological graph is not refined.

Profile nodes

The test for finding the profile nodes uses the distance transform (DT) in a similar way as the test for finding the curve nodes. The DT is known at each voxel node and can be highly variable. If a linear profile is assumed along an edge, the actual values of the corresponding voxels will differ from this assumption. The profile test finds the voxel with the maximum distance to this linear profile and tests this distance to a threshold T_{prof} (see Figure 6). A new profile node is inserted when the threshold is exceeded and the two resulting edges are tested recursively.

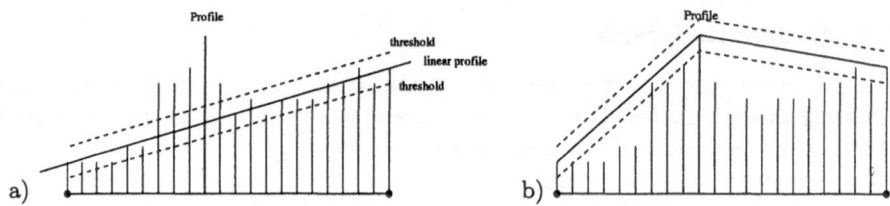

Fig. 6. Finding the profile nodes recursively: a) first step, and b) second step.

Again, the threshold is a measure for shape approximation of the original voxel graph, but this time for the profile of the surface of the object. Also, with a zero threshold the geometric graph is equal to the voxel graph, and with a very large threshold the topological graph remains unchanged.

Figure 7 shows the skeleton graph which results after graph simplification. In the figure the original segmentation is visualized with a transparent iso-surface, and the skeleton graph is visualized with spheres connected by lines. The spheres are located at the node positions and have a radius equal to the DT. In general the skeleton graph gives a good approximation of the shape. However, some objects are cut by the system boundary which results in a flat shape, in these cases the skeleton graph gives a less accurate approximation. The cross-section of a flat object has an elliptical contour, while we assume a circular contour with a radius equal to the distance transform.

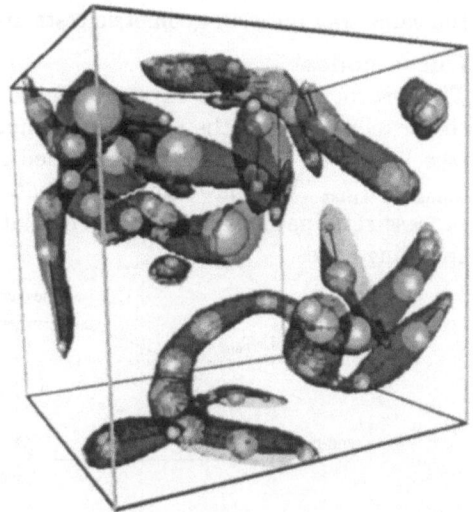

Fig. 7. A visualization of the final skeleton graph. The nodes and edges are shown by spheres and small tubes, and the iso-surface of the object is shown transparently.

5 Shape Reconstruction

The shape of the original object can be approximated by adding a volume to every edge in the skeleton graph. The resulting 3D shapes can be used for an iconic visualization of the features [8]. Several types of geometric objects can be used for "fleshing out" the skeleton graph. The topology is viewed best with an open geometry with for instance lines and spheres, such as shown in Figure 7. A solid structure of spherical and conical volumes gives a clear visualization of the volume of the objects, see Figure 8a). Some variations and intermediate representations to this are possible, one is shown in Figure 8b).

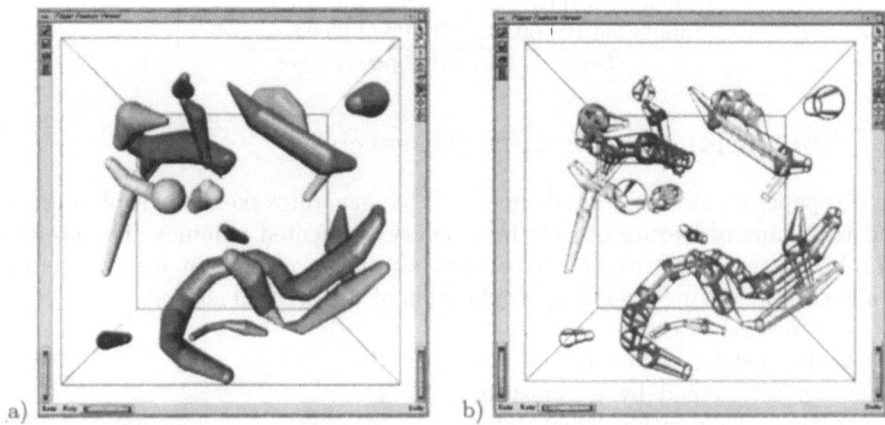

a)   b)

Fig. 8. Skeleton surface reconstruction using solid spheres and cones, or a more open geometry.

In another approach a cubic Hermite interpolating tube is calculated through each path (from end node to end node) in the geometric graph. The interpolation results in a smooth tube, see Figure 9. The thickness of the tube is taken equal to the distance information at the nodes. This way, a smooth surface is drawn representing the feature objects.

6 Applications

The skeleton generation procedure was applied to a dataset with turbulent vortex structures, obtained from a fluid dynamics simulation[1]. The dataset consists of 128^3 grid nodes with vorticity data. After segmentation, 18 objects were found with a total of 119262 voxels. Skeletonization reduced this number of voxels to 966, and after graph simplification only 87 nodes remained. The file size of the skeleton graph is about 2.3 Kb, while the original data file was 8.0 Mb, a reduction factor in the order of 1000. Figures 7, 8, and 9 show different visualizations of this application.

---

[1] Data courtesy D. Silver and X. Wang of Rutgers University.

We also applied the skeleton reconstruction to similar simulations of turbulent vortex structures with a higher Reynolds number. In total three similations were obtained with 128^3 grid nodes with vorticity data. A higher Reynolds number results in more complex and smaller vortex structures, and therefore the skeleton data is somewhat larger. Table 1 shows the reduction in percentages, from segmented volume (V) to voxel nodes (VN), from voxel nodes to geometric nodes(GN) and from volume to geometric nodes, the last column shows the file size of the skeleton file (SF) in Kb (N.B. the file size of the raw data is 8.0 Mb).

dataset	V→VN	VN→GN	V→GN	SF (Kb)
Simulation I	74.3%	81.0%	95.1%	12.3
Simulation II	79.7%	76.5%	95.2%	2.7
Simulation III	89.1%	91.1%	99.9%	2.3

Table 1. Reduction percentages.

7 Conclusions and Future Research

In this paper we have presented a method that generates skeleton graphs describing the shape of feature objects in a binary segmented volumes. The skeleton graph is a set of feature attributes that signifies a significant data reduction, while still preserving a good approximation of the original shapes.

The method works in four stages. First, the skeleton voxels of the objects are determined by an existing skeletonization algorithm. Then, neighboring skeleton voxels are connected into a voxel graph representation with nodes and edges. The nodes represent the voxels and the edges represent the connectivity between voxels. Using the distance transform, also the minimal distance to the surface is known in each voxel. Then, the voxel graph is simplified by recognizing topological nodes (junction nodes, end nodes, and loop nodes) and geometric nodes (curve nodes and profile nodes). The simplification can be controlled by two thresh-old variables, controlling the precision

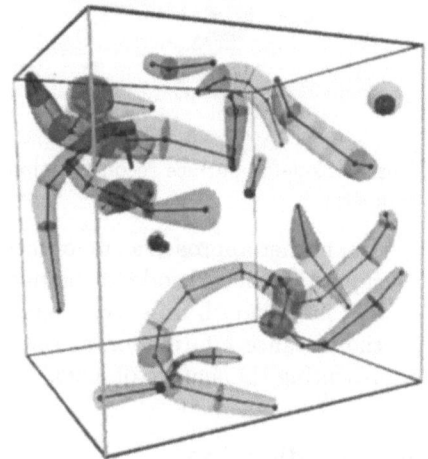

Fig. 9. Skeleton surface reconstruction using hermite tube icons.

of the skeleton graph approximation. Finally, the skeleton graph is used to reconstruct a geometry that approximates the surface of the segmented objects.

The applications with the turbulent vortex structures showed that this method approximates the original surface very well. The skeleton attributes provide a good topological and shape description in a very condensed way. Data reductions in the order of a 1000 were obtained. Figures 7, 8, and 9 show some nice visualizations of this application.

The turbulent vortex structures normally have a worm-like shape which can be very well approximated by this method, however sometimes the method fails. The cross-sections of the worms have a circular contour with a minimal radius equal to the DT. When the shape is more flat the contour has an elliptical shape and the circular approximation can be poor. Two solutions can overcome this problem: 1) for each skeleton voxel create an elliptical approximation, 2) use the surface skeletons (Figure 2b) and do a similar simplification as described in this paper. The implementation of these two solutions is a topic for future research.

Another topic for future research is using the skeleton graph information for feature comparison and tracking [7] in time-dependent data. The skeleton graph can be collected in successive time steps and stored as feature data. For tracking purposes, a metric needs to be defined for the correspondence between two skeleton graphs.

Acknowledgments

This work is supported by the Netherlands Computer Science Research Foundation (SION), with financial support of the Netherlands Organization for Scientific Research (NWO).

References

1. G. Borgefors. Distance Transformations in Arbitrary Dimensions. *Computer Vision, Graphics, and Image Processing*, 27(3):321–345, 1984.
2. P. E. Danielsson. Euclidean Distance Mapping. *Computer Graphics and Image Processing*, 14:227–248, 1980.
3. P. P. Jonker and A. M. Vossepoel. On Skeletonization Algorithms for 2, 3,.. N Dimensional Images. In D. Dori and A. Bruckstein, editors, *Proc. Shape, Structure and Pattern Recognition '94*, pages 71–80, Nahariya, Israel, Oct. 4-6 1995. World Scientific Singapore.
4. T. C. Lee, R. L. Kashyap, and C. N. Chu. Building Skeleton Models via 3-D Medial Surface/Axis Thinning Algorithms. *CVGIP: Graphical Models and Image Processing*, 56(6):462–478, November 1994.
5. F. Leymarie and M. D. Levine. Simulating the Grassfire Transform using an Active Contour Model. *IEEE Transactions on Pattern Analysis and Machine Intelligence*, 14(1):56–75, January 1992.
6. S. Logbregt, P. W. Verbeek, and F. C. A. Groen. Three-Dimensional Skeletonzation: Principle and Algorithm. *IEEE Transactions on Pattern Analysis and Machine Intelligence*, 2:75–77, January 1980.
7. F. Reinders, F.H. Post, and H.J.W. Spoelder. Attribute-Based Feature Tracking. In E. Gröller, H. Löffelmann, and W. Ribarsky, editors, *Data Visualization '99*, pages 63–72. Springer Verlag, 1999.
8. T. van Walsum, F.H. Post, D. Silver, and F.J. Post. Feature Extraction and Iconic Visualization. *Trans. on Visualization and Computer Graphics*, 2(2):111–119, 1996.
9. Y. Xia. Skeletonization via Realization of the Fire Front's Propagation and Extinction in Digital Binary Shapes. *IEEE Transactions on Pattern Recognition and Machine Intelligence*, 11(10):1076–1086, October 1989.

Editors' Note: see Appendix, p. 285 for colored figures of this paper

Progressive Volume Models for Rectilinear Data using Tetrahedral Coons Volumes

David J. Holliday and Gregory M. Nielson

Arizona State University, Tempe AZ 85287-5406, USA
holliday|nielson@asu.edu

Abstract. We present a new technique for modeling rectilinear volume data. The algorithm produces a trivariate model, $F(x, y, z)$, which is piecewise defined over tetrahedra that fits the volume data to within a user specified tolerance. The technique is adaptive leading to an efficient model that is more complex where the data demands it. The novelty of the present technique is that a valid tetrahedrization is not required. Tetrahedral cells are subdivided as required by the error condition only. This type of cellular decomposition leads to a continuous model by the use of a tetrahedral Coons volume which has the ability to interpolate to arbitrary boundary data.

1 Introduction

Visualization of a volume data set is used to gain some insight into the data. Most scientific visualization algorithms produce a graphical image (volume rendering [2]) or an entity that can then be rendered (isosurfacing [5]). In the case of volume rendering a useful image is generated but subsequent analysis of the original data is difficult using such output. Geometry, in the form of triangle meshes, is created as output from typical isosurfacing algorithms. This type of output is more amenable to analysis but it does not give a complete picture of the original volume data.

Much can be gained from modeling discrete volume data, i.e., generating an underlying mathematical representation for the data and using that representation for visualization tasks. Benefits of modeling volume data include the compression of large data sets, the application of visualization algorithms, and the ability to perform analyses or simulations.

1.1 Adaptive Approximations

Adaptive approximative models are useful because they can closely approximate portions of data that have large local variations without using too much information to also represent areas where the data is relatively smooth. One way of forming adaptive approximations to volume data is to define an initial, coarse approximation and apply successive refinements until the model closely approximates the data of interest. A model that consists of piecewise linear functions defined over a tetrahedrization is a popular choice for approximating volume data sets. Each vertex in the tetrahedrization has an associated scalar value, or weight, from which a continuous function inside each tetrahedron can be constructed.

There are several methods for performing local refinements on a tetrahedrization [1, 6]. Both algorithms maintain a valid tetrahedrization at all times. (See [9] for definitions and general background material for tetrahedral decompositions.) A tetrahedron is first selected for refinement, generally based on some local error criterion, and then a recursive rule is applied to also refine neighboring tetrahedra. This helps to ensure that two neighboring tetrahedra will always share a common face. Figure 1(a) shows an example of a tetrahedrization that has undergone several refinements using the algorithm presented in [1].

One problem with such a recursive closure scheme is that many tetrahedra that already closely approximate the data may be refined. Many more tetrahedra than necessary may be generated to construct an adequate approximation to the data. It would be advantageous to only subdivide those tetrahedra in which the error is large instead of also having to subdivide neighboring tetrahedra in order to maintain a valid tetrahedrization. Figure 1(b) shows an example of a tetrahedrization where no recursive closure rule is applied to neighbors after a tetrahedron has been refined. Each refined tetrahedron was split into eight sub-tetrahedra.

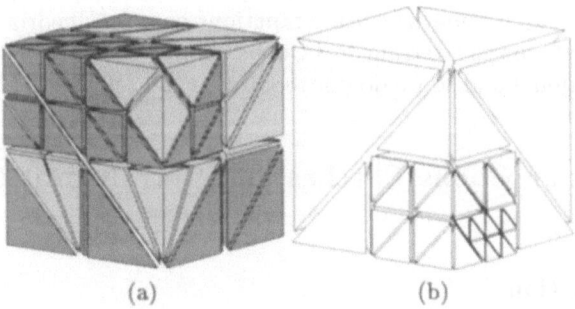

(a)   (b)

Fig. 1. Adaptively refined tetrahedrizations constructed (a) using the red-green algorithm and (b) by not refining neighbors.

If we switch to a refinement strategy which does not maintain a valid tetrahedrization then each tetrahedron cannot have a linear function defined over it and still result in a continuous function across the tetrahedrization. We need to use a different method for defining the volume model over this type of decomposition. In this work we propose the use of a tetrahedral Coons volume defined over each tetrahedron. This will allow us to ignore the incompatibilities between tetrahedra and refine only those tetrahedra that need to be refined based on local error estimates.

This work builds on that presented in [10]. The authors presented a new method for adaptively approximating terrain data and adaptively tessellating parametric surfaces using Coons patches. The idea was to adaptively refine triangles without worrying about also refining neighboring triangles. The approximation to the data then consisted of a collection of Coons patches defined over the triangulation.

1.2   Previous Work

Several different methods of approximating volume data have appeared in the literature. Tensor-product wavelets have been applied to regular volume data to generate multiresolution approximations [7, 8, 4]. Another approach, used by Grosso et. al. [3], is to generate coarse-to-fine approximations using piecewise linear functions over tetrahedrizations. This approach requires a valid tetrahedrization (i.e., one with no T-vertices, see [9]) which is obtained by using the red-green algorithm of [1]. A sequence of approximations of regular volume data can also be generated in a "bottom-up" manner as done in [13]. They start with a tetrahedrization of regular data and merge tetrahedra based on local error estimates. For a different approach to volume models over nonconforming meshes based upon projection operators, we alert the reader to the work of Ohlberger and Rumpf [12].

Our work bears some similarity to that presented in [3]. We generate coarse-to-fine adaptive approximations using functions over tetrahedrizations. The differences are that we use a tetrahedral Coons volume over each tetrahedron (versus linear functions) and so we do not require a valid tetrahedral decomposition.

2   Triangular Coons Patches and Tetrahedral Coons Volumes

2.1   Introduction

Coons patches and volumes have been used in geometric modeling to define patches or volumes that interpolate prescribed boundary curves or functions [9, 11]. The methods generate a smooth surface or volume by blending data from the boundaries in a systematic way. We will first cover triangular Coons patches and then present their extension to tetrahedral Coons volumes. These will then be used in tandem to form approximations to volume data.

2.2   Triangular Coons Patches

The domain of a triangular Coons patch is a triangle so it will be convenient to write points on the surface using barycentric coordinates. The barycentric coordinates of (u_0, u_1, u_2) of a point (x, y) are defined by the relation

$$\begin{bmatrix} x \\ y \end{bmatrix} = u_0 \mathbf{P_0} + u_1 \mathbf{P_1} + u_2 \mathbf{P_2} \tag{1}$$

$$1 = u_0 + u_1 + u_2 \tag{2}$$

where $\mathbf{P_0}$, $\mathbf{P_1}$, and $\mathbf{P_2}$ are vertices of the domain in $I\!\!E^2$.

In order to create a triangular Coons patch we require three compatible boundary curves. We make no requirements on the nature of the boundary curves so we will refer to them as functions defined in terms of boundary edges on the domain triangle. The three boundary curves are denoted as $\mathbf{F_0}(s)$, $\mathbf{F_1}(s)$, and $\mathbf{F_2}(s)$ where $\mathbf{F}$ is an underlying function defined on the boundary of a triangular domain and the parameter s varies from 0 to 1. The boundary curves must be compatible so we require, for example, that $\mathbf{F_0}(0) = \mathbf{F_2}(1)$.

The type of triangular Coons patch used here, the NTW linear/linear patch, was first presented in [11] and more recently appeared in [10]. A surface $\mathbf{S}$ is written as the sum of three components $\mathbf{S_0}(u_1, u_2)$, $\mathbf{S_1}(u_0, u_2)$ and $\mathbf{S_2}(u_0, u_1)$. These components are given by

$$\mathbf{S_0}(u_1, u_2) = \mathbf{F}(u_1 \mathbf{P_1} + (1 - u_1)\mathbf{P_0}) + \mathbf{F}(u_2 \mathbf{P_2} + (1 - u_2)\mathbf{P_0}) - \mathbf{F}(\mathbf{P_0}) \tag{3}$$
$$\mathbf{S_1}(u_0, u_2) = \mathbf{F}(u_0 \mathbf{P_0} + (1 - u_0)\mathbf{P_1}) + \mathbf{F}(u_2 \mathbf{P_2} + (1 - u_2)\mathbf{P_1}) - \mathbf{F}(\mathbf{P_1}) \tag{4}$$
$$\mathbf{S_2}(u_0, u_1) = \mathbf{F}(u_0 \mathbf{P_0} + (1 - u_0)\mathbf{P_2}) + \mathbf{F}(u_1 \mathbf{P_1} + (1 - u_1)\mathbf{P_2}) - \mathbf{F}(\mathbf{P_2}). \tag{5}$$

A point on the patch is written as a barycentric combination of the three components and is given by

$$\mathbf{S}(u_0, u_1, u_2) = u_0 \mathbf{S_0} + u_1 \mathbf{S_1} + u_2 \mathbf{S_2}. \tag{6}$$

Figure 2 illustrates a patch with piecewise linear boundaries.

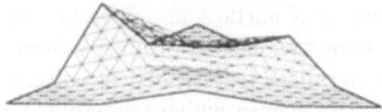

Fig. 2. NTW linear/linear patch that interpolates to piecewise linear boundaries.

2.3 Tetrahedral Coons Volumes

We would like to extend the idea of a triangular Coons patch to define a scalar-valued tetrahedral Coons volume. The volume will be written using barycentric coordinates over a tetrahedral domain and it will be constructed in such a way that it interpolates to four compatible scalar-valued boundary functions.

The tetrahedral Coons volume requires four compatible scalar-valued boundary functions. Each function is defined over a triangular domain and associates a scalar value with a point in its domain. If we denote the vertices of a tetrahedral domain as $\mathbf{P_0}$, $\mathbf{P_1}$, $\mathbf{P_2}$, and $\mathbf{P_3}$ then one compatibility requirement necessitates that $F(\mathbf{P_1}, \mathbf{P_2 P_3})$ and $F(\mathbf{P_0}, \mathbf{P_2 P_3})$ must have the same values along the edge $\mathbf{P_2}$ to $\mathbf{P_3}$ in the domain.

The tetrahedral Coons volume used here is an extension of the NTW linear/linear patch. A volume $V(u_0, u_1, u_2, u_3)$ is written as the sum of four components $V_0(u_1, u_2, u_3)$, $V_1(u_0, u_2, u_3)$, $V_2(u_0, u_1, u_3)$, and $V_3(u_0, u_1, u_2)$. The expression for $V_0(u_1, u_2, u_3)$ is given as follows

$$
\begin{aligned}
V_0(u_1, u_2, u_3) = {} & F(u_1 \mathbf{P_1} + u_2 \mathbf{P_2} + (1 - u_1 - u_2)\mathbf{P_0}) + \\
& F(u_2 \mathbf{P_2} + u_3 \mathbf{P_3} + (1 - u_2 - u_3)\mathbf{P_0}) + \\
& F(u_3 \mathbf{P_3} + u_1 \mathbf{P_1} + (1 - u_3 - u_1)\mathbf{P_0}) - \\
& F(u_1 \mathbf{P_1} + (1 - u_1)\mathbf{P_0}) - \\
& F(u_2 \mathbf{P_2} + (1 - u_2)\mathbf{P_0}) - \\
& F(u_3 \mathbf{P_3} + (1 - u_3)\mathbf{P_0}) + \\
& F(\mathbf{P_0}).
\end{aligned}
\tag{7}
$$

In a manner similar to the triangular Coons patch, the value of the volume corresponding to barycentric coordinates (u_0, u_1, u_2, u_3) is written as a convex combination of the four components and is given by

$$
V(u_0, u_1, u_2, u_3) = u_0 V_0 + u_1 V_1 + u_2 V_2 + u_3 V_3.
\tag{8}
$$

There are other types of triangular Coons surfaces which differ in the way in which the components of a surface are defined and combined [9, 11]. The transfinite scheme used here was chosen primarily because of its simplicity and the ease with which it generalizes to tetrahedral volumes.

We will next describe the way in which triangular Coons patches and tetrahedral Coons volumes will be used to approximate volume data. Using the ideas presented in this section we will be able to build models of volume data in which the need for maintaining a valid tetrahedrization is eliminated.

3 Adaptive Approximations using Tetrahedral Coons Volumes

3.1 Algorithm

The algorithm that we present will be used to adaptively approximate regular volume data. We require that the data set is of size $(2^n + 1) \times (2^n + 1) \times (2^n + 1)$.

The reason for this is that vertices in the tetrahedrization will be associated with points in the input. As tetrahedra are refined, new vertices will continue to correspond to data points. In particular, a tetrahedron will be refined by splitting each edge at its midpoint and joining those to form new tetrahedra. Because of the special size of the data set, the midpoint of each edge will also correspond to a data point.

We will now describe the face functions that will be used in order to define tetrahedral Coons volumes over a tetrahedrization like that shown in Figure 1(b). Each face function is scalar-valued and is defined over a triangular domain. Because a tetrahedron may have one of its faces shared by many other tetrahedra, the domain of a face function can be thought of as an adaptive triangular decomposition.

In order to define a continuous function across an adaptively refined domain we define a triangular Coons patch over each triangle in the domain of a face function. The portions of the function that correspond to triangles in the domain without T-vertices are simply planar triangles. If all the face functions for a tetrahedron are single triangles without adaptive refinements then the Coons volume evaluates to a linear function over the tetrahedron.

The algorithm to adaptively approximate volume data is shown in Figure 3. Since the original data is being approximated, we require a user-specified error tolerance, ϵ, for the fitting process. The initial tetrahedrization for the approximation consists of the unit cube tetrahedrized into six tetrahedra where each tetrahedron shares the main diagonal of the cube. We also define a maximum level that prevents the tetrahedra from being subdivided too many times.

There are several strategies for determining if a tetrahedron requires refinement. We refine a tetrahedron if the difference between the weight associated with any data point inside it and the value of the Coons volume evaluated at the data point's location exceeds a tolerance. Different strategies might include comparing the average or the median of the differences to a threshold.

3.2 Results

We demonstrate our adaptive approximation method on two data sets. The first is a synthetic data set of size 33^3 where the dependent values, w, have been computed by

$$w = f(x, y, z) = \frac{1}{2}e^{(-10((x-0.25)^2+(y-0.25)^2))}$$

$$+\frac{3}{4}e^{(-16((x-0.25)^2+(y-0.25)^2+(z-0.25)^2))}$$

$$+\frac{1}{2}e^{(-10((x-0.75)^2+(y-0.125)^2+(z-0.5)^2))}$$

$$-\frac{1}{4}e^{(-20((x-0.75)^2+(y-0.75)^2))}. \tag{9}$$

The following table summarizes the fitting process for this data set. The user-specified tolerance is given followed by the number of vertices and tetrahedra

```
repeat
{
    for each tet Tᵢ ∈ 𝒯
    {
        for each data point P = (x, y, z; w) inside Tᵢ
        {
            // convert (x,y,z) to barycentric coordinates with respect to Tᵢ
            // evaluate point using Coons volume associated with tet Tᵢ
            // see equation (8)
            ŵ := V_{Tᵢ}(u₀, u₁, u₂, u₃)
            if (|w − ŵ| > ε)
            {
                mark Tᵢ for refinement
            }
        }
    }
    for each tet T marked for refinement
    {
        if (T not at maximum level)
        {
            split T into eight sub-tetrahedra
        }
    }
} until no tets refined
```

Fig. 3. Algorithm for adaptively approximating regular volume data.

in the resulting tetrahedrization. The number of volumes indicates the number of tetrahedra that have non-trivial face functions and must be evaluated as tetrahedral Coons volumes.

tolerance	vertices	tetrahedra	volumes	rms error
0.05	756	1455	558	0.0283
0.02	3110	6173	2220	0.00914
0.01	7659	16141	4934	0.00421

Table 1. Statistics for the approximations shown in Figure 4.

In order to visualize the approximations we performed a regular sampling of the Coons volumes and then use marching cubes to generate isosurfaces and normals for shading. The sampled data was also used to compute the rms errors. They were computed by taking the sum of the squares of the differences between the weight associated with a data point and the value of the Coons volume evaluated at that location.

Figure 4 shows the results of the fitting process for this data set. The top row of the figure shows isosurfaces from the original data set (the left is threshold 0.21 and the right is 0.50). The next three rows are the approximations (tetrahedrizations and isosurfaces) using tolerances 0.05, 0.02, and 0.01, respectively.

The second data set is a 33^3 data set from a MRI scan. The top row of Figure 5 shows an isosurface computed from the data set (isosurface threshold 0.095). The following table summarizes the statistics for several approximations of this data set.

tolerance	vertices	tetrahedra	volumes	rms error
0.07	18493	35452	11181	0.0232
0.03	26081	63712	10432	0.00398

Table 2. Statistics for the approximations shown in Figure 5.

Figure 5 shows results of applying our adaptive approximation algorithm to this data set. The top row is an isosurface computed from the original data. Each row thereafter shows an approximation in the form of a tetrahedrization and an isosurface for tolerances 0.07 and 0.03. The isosurface threshold is the same as that used on the original data set.

4 Summary

We have presented a method of performing adaptive approximations of regular volume data using tetrahedral Coons volumes. The advantage of using Coons volumes over existing approaches is that a valid tetrahedrization does not need to be maintained. Only those tetrahedra in which the error is large (i.e., areas where the model does not adequately approximate the data) need to be refined instead of also needing to refine neighboring tetrahedra like existing local refinement algorithms.

Future work includes applying a least squares approach to the fitting process to approximate data that does not meet the size requirements as given in Section 3.1.

5 Acknowledgments

We wish to acknowledge the support of the Office of Naval Research under grant N00014-97-1-0243, the support of the National Aeronautical and Space Administration under NASA-Ames Grant NAG 2-990, and the support of the National Science Foundation under grant IIS 9980166.

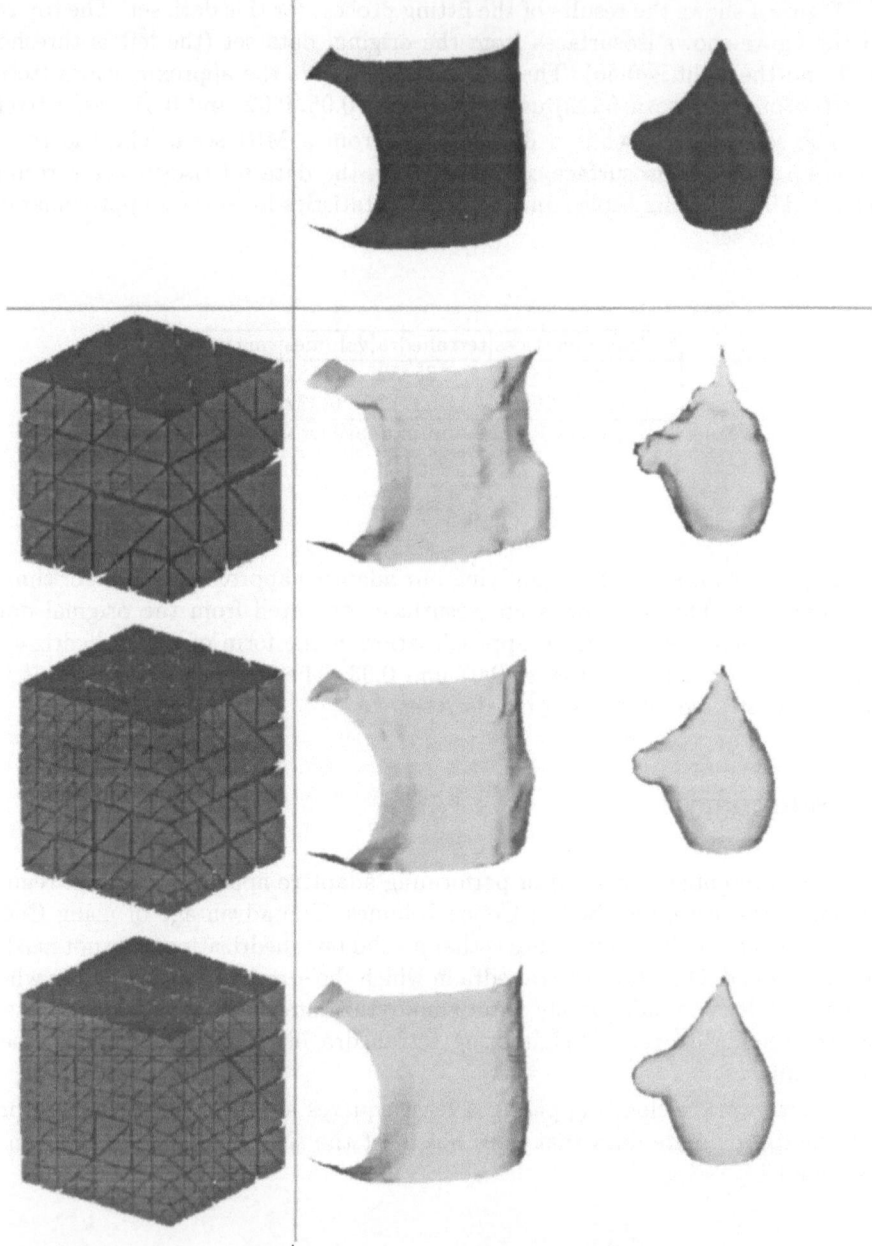

Fig. 4. An adaptively approximated approximated synthetic data set. The top row shows isosurfaces of the original data set. The next rows show the tetrahedrization resulting from a fit and the isosurfaces computed from the model.

References

1. Jürgen Bey. Tetrahedral mesh refinement. *Computing*, 55(4):355–378, 1995.
2. Robert A. Drebin, Loren Carpenter, and Pat Hanrahan. Volume rendering. In John Dill, editor, *SIGGRAPH 88 Conference Proceedings*, Annual Conference Series, pages 65–74. ACM SIGGRAPH, Addison Wesley, August 1988.
3. Roberto Grosso, Christoph Lürig, and Thomas Ertl. The multilevel finite element method for adaptive mesh optimization and visualization of volume data. In Roni Yagel and Hans Hagen, editors, *Proceedings of Visualization '97*, pages 387–394. IEEE Computer Society Press, Novemeber 1997.
4. Insung Ihm and Sanghun Park. Wavelet-based 3D compression scheme for interactive visualization of very large volume data. *Computer Graphics Forum*, 18(1):3–15, March 1999.
5. William E. Lorensen and Harvey E. Cline. Marching cubes: A high resolution 3D surface construction algorithm. In Maureen C. Stone, editor, *SIGGRAPH 87 Conference Proceedings*, Annual Conference Series, pages 163–169. ACM SIGGRAPH, Addison Wesley, July 1987.
6. Joseph M. Maubach. Local bisection refinement for N-simplicial grids generated by reflection. *SIAM Journal on Scientific Computing*, 16(1):210–227, January 1995.
7. Shigeru Muraki. Approximation and rendering of volume data using wavelet transforms. In Arie E. Kaufman and Gregory M. Nielson, editors, *Proceedings of Visualization '92*, pages 21–28. IEEE Computer Society Press, Oct 1992.
8. Shigeru Muraki. Volume data and wavelet transforms. *IEEE Computer Graphics and Applications*, 13(4):50–56, July 1993.
9. Gregory M. Nielson, Hans Hagen, and Heinrich Müller. *Scientific Visualization: Overview, Methodologies, and Techniques*, chapter Tools for Triangulations and Tetrahedrizations, pages 429–525. IEEE Computer Society, Los Alamitos, CA, 1997.
10. Gregory M. Nielson, David J. Holliday, and Tom Roxborough. Cracking the cracking problem with Coons patches. In David Ebert, Markus Gross, and Bernd Hamann, editors, *Proceedings of Visualization '99*. IEEE Computer Society Press, Novemeber 1999.
11. Gregory M. Nielson, Donald H. Thomas, and James A. Wixom. Boundary data interpolation on triangular domains. Technical Report GMR-2834, General Motors Research Laboratories, 1978.
12. Mario Ohlberger and Martin Rumpf. Adaptive projection operators in multiresolution scientific visualization. *IEEE Transaction on Visualization and Graphics*, 4(4):344–364, 1998.
13. Yong Zhou, Baoquan Chen, and Arie Kaufman. Multiresolution tetrahedral framework for visualizing regular volume data. In Roni Yagel and Hans Hagen, editors, *Proceedings of Visualization '97*, pages 135–142, October 1997.

Hardware Accelerated Wavelet Transformations

Matthias Hopf   Thomas Ertl

{hopf,ertl}@informatik.uni-stuttgart.de
http://wwwvis.informatik.uni-stuttgart.de/

Visualization and Interactive Systems Group, University of Stuttgart

Abstract. Wavelets and related multiscale representations are important means for edge detection and processing as well as for segmentation and registration. Due to the computational complexity of these approaches no interactive visualization of the extraction process is possible nowadays. By using the hardware of modern graphics workstations for accelerating wavelet decomposition and reconstruction we realize a first important step for removing lags in the visualization cycle.

1   Introduction

Feature extraction has been proven to be a useful utility for segmentation and registration in volume visualization [7, 13]. Many edge detection algorithms used in this step employ wavelets or related basis functions for the internal representation of the volume. Additionally, wavelets can be used for fast volume visualization [5] using the Fourier rendering approach [8, 12].

Wavelet analysis is a mainly memory bound problem. Graphics hardware on the other hand regularly has memory systems that can be addressed extremely fast. As modern graphics hardware of several vendors, for instance Silicon Graphics [9], has support for two dimensional convolution and the ability to scale bitmaps by arbitrary factors, all necessary steps needed for wavelet decomposition and reconstruction are available.

Additionally, three dimensional convolution with separable filter kernels can be implemented by using these hardware supported convolution filters along with volume textures [3], paving the way to 3D wavelet analysis, which will benefit from the high memory bandwidth of the graphics hardware even more.

However, there are still several pitfalls to be circumvented, which are addressed in our previous paper about the first steps to hardware based wavelet analysis [4]. In this paper, we will emphasize new algorithmic aspects of the acceleration process by utilizing special OpenGL features.

2   Wavelets

In the past two decades, wavelet analysis has grown from a mathematical curiosity into a major source of new basis decomposition and signal processing algorithms [10, 14]. The importance of orthonormal basis of wavelets and multi-resolution analysis resides in their hierarchical nature, which offers a mathematical framework for describing functions at different levels of resolution. Using basis functions with good approximation properties, i.e. with many vanishing moments, one can represent functions by keeping only the important coefficients (regularly called *features*) and discarding all others. This sections gives a short introduction into the basics of wavelet theory. Details on the theory can be found in [1, 2, 6].

A multi-resolution analysis can be thought of as a ladder of approximating closed subspaces $(V_j)_{j \in \mathbf{Z}}$ of $L^2(\mathbf{R})$. The functions in these subspaces have well defined scaling and translation properties. Furthermore, there exists a function $\phi \in V_0$ such that $\{\phi_{0,n}; j, n \in \mathbf{Z}\}$ with $\phi_{j,n} = 2^{j/2}\phi(2^j x - n)$ is an orthonormal basis of V_0. Under these conditions one can construct an orthonormal wavelet basis $\{\psi_{j,n}; j, n \in \mathbf{Z}\}$ with $\psi_{j,n} = 2^{j/2}\psi(2^j x - n)$, such that for any function f in $L^2(\mathbf{R})$

$$P_j f = P_{j-1} f + Q_{j-1} f , \tag{1}$$

where P_j and Q_j are the orthogonal projections onto V_j and W_j, respectively:

$$P_j f = \sum_{n \in \mathbf{Z}} < f, \phi_{j,n} > \phi_{j,n} , \quad Q_j f = \sum_{n \in \mathbf{Z}} < f, \psi_{j,n} > \psi_{j,n} .$$

The function ψ is sometimes called the *mother* wavelet. The projection $P_j f$ onto the subspaces V_j corresponds to the different resolution levels in which the function f can be decomposed. These projections contain the *smooth* information of f at a given level of resolution. The projections $Q_j f$ onto the subspaces W_j spanned by the $\psi_{j,n}$ represent the *detail* information of f required to move from one resolution approximation subspace to the next finer one. Equation (1) is the wavelet decomposition of the function f. The *scaling* function ϕ satisfies the *two-scale* relation

$$\phi = \sum_n h_n \phi_{1,n} , \tag{2}$$

which is a discrete *low-pass filter* operation with the filter $\{h_n\}_{n \in \mathbf{Z}}$.

Now we start with a scale approximation $f^{j+1} = P_{j+1} f$ of a function f in V_{j+1} and decompose it into a coarser approximation in V_j. Due to the fact that $V_{j+1} = V_j \oplus W_j$, we have $f^{j+1} = f^j + \delta^j$, where $\delta^j = Q_j f$. In terms of the orthonormal bases $\{\phi_{j,n}\}_{n \in \mathbf{Z}}$ and $\{\psi_{j,n}\}_{n \in \mathbf{Z}}$, we have

$$f^j = \sum_n c_n^j \phi_{j,n} , \quad \delta^j = \sum_n d_n^j \psi_{j,n} ,$$

where the relation between the coefficients of the two levels of resolution is given by

$$c_n^{j-1} = \sum_k h_{k-2n} c_k^j , \quad d_n^{j-1} = \sum_k g_{k-2n} c_k^j \tag{3}$$

and $g_n = (-1)^n h_{1-n}$. h and g are the low-pass and high-pass filters, respectively. The decimation by a factor 2 corresponds to a down-sampling when going from one level to the next coarser one. This decomposition can be continued using the relation $V_{j+1} = V_j \oplus W_j$ and so on until a given level $J < j$, obtaining the following approximation for f:

$$f^{j+1} = \delta^j + \cdots + \delta^{J+1} + \delta^J + f^J$$

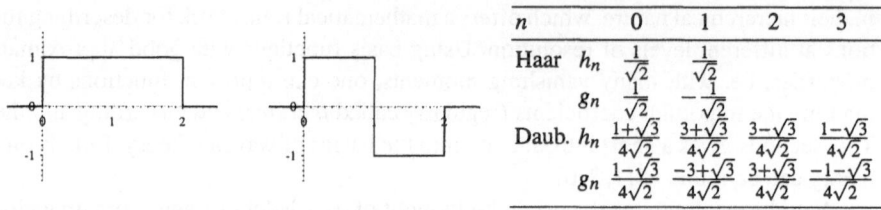

n	0	1	2	3
Haar h_n	$\frac{1}{\sqrt{2}}$	$\frac{1}{\sqrt{2}}$		
g_n	$\frac{1}{\sqrt{2}}$	$-\frac{1}{\sqrt{2}}$		
Daub. h_n	$\frac{1+\sqrt{3}}{4\sqrt{2}}$	$\frac{3+\sqrt{3}}{4\sqrt{2}}$	$\frac{3-\sqrt{3}}{4\sqrt{2}}$	$\frac{1-\sqrt{3}}{4\sqrt{2}}$
g_n	$\frac{1-\sqrt{3}}{4\sqrt{2}}$	$\frac{-3+\sqrt{3}}{4\sqrt{2}}$	$\frac{3+\sqrt{3}}{4\sqrt{2}}$	$\frac{-1-\sqrt{3}}{4\sqrt{2}}$

Fig. 1. The Haar scaling function, wavelet, and filter coefficients for Haar and Daubechies (4)

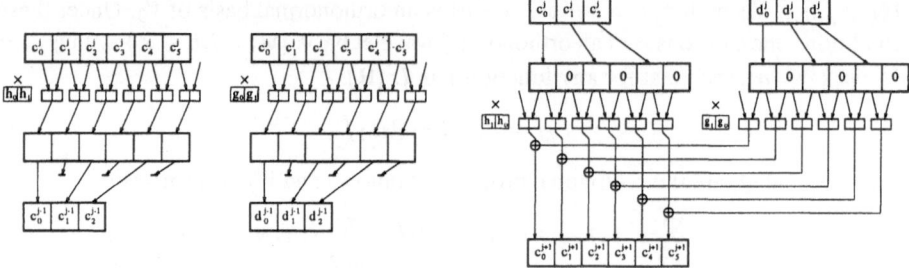

Fig. 2. Decomposition using Haar wavelets   **Fig. 3.** Reconstruction using Haar wavelets

The inverse operation, the reconstruction of f^{j+1} from f^j and δ^j, is simply given by:

$$c_k^{j+1} = \sum_n (h_{k-2n}\, c_n^j + g_{k-2n}\, d_n^j) \qquad (4)$$

Now let us take a look at an example. The simplest possible wavelet is the *Haar* wavelet. Figure 1 depicts its scaling function and the mother wavelet together with the filter coefficients.

We will now decompose a set of coefficients c_k^j into the c_k^{j-1} of the next coarser level. In Figure 2 the decomposition process is explained. The input data are convolved with the filter kernels h_n and g_n and down-sampled by a factor of 2. This process can be continued with the low-pass filtered coefficients c_k^{j-1}, until only one coefficient is left.

In order to reconstruct the original signal, the low- and high-pass filtered coefficients are processed as shown in Figure 3. The coefficients are up-sampled and then convolved with the reverted filter kernels according to (4).

So far we have only dealt with one-dimensional data. For higher dimensions bases which are tensor products of the one-dimensional case are used. There exist other approaches for selecting orthogonal basis functions, but tensor product wavelets are easier to understand and faster to compute.

3 The Rendering Pipeline

As it can be directly derived from Equations (3) and (4), wavelet decomposition is practically done by an input signal filtering and a down-sampling step. Reconstruction on the other hand is performed by first up-sampling and filtering afterwards. Modern

96

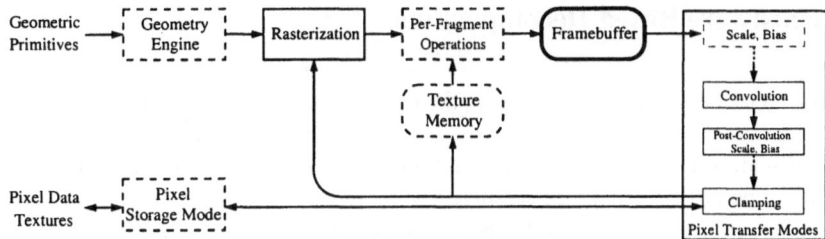

Fig. 4. The OpenGL graphics pipeline

graphics hardware supports filtering and scaling (resampling) for image transfer operations, which we will utilize for hardware based wavelet decomposition and reconstruction. The relevant part of the the OpenGL graphics pipeline is depicted in Figure 4.

In order to map the wavelet transformation onto the graphics hardware, we will use a mathematical specification of the graphics pipe. A more elaborated model has been derived in [4]. Let us consider the relevant parts of the graphics pipeline for image data. When a rectangular part of the frame buffer is to be copied from a source area, its color values are piped through the pixel transfer system and the rasterizer, before they are written to the destination area. Pixel transfer includes scaling and biasing of the color values, convolution with a prior defined filter kernel and clamping to the usual color value range $[0, 1)$. The rasterizer transposes the input image to the designated destination area while zooming it with arbitrary zoom factors, in other words, it performs up- and down-sampling.

Now let p^{n+1} be the pixel data that results from a graphical operation on p^n. For simplification we will assume that p^n is one-dimensional. A first approximation of the relevant part of the graphics pipeline can be written as a composition of a convolution (co), a clamping step (cl), a transposition (tr), and the scaling step (sc):

$$p^{n+1} = \text{sc} \circ \text{tr} \circ \text{cl} \circ \text{co}\,(p^n) \tag{5}$$

$$\text{sc}(p_i) = p_{\lfloor zi \rfloor} \tag{6}$$

$$\text{tr}(p_i) = p_{i-x_s+x_d} \tag{7}$$

$$\text{cl}(p_i) = \max(0, \min(1, p_i)) \tag{8}$$

$$\text{co}(p_i) = s \cdot \sum_{j=0}^{m} k_j\, p_{i+j} + b\,, \tag{9}$$

with zoom z, source x_s and destination x_d position, scaling s, and bias b parameters, and with a convolution kernel k of size m. As explained above, (co) and (cl) are performed in the pixel transfer system, while (tr) and (sc) describe the task of the rasterizer.

These equations are applied to pixels p_i^{n+1} of the destination area $i \in [x_d, (x_d + w + 1 - m) \cdot z)$, with w being the image size. The remaining pixels stick to their old values, that is, they are equal to p_i^n.

As we now have a mathematical model of the rendering pipeline, we can address the problem of mapping wavelet transformations onto the hardware as the next logical step.

4 Hardware Based Decomposition

Compared to the order of operations in the graphics pipeline, of which the relevant part is depicted in Figure 4, wavelet decomposition fits neatly into its scheme. Remembering that scaling is a part of the rasterization process, convolution is performed in the graphics pipe just before image scaling.

When we write the wavelet decomposition (3) as

$$\check{c}_n^{j-1} = \sum_i h_i c_{n+i}^j, \quad \check{d}_n^{j-1} = \sum_i g_i c_{n+i}^j, \tag{10}$$

$$c_n^j = \check{c}_{2n}^j, \quad d_n^j = \check{d}_{2n}^j \tag{11}$$

and compare it to Equations (5) to (9), we see that each of the wavelet decomposition filter steps matches the calculations of the OpenGL graphics pipe perfectly, except for the clamping steps. Clamping introduces several problems to these algorithms, that have to be addressed by using arbitrary scale and bias parameters. This aspect is discussed in detail in [4]. (6) implements the down-scaling in (11) and (10) can be expressed with the convolution filters (9).

One thing to note is that the image data p_j^n as well as the filter kernel k_j are only defined for $j \geq 0$. The filter kernel size is further limited by hardware specific constants, which are rather small. Thus it is necessary to displace the filter kernel and the input and output image specifications before invocation. Of course, the displacement has to be compensated in the final convolution step.

The input data have to be convolved using two different filters, so either the resulting images have to be written to another part of the frame buffer, just like in our earlier approach, or they have to be done together in one step. Now remember that we are actually dealing with 2D images. When we combine both tensor product steps with the two different filters, we get a total of four filters that have to be applied to the data.

As the graphics pipeline works on RGBA images nevertheless, it seems to be straightforward to use RGBA convolution filters instead of luminance only filters to combine these four steps into one as depicted in Figure 5. This will speed up the decomposition significantly, as the raster manager needs to address only one fourth of the number of pixels of the previous mentioned approach, and the convolution pipeline is implemented for color filters anyway. Additionally, we do not have to copy the source image in order to save it for the second filter, which makes for another factor of two.

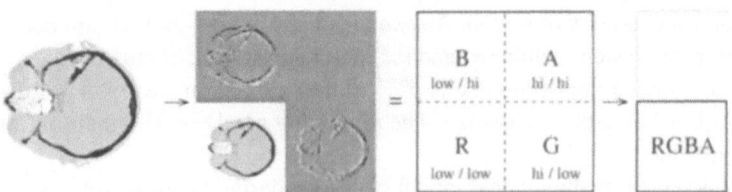

Fig. 5. Using one RGBA convolution instead of four different luminance only convolutions

Create convolution filter: $\tilde{h}_j = h_{j+\alpha}$, $\tilde{g}_j = g_{j+\alpha}$,

$\mathbf{f}^R_{j,k} = \tilde{h}_j \cdot \tilde{h}_k$, $\mathbf{f}^G_{j,k} = \tilde{g}_j \cdot \tilde{h}_k$, $\mathbf{f}^B_{j,k} = \tilde{h}_j \cdot \tilde{g}_k$, $\mathbf{f}^A_{j,k} = \tilde{g}_j \cdot \tilde{g}_k$ $\forall j, k$.

Calculate scaling **s** and bias **b** .   Set post-convolution scaling to **s** .

Set post-convolution bias to **b** .

Set pixel zoom to 1.0×1.0 .   Set color matrix to $\begin{pmatrix} 1 & 1 & 1 & 1 \\ 0 & 0 & 0 & 0 \\ 0 & 0 & 0 & 0 \\ 0 & 0 & 0 & 0 \end{pmatrix}$.

Copy area $[\delta_x + \alpha + i_x , \delta_x + \alpha + i_x + w_x + \Delta - 1) \times [\delta_y + \alpha + i_y , \delta_y + \alpha + i_y + w_y + \Delta - 1)$ to $[o_x , o_x + w_x + \Delta - 1) \times [o_y , o_y + w_y + \Delta - 1)$.

Set pixel zoom to 0.5×0.5 .   Disable color matrix.

Copy area $[o_x , o_x + w_x + \Delta - 1) \times [o_y , o_y + w_y + \Delta - 1)$ to $[o_x , o_x + \frac{1}{2} w_x) \times [o_y , o_y + \frac{1}{2} w_y)$, using convolution filter **f** (size Δ^2).

---

h_j, g_j Low- and high-pass filters, respectively
α Index of first non-zero element of both filters
Δ Size of filters
δ Shift offset (see text)
i, w Input image offset and size
o Output image offsets

Fig. 6. Implementation sequence for wavelet decomposition in hardware

However, it turns out that we still have to copy the source image, because OpenGL does not provide a pre-convolution color matrix, which would be necessary to provide the same information to the four different filters. As we want to address only the low-pass filtered data of the previous step, which is stored in the red component of the calculated image, we have to spread this information to all four color channels using SGI's color matrix OpenGL extension before invoking the convolution filter. Still, we have the advantage of better utilization of the graphics pipe.

Unfortunately, OpenGL is no pixel exact specification. In particular, zooming is only well defined according to (6) for up-sampling, that is for zoom factors greater than one. When images are scaled down, it is up to the implementation which pixels to transfer. We have found that even the implementations of one vendor — Silicon Graphics in our case — vary from architecture to architecture. In order to address this problem, a so-called *shift offset* δ is determined. When added to the specification of the source image's left edge, it corrects the internal pixel offset. Currently the only way to determine the shift offset is to draw a scaled-down version of a well-known image for several different shift values and to read it back afterwards for comparison with the desired result.

Additionally, care has to be taken at the borders of the input image. Several strategies have already been discussed, with blanking being the easiest and input mirroring being one of the best methods in order to suppress high frequencies that are not part of the image, but introduced by aliasing effects.

Finally, Figure 6 shows the implementation sequence for wavelet decomposition using graphics hardware. The calculation of the scaling and bias values, which is left out here for clarity, is discussed in detail for the one dimensional case in [4].

5 Hardware Based Reconstruction

In contrast to the decomposition algorithm, wavelet reconstruction is much more complicated, because according to Equation (4) scaling and convolution is to be performed in inverse order compared to the rendering pipeline (Figure 4). Either scaling and convolution have to be performed in separate rendering steps, or the filters have to be split and special care has to be taken in order to render even and odd pixel positions separately. Either way, reconstruction is more complicated than decomposition.

Moreover, due to different scaling and bias values for odd and even pixels, using separate rendering steps is not a feasible option. Therefore, we will concentrate on the second possibility of splitting the filters.

Now we examine the wavelet reconstruction (4). In order to simplify the expression, we have to distinguish between k being even and odd. For even k we substitute h_{k-2n} using $h_n^{ev} = h_{-2n}$ (g accordingly) and get

$$\check{c}_n^{j+1} = \sum_i (h_i^{ev} c_{i+n}^j + g_i^{ev} d_{i+n}^j) \,, \tag{12}$$

$$c_k^{j+1} = c_{2n}^{j+1} = \check{c}_n^{j+1} \,. \tag{13}$$

For odd k we use $h^{od} = h_{1-2n}$, which results in

$$\hat{c}_n^{j+1} = \sum_i (h_i^{od} c_{i+n}^j + g_i^{od} d_{i+n}^j) \,, \tag{14}$$

$$c_k^{j+1} = c_{2n+1}^{j+1} = \hat{c}_n^{j+1} \,. \tag{15}$$

Again, we will concentrate on the low pass filtered data first and simply neglect g in the terms above. We can see that (13) and (15) can be performed by setting according zoom factors in (6). (12) and (14) can be implemented in (9) by choosing h^{ev} and h^{od} as filter kernels, respectively.

Of course, when rendering the odd coefficients, we have to make sure that we do not overwrite the previously rendered even coefficients. OpenGL knows about a so-called *stencil* buffer, which provides masking tests in the per-fragment operation part of the graphics pipeline. The stencil buffer has to be initialized with a striped pattern only once, after that the stencil test can be set to render even or odd pixels only. We activate the test for rendering odd pixels only due to speed reasons, as each activated test can slow down the rendering process.

Up to now we have only dealt with the low-pass filtered coefficients c_n^j. As we have the necessary hi-pass filtered coefficients d_n^j stored as another component of the same pixels, we can use SGI's color matrix extension to combine them. Again, we use all four red, green, blue, and alpha components in order to work on 2D tensor product wavelets in one step. This is different to our previous approach, where we treated the different coefficients in separate steps. The new approach is not only faster, but even more accurate, because color matrix operations are performed with higher precision than blending operations in the frame buffer, and we do not have to deal with clamping artifacts in this case either. We disable rendering to the green, blue, and alpha channels in order to not overwrite the hi-pass filtered coefficients there, which will be needed in the next reconstruction step.

Create convolution filters:

$$\tilde{h}_j^{ev} = h_{2\lfloor \frac{\alpha_h + \Delta_h}{2} \rfloor - 2j} , \quad \tilde{h}_j^{od} = h_{2\lceil \frac{\alpha_h + \Delta_h}{2} \rceil - 2j+1} , \quad \tilde{g}_j^{ev} = g_{2\lfloor \frac{\alpha_g + \Delta_g}{2} \rfloor - 2j} , \quad \tilde{g}_j^{od} = g_{2\lceil \frac{\alpha_g + \Delta_g}{2} \rceil - 2j+1} .$$

$$\mathbf{f}_{j,k}^{x,y,R} = \tilde{h}_j^x \cdot \tilde{h}_k^y , \quad \mathbf{f}_{j,k}^{x,y,G} = \tilde{g}_j^x \cdot \tilde{h}_k^y , \quad \mathbf{f}_{j,k}^{x,y,B} = \tilde{h}_j^x \cdot \tilde{g}_k^y , \quad \mathbf{f}_{j,k}^{x,y,A} = \tilde{g}_j^x \cdot \tilde{g}_k^y \quad \forall j,k, \forall x,y \in \{ev, od\} .$$

$$\delta^{ev} = -\lfloor \tfrac{\alpha + \Delta - 1}{2} \rfloor , \quad \delta^{od} = 1 - \lceil \tfrac{\alpha + \Delta - 1}{2} \rceil , \quad \Delta^{ev} = -\delta^{ev} - \lceil \tfrac{\alpha}{2} \rceil + 1 , \quad \Delta^{od} = -\delta^{od} - \lfloor \tfrac{\alpha}{2} \rfloor + 1 .$$

Calculate scaling $\mathbf{s}$ and bias $\mathbf{b}^{x,y}, x,y \in \{ev, od\}$.

Set pixel zoom to 2.0×2.0 . Enable rendering to R only, disable rendering to G, B, and A.

Set color matrix to $\begin{pmatrix} 1 & 0 & 0 & 0 \\ 1 & 0 & 0 & 0 \\ 1 & 0 & 0 & 0 \\ 1 & 0 & 0 & 0 \end{pmatrix}$. Initialize stencil buffer with $\begin{cases} 0 & x \text{ even}, y \text{ even} \\ 1 & x \text{ odd}, y \text{ even} \\ 2 & x \text{ even}, y \text{ odd} \\ 3 & x \text{ odd}, y \text{ odd} \end{cases}$.

Disable stencil test. Set post-convolution scaling and bias to $\bar{\mathbf{s}}$ and $\bar{\mathbf{b}}^{ev,ev}$.

Copy area $[i_x + \delta^{ev} , i_x + \delta^{ev} + w_x + \Delta^{ev} - 1) \times [i_y + \delta^{ev} , i_y + \delta^{ev} + w_y + \Delta^{ev} - 1)$

to $[o_x , o_x + \tfrac{1}{2} w_x) \times [o_y , o_y + \tfrac{1}{2} w_y)$, using convolution filter $\mathbf{f}^{ev,ev}$ (size $\Delta^{ev} \times \Delta^{ev}$) .

Do $\forall x,y \in \{ev, od\}$:

Enable stencil test, render only pixels with stencil value $\begin{cases} 1 & x = od, y = ev \\ 2 & x = ev, y = od \\ 3 & x = od, y = od \end{cases}$.

Set post-convolution bias to $\bar{\mathbf{b}}^{x,y}$.

Copy area $[i_x + \delta^x , i_x + \delta^x + w_x + \Delta^x - 1) \times [i_y + \delta^y , i_y + \delta^y + w_y + \Delta^y - 1)$

to $[o_x , o_x + \tfrac{1}{2} w_x) \times [o_y , o_y + \tfrac{1}{2} w_y)$, using convolution filter $\mathbf{f}^{x,y}$ (size $\Delta^x \times \Delta^y$) .

h_j, g_j	Low- and high-pass filters, respectively
α	Index of first non-zero element of both filters
Δ	Size of both filters
o_c, o_d, w	Input image offsets and size
o_o	Output image offset

Fig. 7. Implementation sequence for wavelet reconstruction in hardware

As we are up-sampling during reconstruction, we do not have to care about any shift offsets during zooming, as the OpenGL specification is pixel exact in this case. However, we have to care about the fact that hardware filter kernels h_k are only to be specified for non-negative k. Together with the problem of odd sized filter kernels this leads to quite horrible filter kernel specifications, which can be noted in the implementation sequence in Figure 7. Again, the scaling and bias values that have to be computed here have been discussed in detail in our previous paper. Care has to be taken about image borders as well. The policy here depends heavily on the policy taken during the decomposition step. Note that Haar wavelets are quite uncomplicated, as the reconstruction filters have the size 1, which is a mere scaling.

6 Results

Table 1 reveals that Hardware based wavelet filtering is much faster than a well tuned software implementation. Only for very small images the software system outperforms the OpenGL hardware. Scaling and bias computation as well as filter kernel download adds an almost constant overhead which unsurprisingly leads to bad times for small images. On the other hand, performance analysis shows that the filter operations of current

Size	Haar wavelet					Daubechies (4) wavelet				
	32^2	64^2	128^2	256^2	512^2	32^2	64^2	128^2	256^2	512^2
Software decomp.	0.50	2.0	7.8	31	150	0.70	2.8	11	45	209
Hardware decomp.	0.65	1.4	4.5	16	62	0.70	1.8	5.5	19	74
Factor	0.77	1.4	1.7	1.9	2.4	1.0	1.6	2.0	2.4	2.8
Software recons.	0.80	3.6	14	55	240	1.2	5.0	19	78	340
Hardware recons.	1.4	2.0	5.0	18	66	1.4	2.0	5.1	18	66
Factor	0.57	1.8	2.8	3.1	3.6	0.86	2.5	3.7	4.3	5.2

Table 1. Filter times in ms per 2D wavelet step

graphics hardware are still not optimized and in the future much higher throughput can be expected.

All times have been measured on a Silicon Graphics Octane with R10000 195MHz processor and a MXE graphics pipe. We will add performance figures for the Intergraph Wildcat as well as soon as possible.

As hardware based wavelet filtering uses the frame buffer for its computations, which has only a limited depth, the accuracy of the computations cannot be as good as with software based techniques, which in contrast only have to tolerate the typically small floating point errors. On the other hand, when using a frame buffer with a depth of 12 bits per base color, only single bit errors can be found in images of size 512^2 after complete wavelet decomposition and reconstruction, as it can be seen on the color plate in Figures 8 to 11. Note that the difference images have been enhanced so that one bit differences are visible.

On the other hand, frame buffers with only eight bits per base color yield less pleasing results. Figures 12 to 13 reveal the differences after complete decomposition and reconstruction. Again, the last image has been enhanced in order to reveal the differences. The maximum absolute difference between the original image and the wavelet decomposed image is 13, that is about 5% of the total 8 bit color range.

7 Conclusion

We have introduced a wavelet decomposition and reconstruction algorithm, that directly works on the graphics hardware of modern OpenGL capable workstations and accelerates the time consuming filtering steps a lot. By using the convolution and color matrix extensions together with OpenGL's facilities to scale images during copy instructions, we are able to perform all necessary steps of 2D tensor product wavelet filtering without copying data from or to the machine's main memory, thus avoiding typical bottlenecks in the visualization cycle. Different possibilities to use hardware based wavelets for enhanced feature detection are currently subject of further investigations.

Using the frame buffer for mathematical operations is usually problematic in terms of accuracy [11] due to the limited depth of the frame buffer. However, wavelet decomposition and reconstruction have proven to be relatively robust. Only single-bit differences between software and hardware decomposed data can be detected when rendering intermediate images to 12 bit accurate frame buffers.

8 Acknowledgments

We would like to thank our colleague Rüdiger Westermann for his helpful discussion regarding wavelet basis and hardware implementation issues. Additionally, we would like to thank our former colleague Christoph Lürig for giving us some ideas about how to accelerate hardware based wavelet transformations even more.

References

1. C. K. Chui. *An Introduction to Wavelets*. Academic Press, Inc., San Diego, 1992.
2. I. Daubechies. *Ten Lectures on Wavelets*. Number 61 in CBMS-NSF Series in Applied Mathematics. SIAM, Philadelphia, 1992.
3. M. Hopf and T. Ertl. Accelerating 3D Convolution using Graphics Hardware. In D. Ebert, M. Gross, and B. Hamann, editors, *Visualization '99*, pages 471–474, San Francisco, CA, 1999. IEEE Computer Society, IEEE Computer Society Press.
4. M. Hopf and T. Ertl. Hardware Based Wavelet Transformations. In B. Girod, H. Niemann, and H.-P. Seidel, editors, *Vision, Modeling, and Visualization '99*, pages 317–328, Erlangen, Germany, November 1999. SFB 603, Graduate Research Center, IEEE, and GI, Infix Press.
5. L. Lippert, M. H. Gross, and C. Kurmann. Compression Domain Volume Rendering for Distributed Environments. In D. Fellner and L. Szirmay-Kalos, editors, *EUROGRAPHICS '97*, volume 14, pages C95–C107. Eurographics Association, Blackwell Publishers, 1997.
6. A. K. Louis, P. Maass, and A. Rieder. *Wavelets*. B. G. Teubner Stuttgart, Germany, 1994.
7. C. Lürig, R. Grosso, and T. Ertl. Combining Wavelet Transform and Graph Theory for Feature Extraction and Visualization. In *Proc. 8th Eurographics Workshop on Visualization in Scientific Computing*, pages 137–144. Eurographics Association, 1997.
8. T. Malzbender. Fourier-Volume-Rendering. *ACM Transactions on Graphics*, 12(3):233–250, July 1993.
9. SGI. *OpenGL on Silicon Graphics Systems*. Silicon Graphics Inc., Mountain View, California, 1996.
10. G. Strang and T. Nguyen. *Wavelets and Filter Banks*. Wellesley-Cambridge Press, Wellesley, Massachusetts, 1996.
11. C. Teitzel, M. Hopf, R. Grosso, and T. Ertl. Volume Visualization on Sparse Grids. Technical Report 8/1998, Universität Erlangen-Nürnberg, Lehrstuhl für Graphische Datenverarbeitung (IMMD IX), Erlangen, July 1998. Accepted for publication in *Computing and Visualization in Science*, Springer-Verlag, Heidelberg.
12. T. Totsuka and M. Levoy. Frequency Domain Volume Rendering. *Computer Graphics*, 27(4):271–78, August 1993.
13. R. Westermann and T. Ertl. A Multiscale Approach to Integrated Volume Segmentation and Rendering. In *Computer Graphics Forum 16(3) (Proc. EUROGRAPHICS '97)*, pages 117–129. Blackwell, 1997.
14. M. V. Wickerhauser. *Adapted Wavelet Analysis from Theory to Software*. IEEE Press, New York, 1994.

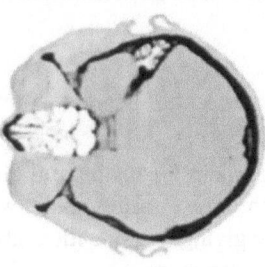

Fig. 8. The head data set

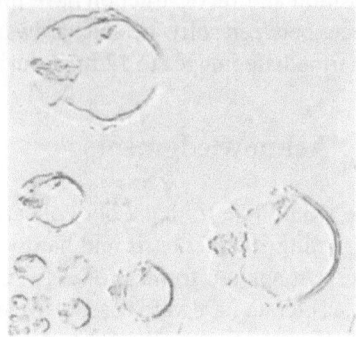

Fig. 9. Haar wavelet decomposition

Fig. 10. 1-bit differences after full Haar decomposition and reconstruction using a frame buffer with 12 bits per color

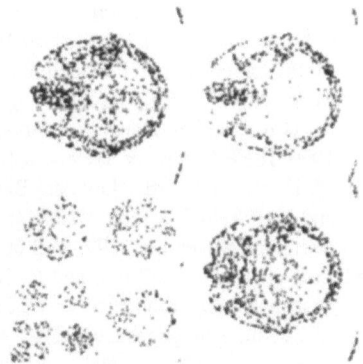

Fig. 11. 1-bit differences between software and hardware Haar decomposition using a frame buffer with 12 bits per color

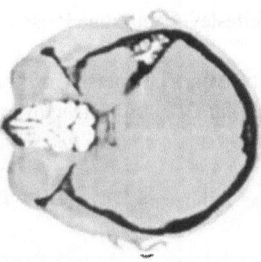

Fig. 12. Reconstructed image using a frame buffer with 8 bits per color

Fig. 13. Enhanced differences after full Haar decomposition and reconstruction using a frame buffer with 8 bits per color

Multiresolution Techniques for Interactive Texture-based Rendering of Arbitrarily Oriented Cutting Planes

Eric LaMar

Mark A. Duchaineau, Bernd Hamann, Kenneth I. Joy

Center for Image Processing and Integrated Computing
Department of Computer Science
University of California, Davis, CA 95616-8562, USA

Center for Applied Scientific Computing
Lawrence Livermore National Laboratory
Box 808, L-561, Livermore, CA 94551, USA

Abstract. We present a multiresolution technique for interactive texture-based rendering of arbitrarily oriented cutting planes for very large data sets. This method uses an adaptive scheme that renders the data along a cutting plane at different resolutions: higher resolution near the point-of-interest and lower resolution away from the point-of-interest. The algorithm is based on the approximation of a tri-tree space hierarchy, where the leaves of the tree define the original data and the first and finer nodes of the tree-resolution version. Rendering is done adaptively by selecting high-resolution cells close to a center-of-attention and lower-resolution cells away from it. We limit the artifacts introduced by this method by blending between different levels of resolution to produce a smooth image. This technique can be used to produce view-point-dependent renderings.

1. Introduction

Computing technology has steadily improved for more than four decades, and
visualization techniques equally. These increased computing capabilities have
scaled applications to scale accordingly. In turn all throughput and resulting data
sets have. However, current visualization techniques break down when operating
in this environment due to the massive size of the data sets. New techniques are
necessary to enable exploration of large multi-dimensional data sets.

In this paper, we combine hardware-assisted texture mapping and multiresolution methods for rendering cutting planes of large volumetric data sets. The
general idea is to assist prioritizer to different regions of the volume and to render
the high-priority region with highest accuracy while low-complexity regions are
rendered in progressively less accuracy, and progressively faster.

Eric.LaMar@vis.ucla.edu, duchaineau1@llnl.gov, {hamann, joy}@cs.ucdavis.edu

Multiresolution Techniques for Interactive Texture-based Rendering of Arbitrarily Oriented Cutting Planes

Eric LaMar*
Mark A. Duchaineau*, Bernd Hamann*, Kenneth I. Joy*

Center for Image Processing and Integrated Computing
Department of Computer Science
University of California, Davis, CA 95616-8562, USA

Center for Applied Scientific Computing
Lawrence Livermore National Laboratory
Box 808, L-561   Livermore, CA 94551, USA

Abstract. We present a multiresolution technique for interactive texture based rendering of arbitrarily oriented cutting planes for very large data sets. This method uses an adaptive scheme that renders the data along a cutting plane at different resolutions: higher resolution near the point-of-interest and lower resolution away from the point-of-interest. The algorithm is based on the segmentation of texture space into an octree, where the leaves of the tree define the original data and the internal nodes define lower-resolution versions. Rendering is done adaptively by selecting high-resolution cells close to a center of attention and low-resolution cells away from it. We limit the artifacts introduced by this method by blending between different levels of resolution to produce a smooth image. This technique can be used to produce viewpoint-dependent renderings.

1   Introduction

Computing technology has steadily improved for more than four decades and continues to improve rapidly. These increased computing capabilities have enabled applications to scale accordingly in overall throughput and resulting data set sizes. However, current visualization techniques break down when operating in this environment due to the massive size of the data sets. New techniques are necessary to enable exploration of large multidimensional data sets.

In this paper, we combine hardware-assisted texture mapping and multiresolution methods for rendering cutting planes of large volumetric data sets. The general idea is to assign priorities to different regions of the volume and to render the high-priority regions with highest accuracy, while lower-priority regions are rendered with progressively less accuracy, and progressively faster.

---

* eclamar@cipic.ucdavis.edu, duchaine@llnl.gov, {hamann,joy}@cs.ucdavis.edu

We use an octree to decompose texture space producing several coarser levels of the original data set. Each level is associated with a level in the octree and each level is half the resolution of the next level. The leaf nodes are associated with the original resolution, and the root node is associated with the coarsest resolution. Interior nodes are created by subsampling the eight child nodes. Each node contains two texture tiles, called *high* and *low*. The *high* tile stores the node's copy of the data; the *low* tile stores portión of the parent's *high* tile that covers the same area as the node.

Rendering a cutting plane involves traversing the octree and applying a selection filter to each node, building a selected node tree. Three results are possible: (1) the node (and its children) are skipped entirely; (2) the node is skipped, but its children are visited; or (3) the node is rendered and the children are skipped. The selected node tree forms an incomplete octree with the leaves being the nodes selected for rendering. The second step is to balance the selected node tree: all adjacent nodes must differ by no more than one level of resolution. The final step is to render each node, blending the *high* and *low* tiles when the node is adjacent to a lower-resolution node.

This technique reduces the amount of data accessed to produce a rendering. This is important in data mining or visual steering applications, where a user does not know the point-of-interest or would just like to browse the data. Another application is progressive visualization: often, a data set is too large to be placed on one computer system, and portions are distributed across a network of machines. It is not always practical to wait for all systems to finish rendering. With our technique, an initial approximation is first rendered. As higher-resolution data is received, a higher-quality approximation is rendered. This continues until all the data is received or the user changes viewing parameters.

Section 2 contains a survey of related work. Section 3 discusses construction of the texture hierarchy, and Section 4 covers how to process and render the texture hierarchy. Section 5 shows results for two data sets and provides performance results. Conclusions and future work are presented in Section 6.

2 Related Work

High-performance computer graphics systems are evolving rapidly. Silicon Graphics, Inc. (SGI) has been a primary developer of rendering technology, introducing the RealityEngine graphics system [1] in 1994 and the InfiniteReality graphics system [8] in 1998. SGI has also provided extensions to OpenGL [9], [7] that allow taking advantage of this hardware.

Cabral et al. [2] show that volume rendering and reconstruction integrals are generalizations of the Radon and inverse Radon transforms. They show that the Radon and inverse Radon transforms have similar mathematical forms and, by developing this relationship, show that both volume rendering and volume reconstruction can be implemented with hardware-accelerated textures. Cullip and Neumann [3] discuss general implementation issues for hardware-assisted texture-based volume visualization and illustrate the superiority of viewport-versus object-aligned sampling planes. Wilson et al. [13] and Van Gelder and Kim [11] develop the mathematical foundation for generating texture coordinates. Van Gelder and Kim also introduce a quantized gradient method for interactive shading for volume visualization. Westermann et al. [12] show how to visualize isosurfaces using fragment testing and discuss a technique to shade the texture-based isosurfaces. Grzeszczuk et al. [5] enumerate many methods using hardware-accelerated texturing to provide interactive volume visualization, and they introduce a library for texture-based rendering called *Volumizer* [4].

LaMar et al. [6] discuss techniques on which this work is based. This paper [6] shows that multiresolution techniques, when applied to large data sets and used for volume rendering applications, are a reasonable approach to reducing both rendering time and amount of data rendered. Shen et al. [10] discuss a temporally based multiresolution scheme for volume visualization of unsteady data sets.

Our method differs from these prior approaches in that we allow adaptive rendering of a cutting plane. Prior algorithms assume that a data set is "uniformly complex" or "uniformly important." This is not the case in an immersive environment, where data closer to the viewer has more visual importance than data far away. Our method of rendering tiles at different resolutions enables us to treat quality as a "tunable" parameter. Artifacts that may appear are removed by blending higher-resolution nodes into lower-resolution nodes.

3 Generating The Texture Hierarchy

3.1 *High/Low* Texture Tiles

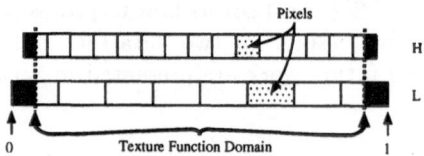

Fig. 1. A node with one-dimensional tiles, *high(H)* and *low(L)*.

In hardware texturing, linear interpolation is used to interpolate the values at the centers of adjacent texels. To allow for blending within a node, each node contains two texture map tiles (Figure 1). The *high* tile is the normal data associated with that node. The *low* tile is that part of the parent's *high* tile that is covered by the child node. The size ratio *high* to *low* is defined as $|high| = |low| * 2 - 1$. Thus one of the tiles must have odd size. If the size of a texture tile must be a power of two, then this relationship will incur some memory overhead. Our system uses a power-of-two size for the *low* tile, and the size for the *high* tile is calculated accordingly.

3.2 The Multiresolution Texture Hierarchy

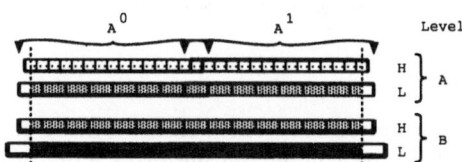

Fig. 2. A texture hierarchy of two levels.

Figure 2 shows a texture hierarchy consisting of two levels: The higher-resolution level is denoted as level A, with nodes A^0 and A^1, and the lower-resolution level as B. The image represented by A can be approximated by B. The *high* and *low* tiles in B are the same size as the *high* and *low* tiles in A^0 or A^1, and half the total size of the *high* and *low* tiles in A. We note that the natural relationship for two textures whose resolutions differ by a factor of two is using texel-center alignment. In the binary tree arrangement defined by this one-dimensional texture, B is the parent of A^0 and A^1. Also, note the correspondence between the *low* tile of the children to the *high* tile of the parent.

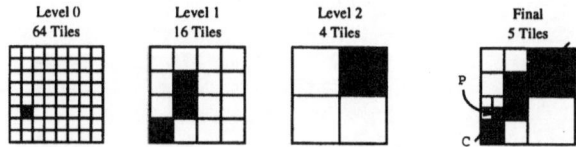

Fig. 3. Selecting a set of tiles from a 2D hierarchy of four levels (level 3 not shown).

Figure 3 shows a two-dimensional quadtree example. The original texture, level 0, contains 64 nodes. The dark regions show the portion of the level used in rendering the cutting plane. Nodes are selected when the distance from the center of the node to the point p is greater than the diagonal length of the node, and when the node intersects the cutting plane c. The selected nodes are shaded. The original texture, divided into 64 nodes, requires 64 time units to transfer. The multiresolution rendering uses five nodes, requiring five time units which implies a speed-up factor of about 13.

This technique extends to three-dimensional textures. Approximations are generated by subsampling the textures. The amount of memory "wasted" over the prior technique [6] is the storage of the *low* tile with each node; since each *low* tile is $\frac{1}{8}$ the size of the *high* tile, the additional memory overhead is $\frac{1}{8}$.

4   Rendering

The rendering phase is divided into the following steps: (1) selecting nodes to be rendered and building the selected node tree; (2) balancing the selected node tree; (3) computing the blending ratios; and (4) rendering the nodes.

4.1   Selecting Nodes

The first rendering step determines which nodes will be rendered. The general filtering logic starts at the root node and performs a depth-first traversal of the octree. For each node, we evaluate a selection filter, which returns one of three possible responses:

- Ignore this node and all of its children. This response is used to cull the tree. For example, if a node is not in the view frustum, then we can ignore the node and its children.
- The node satisfies all criteria. Render the node and do not consider the children.
- The node does not satisfy the criteria. Check the children.

Our primary selection filter is based on one of these two criteria:

- Cutting Plane. This filter selects a node when it intersects the cutting plane.
- Multiresolution Cutting Plane. This filter selects a node when it intersects the cutting plane and the distance from the node center to the point-of-interest (on the cutting plane) is smaller than the diagonal length of the node.

4.2   Blending

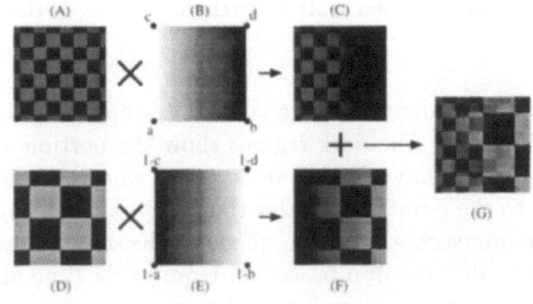

Fig. 4. Blending red and green checker board patterns.

Texturing is performed by modulating the color of the proxy geometry by the texture; the color is white and constant across a polygon. However, to blend

110

two images, we can change the polygon color to implement bilinear filtering. In Figure 4, image (G) is created by performing a per-pixel affine combination of images (A) and (D). Image (B), with ratios of $a = c = 1$ and $b = d = 0$, multiplies (A) and produces (C). Image (E) multiplies (D) and produces (F). Images (B) and (E) sum to unity. Adding (C) and (F) produces (G): a transition from red checks on the left to green checks on the right. We obtain a smooth transition provided (A) and (D) are two different resolutions of the same image.

4.3 Neighborhoods and "Balancing"

The blending algorithm described in section 4.2 requires that all selected nodes in a 26-neighborhood (across node faces, edges, and corners) have resolutions that differ by at most one level in the octree. Blending within a node can only blend between two texture resolutions: the high-resolution texture is blended into the low-resolution texture. Nodes have two textures tiles, *high* and *low*, so that a pair of nodes that differ by one level in the tree can be blended. Those that differ by two or more levels do not share any textures and cannot be blended.

After balancing the tree, we examine the neighbors of all selected nodes. The nodes adjacent to a node of lower-resolution must be blended such that the textures match. For each corner of a given node, when any of the seven adjacent nodes exist and have a lower-resolution, that corner must blend to the *low* tile; otherwise, it must use the *high* tile.

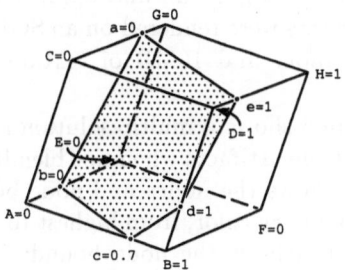

Fig. 5. Cutting plane clipped to an intersecting node.

Figure 5 shows a cutting plane clipped to an intersecting node. A to H are the blend ratios associated with the node: corners B, E, and H are adjacent to lower-resolution nodes, so that the blend ratio is one; the other corners have a blend ratio of zero, selecting the *low* and *high* tile of the node, respectively. The values a to e are the blend ratios associated with the clipped cutting planes vertices. Ratios on an edge are linear combinations of the ratios at the ends of that edge, and are proportional to the position of the point along the edge.

For rendering, we first define the RGB value for each clipped cutting plane vertex to the ratio (a to e in Figure 5), download the *low* texture tile, and draw the polygon. The color values will be interpolated across the polygon, multiplying the texture and producing the first weighted image. Next, we download the *high* texture tile, define the RGB value for each clipped cutting plane vertex to one minus the ratio, and draw the polygon, producing the second weighted image. Finally, by adding the first and second images, we produce the blended result.

5 Results

	Mandrill (Fig. 6)		Visible Female (Fig. 7)	
Data set resolution	$256^2 * RGB$ (2D)		$500^2 * 250 * RGBA$ (3D)	
Data set size	192K		238MB	
Tile resolution (high/low)	$15^2/8^2$		$32^3/16^3$	
Tile size (high/low)	1024/256 bytes		128K/16K bytes	
Level 0 nodes	324		2601	
Rendered nodes: fixed/MR	324	41	443	50
Bytes transmitted	405K	51K	56MB	7MB
Rendering time	-	-	2.0 sec.	0.37 sec.

Table 1. Timing results for Mandrill and Visible Female data sets.

We have implemented the algorithm and applied it to parts of the Visual Female data set. The data sets were rendered on an SGI Onyx2 computer system with 512MB of main memory and 16MB of texture memory, using a single 195MHz R10K processor.

For comparison, Figure 6 shows a multiresolution image of a Mandrill. This image is used to point out the artifacts when not blending across different levels of resolution. Image 6(b) shows the nodes and node boundaries: the resolution is shown by the node's boundary color, from highest to lowest: black, red, green, and yellow; notice the artifacts at the node boundaries in image 6(a). Image 6(c) shows the blending result, with nearest-neighbor filtering; notice that the pixel sizes blend smoothly across the nodes. Image 6(d) shows the final result; notice how the image is free of the boundary artifacts and smoothly blends high resolution nodes to low resolution nodes.

Figure 7 shows a multiresolution view of the Visible Female data set. The 443 nodes of the Visible Female represent the highest-resolution nodes that intersect the cutting plane (the other 2158 are never considered). The performance results shown in Table 1 are for a single frame; at 20 frames per second. The 1.1GB/sec required for the non-multiresolution approach exceeds the SGI InfiniteReality Engine's maximum transfer rate for textures of 320MB/sec by a factor of about 3.5, while the 140MB for the multiresolution approach has capacity to spare. The selection criteria are flexible and under user control. When the bandwidth is very low (e.g., over a modem), even fewer nodes can be selected.

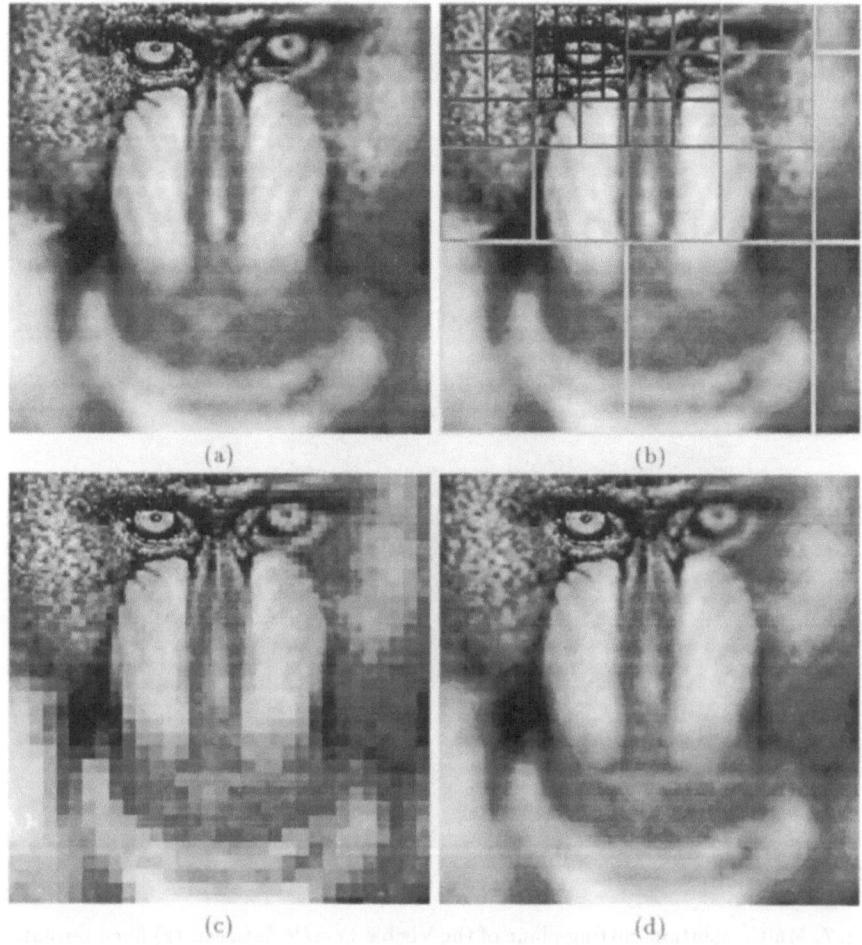

Fig. 6. Multiresolution Mandrill: (a) without blending; (b) node boundaries highlighted; (c) blended nearest-neighbor; and (d) blended, bilinear filtering.

6   Conclusions

We have presented an algorithm for interactive rendering of multiresolution cutting planes. We use hardware-based texturing, multiresolution techniques, and image blending to render a smooth approximation of a cutting plane. We have shown that our algorithm can produce a reasonable approximation while using less data. Despite the fact that our overall system is limited by the amount of available texture memory, the algorithm produces very good results, and we expect that this approach will have a major impact on the exploration of massive volumetric data sets that are currently generated in numerous applications.

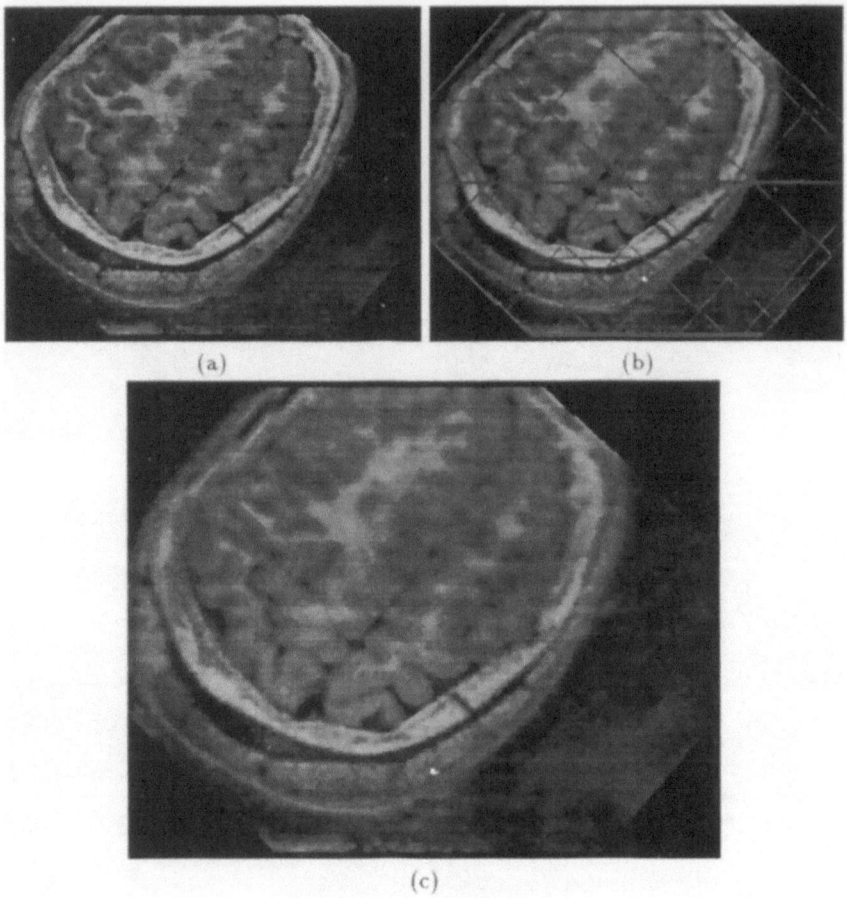

Fig. 7. Multiresolution cutting plane of the Visible Female data set: (a) fixed resolution; (b) blending with node boundaries high-lighted; and (c) MR, blending.

Future work includes error analysis. We will implement this technique in our multiresolution volume visualization system and extend it to visualizing vector fields.

7 Acknowledgments

This work was supported by the National Science Foundation under contract ACI 9624034 (CAREER Award) and through the National Partnership for Advanced Computational Infrastructure (NPACI);the Office of Naval Research under contract N00014-97-1-0222; the Army Research Office under contract ARO 36598-MA-RIP; the NASA Ames Research Center through an NRA award under contract NAG2-1216; the Lawrence Livermore National Laboratory under

ASCI ASAP Level-2 Memorandum Agreement B347878 and under Memorandum Agreement B503159; and the North Atlantic Treaty Organization (NATO) under contract CRG.971628 awarded to the University of California, Davis. We also acknowledge the support of ALSTOM Schilling Robotics, Chevron, Silicon Graphics, Inc. and ST Microelectronics, Inc. We thank the members of the Visualization Thrust at the Center for Image Processing and Integrated Computing (CIPIC) at the University of California, Davis.

References

1. Kurt Akeley. RealityEngine graphics. In *Proceedings of Siggraph 93*, pages 109–116. ACM, August 1993.
2. Brian Cabral, Nancy Cam, and Jim Foran. Accelerated Volume Rendering and Tomographic Reconstruction Using Texture Mapping Hardware. In *1994 Symposium on Volume Visualization*, pages 91–98. ACM, October 1994.
3. Timothy J. Cullip and Ulrich Neumann. Accelerating Volume Reconstruction With 3D Texture Hardware. Technical Report TR93-027, Department of Computer Science, University of North Carolina - Chapel Hill, May 1994.
4. George Eckel. *OpenGL Volumizer Programmer's Guide*. SGI, Inc., 1998.
5. Robert Grzeszczuk, Chris Henn, and Roni Yagel. *SIGGRAPH '98 "Advanced Geometric Techniques for Ray Casting Volumes" course notes*. ACM, July 1998.
6. Eric LaMar, Bernd Hamann, and Kenneth I. Joy. Multiresolution Techniques for Interactive Hardware Texturing-based Volume Visualization. In *IEEE Visualization 99*, pages 355–361. IEEE, November 1999.
7. Tom McReynolds and Davis Blythe. *SIGGRAPH '98 "Advanced Graphics Programming Techniques Using OpenGL" course notes*. ACM, July 1998.
8. John S. Montrym, Daniel R. Baum, David L. Dignam, and Christopher J. Migdal. Infinite Reality: a Real-Time Graphics System. In *Proceedings of Siggraph 97*, pages 293–302. ACM, August 1997.
9. Mark Segal and Kurt Akeley. *The OpenGL Graphics System: A Specification (Version 1.2)*. SGI, Inc., 1998.
10. Han-Wei Shen and Kwan-Liu Ma. A Fast Volume Rendering Algorithm for Time-Varying Fields Using A Time-Space Partitioning (TSP) Tree. In *IEEE Visualization 99*, pages 371–377. IEEE, November 1999.
11. Allen Van Gelder and Kwansik Kim. Direct Volume Rendering with Shading via Three-Dimensional Textures. In *Proceesings of 1996 Volume Visualization Symposium*, pages 23–30. IEEE, October 1996.
12. Rüdiger Westermann and Thomas Ertl. Efficiently Using Graphics Hardware In Volume Rendering Applications. In *Proceedings of Siggraph 98*, pages 169–177. ACM, July 1998.
13. Orion Wilson, Allen Van Gelder, and Jane Wilhelms. Direct Volume Rendering via 3D Textures. Technical Report UCSC-CRL-94-19, University of California, Santa Cruz, June 1994.

Editors' Note: see Appendix, p. 287 for colored figures of this paper

Very Large Scale Visualization Methods for Astrophysical Data

Andrew J. Hanson, Chi-Wing Fu, and Eric A. Wernert

Indiana University, Bloomington, IN 47405, USA,
{hanson, cwfu, ewernert}@cs.indiana.edu,
WWW home page: http://www.cs.indiana.edu/

Abstract. We address the problem of interacting with scenes that contain a very large range of scales. Computer graphics environments normally deal with only a limited range of orders of magnitude before numerical error and other anomalies begin to be apparent, and the effects vary widely from environment to environment. Applications such as astrophysics, where a single scene could in principle contain visible objects from the subatomic scale to the intergalactic scale, provide a good proving ground for the multiple scale problem. In this context, we examine methods for interacting continuously with simultaneously active astronomical data sets ranging over 40 or more orders of magnitude. Our approach relies on utilizing a single scale of order 1.0 for the definition of all data sets. Where a single object, like a planet or a galaxy, may require moving in neighborhoods of vastly different scales, we employ multiple scale representations for the single object; normally, these are sparse in all but a few neighborhoods. By keying the changes of scale to the pixel size, we can restrict all data set scaling to roughly four orders of magnitude. Navigation problems are solved by designing constraint spaces that adjust properly to the large scale changes, keeping navigation sensitivity at a relatively constant speed in the user's screen space.

1 Introduction

We study the problem of supporting interactive graphics exploration of datasets spanning dozens of orders of magnitude in space and time.

The physical universe is precisely such a data set, and, as our theoretical and experimental understanding of physics has improved over the years, the development of appropriate techniques for making these vast scales accessible to human comprehension has become increasingly important. From the scale of galactic super-clusters down to the quarks in atomic nuclei, we have approximate sizes ranging from 10^{25} meters down to 10^{-15} meters. To look closely at cosmological scales corresponding to the distance that far-away light has traveled since near the beginning of the Big Bang [10], we may need to go to even larger scales, while to visualize the tightly wound Calabi-Yau spaces intrinsic to the "hidden dimensions" of modern string theory [6,9], we need to delve down to the Planck length of 1.6×10^{-35} meters, giving a potential requirement for 60 or more orders of magnitude.

Large scale data sets are common in visualization, but the focus of the visualization literature has generally been on data sets with massive amounts of information (see, e.g., [13]). As increasingly detailed data have become available on distant astronomical structures (see, e.g., [2, 5]), the combined problems of large scale in the size and large scale in the extent of astronomical data have received increasing attention (see, e.g, [16]).

The concept of creating pictures and animations to expose these vast realms has a long history. As early as 1957, a small book by Kees Boeke entitled "Cosmic View: The Universe in Forty Jumps" [1] had already appeared with the purpose of teaching school children about the scales of the universe. The effort to turn the ideas in this book into a film were pursued by the creative team of Charles and Ray Eames for over a decade, starting with a rough draft completed in 1968 and culminating in the classic film "Powers of 10" in 1977 [3]. Five years later, Philip and Phylis Morrison annotated the movie in a richly illustrated Scientific American book [12]. The Eames Office continues to develop educational materials based on the original film, including a CD-ROM [14] that adds many additional details. Recently, the IMAX corporation sponsored the production of "Cosmic Voyage," a feature-length film [15] that reframes many of the features of "Powers of 10" using modern cosmological concepts combined with current large scale 3D data sets and simulations to provide additional richness and detail.

What all these efforts have in common is that they are relatively static, with no easy way to let the student of the universe ask his or her own questions and wander through the data choosing personal viewpoints. Even given the general availability of many interesting astronomical data sets and the computer power to display portions of them in real time, myriad problems confront the visualizer who wishes to roam freely through the whole range of scales of the physical universe. In this paper we present an elegant approach to solving this problem.

2 Strategies

Our experience with brute force attempts to display data sets of many orders of magnitude is that major flaws in precision occur between scales of 10^{13} and 10^{19}; the problems can actually be worse on more expensive workstations, regardless of the apparent uniformity of the promised hardware support for OpenGL transformation matrices. One can surmise that the onset of problems traces to normalizing vectors by taking the inverse square root of the scale squared; with an available exponent range in single precision arithmetic up to about 10^{38}, our experimental observations are theoretically plausible.

One of our chosen tasks is therefore to find a way that would allow many more orders of magnitude to be represented without significant arithmetic errors. We accomplish this with three major strategies: (1) the replacement of ordinary homogeneous coordinates by "log scale" homogeneous coordinates, with the fourth coordinate representing the log of the current scale in the chosen base, typically 10; (2) the use of a small number of cycling scale frames, covering approximately three orders of magnitude, that are reused so that all data are rendered at or near unit scale, regardless of the actual scale; (3) pixel-level replacement of enormous but distant visible data sets by hierarchically scaled models reducing to environment texture maps whenever the current scale of local user motions would result in negligible screen motions of distant objects.

We then combine these techniques with a scale-intelligent implementation of the constrained navigation framework [7, 17] to keep user navigation response constant in screen space; this is essential to retain an appropriate level of user response universally across scales. A crucial side benefit of constrained navigation is somewhat similar to the advantages of carefully constraining the camera motion in Image-Based Rendering: one allows the viewer a great deal of freedom, but restricts access in such a way that the expense of data retrieval is reduced compared to that needed for full six-degree-of-freedom "flying."

In the following sections, we begin our work by deriving the conditions needed to represent a large range of scales in spatial data. Next, we describe our hierarchical approach to object representation, ways to support many scale magnitudes within a single object, and the concept of variable-representation geometry adapted from classical cartography. Finally, we treat several examples in the context of constrained navigation with intelligent scaling built into the user interface.

3   The Geometry of Scale Uniformization

Navigation through a virtual environment and the placement of objects in the environment utilize the standard six degrees of freedom: three orientation parameters, which are independent of scale, and three position parameters, which must be adapted to our scaling requirements. In order to support local scales near unity, we introduce one additional parameter, the power scale, which is effectively an index into an array of local, mipmap-like unit-scale environments whose scales are related by exponentiating the indices.

3.1   Homogeneous Power Coordinates

We define the homogeneous power coordinate representation of a point as

$$p = (x, y, z, s) \tag{1}$$

where s is typically an integer and the physical coordinates in 3D space corresponding to p are

$$\mathbf{X}(p) = (xk^s, yk^s, zk^s) = \mathbf{x} \times k^s.$$

Here k is the chosen scale base, typically $k = 10$. The homogeneous power coordinates p are thus equivalent to the ordinary homogeneous coordinates $X = (x, y, z, w)$ with

$$s = -\log_k w .$$

Thus if we choose $k = 10$ and units of meters, we see that $s = 0$ corresponds to the human scale of one meter, and $s = 7$ to about the diameter of the Earth (1.3×10^7m). Table 1 gives a rough picture of some typical scales in these units.

We note that s can be generalized to a vector, $\mathbf{s} = (s_x, s_y, s_z)$, such that s_x, s_y and s_z are the power scales applied to the x, y and z-components of p independently.

Object	Power of 10	Object	Power	Object	Power	Object	Power
Planck length	-35	virus	-7	Earth	7	Local galaxies	23
proton	-14	cell	-5	Solar system	14	Super-cluster	25
hydrogen atom	-10	human	0	Milky Way	21	Known Universe	27

Table 1. Base 10 logarithms of scales of typical objects in the physical universe in units of meters. To convert to other common units, note that 1 $au = 1.50 \times 10^{11}$ m, 1 $ly = 9.46 \times 10^{15}$ m, and 1 $pc = 3.26$ $ly = 3.08 \times 10^{16}$ m, where au = astronomical unit, ly = light year, and pc = parsec.

3.2 Homogeneous Power Coordinate Interpolation

Since each coordinate $p = (x, y, z, s)$ may have a different power scale, our first requirement is the formulation of interpolation methods that may be used to transition smoothly among scales. To interpolate between two homogeneous power coordinates p_0 and p_1 using the parameter t, where $t \in [0, 1]$, we may immediately write $p(t) = (1 - t)p_0 + tp_1$. Since $X(p) = x \times k^s$ is the ordinary space representation of p, the interpolation in ordinary space becomes

$$X(p(t)) = [(1 - t)x_0 + tx_1] \times k^{(1-t)s_0 + ts_1} .$$

When $s_0 = s_1$, this reduces to ordinary interpolation within a single scale domain. When we fix $x_0 = x_1$, we can implement a "Powers of [Base k]" journey by interpolating directly in power-scale space alone.

In order to adjust display positions of objects with different scales appearing in the same scene, we need scale-to-scale correspondence rules. Thus, if we let p_0 and p_1 be the positions of a single object expressed with respect to power scales of s_0 and s_1, respectively, we require equivalence of the physical coordinates:

$$X_0(p_0) = X_1(p_1)$$
$$x_0 \times 10^{s_0} = x_1 \times 10^{s_1} .$$

That is, $x_1 = x_0 \times 10^{\delta s}$, where $\delta s = s_1 - s_0$ is the change in the power scale.

3.3 Environment Map Angular Scale Criteria

Next, we need to set up the framework for viewing large scale objects that intrude upon the current scale of the user's display.

Let x_1 and x_2 be the limiting boundaries of a family of $3D$ observation positions spanning a distance $|x_2 - x_1| = d$, and let O be an object at a radial distance r from the mid-point of the line joining x_1 and x_2, as shown in Figure 1. As we move along the line between x_1 and x_2, O will have the greatest angular displacement (as perceived by the viewer) when O is located on the perpendicular bisecting plane of x_1 and x_2. Thus if α is the maximal angular displacement of O with respect to a motion connecting x_1 and x_2, then $\tan(\alpha/2) = d/2r$, so

$$\alpha = 2 \tan^{-1} \frac{d}{2r} . \tag{2}$$

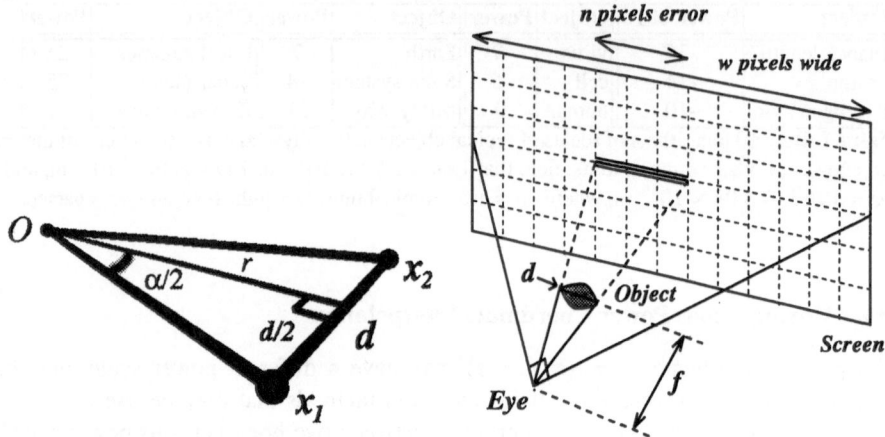

Fig. 1. Geometry for the calculation of an object's maximum subtended angle α.

Fig. 2. Schematic diagram for angular limits deciding whether to omit an object.

Next let w be the maximum resolution of a desktop display screen, or perhaps one wall of a CAVE^{TM}. Assuming a 90-degree field of view, the typical angle subtended by one pixel is $\pi/2w$. We impose the assumption that whenever an object is displaced by fewer than n pixels as we move our viewpoint across its maximal local range, we can represent that object by a static texture map (an *environment map*) without any major perceptual error. The condition becomes

$$\alpha < \frac{n\pi}{2w} \; .$$

Thus we conclude that $\tan^{-1}(d/2r) < n\pi/4w$, so

$$r > \frac{d}{2\tan(n\pi/4w)} \; . \tag{3}$$

Thus we may replace an object O by its pixel-level environment map representation whenever its distance r from the worst-case head-motion displacement axis with $d = |x_2 - x_1|$ satisfies Eq. (3).

Local User Motion Scales. Let s be the scale of our current environment. This means, e.g., that we typically restrict local motions to lie within a virtual cube of side 10^s accessible to the user without a scale change. Assuming the average extreme motion is given simply by the cube edge length, and letting the distance of the object in question be of order $r = 10^l$, we find

$$10^l = \left(\frac{10^s}{2}\right)\frac{1}{\tan\left(\frac{n\pi}{4w}\right)}$$

$$\delta s(n) = l - s = -\log_{10} 2 - \log_{10} \tan\left(\frac{n\pi}{4w}\right) \; .$$

Plugging in typical values for n and taking $w = 1024$, we find the following results:

$$\delta s(1) = 2.81, \quad \delta s(2) = 2.5, \quad \delta s(3) = 2.34, \quad \delta s(4) = 2.21 .$$

Therefore, to represent the Milky Way with individual star data from the Bright Star (BSC) catalog [8] or Hipparcos [4], we need a hierarchical strategy: whenever the current scale differs from that of the star data scale by about 3 units, we can just use an environment map. As we move within the scale of a single local environment, this guarantees that no object position will differ by more than n pixels from the full 3D rendering.

3.4 Object Disappearance Criteria

When will an object's scale be so small that we do not need to draw it? In this section, we derive the estimates needed to decide when to "recycle" a particular scale representation so that we can adhere to our requirements that nothing be represented too far from unit scale in *actual* calls to the graphics hardware. This of course can only be accomplished if the transition between successive scales is smooth and has the appearance of a large scale change. To carry out this illusion, we need to eliminate small scale representations in favor of new unit scale representations as the camera zooms out (and also the reverse).

The Smallest Visible Object. We begin by assuming that any object subtending less than n pixels when projected to the screen or CAVE wall can be ignored and does not need to be drawn.

As before, the resolution of the screen is described by the average angle subtended by one pixel, i.e., $\pi/2w$, where w be the maximum screen resolution in pixels; thus, if the projected size of anything is smaller than $n\pi/2w$, we can ignore it.

Largest Angle Subtended by an Object. Given an object of size d, we need to calculate the worst-case projected size on the screen in order to decide when it can be omitted.

If f is the distance from the viewpoint to the near viewing plane, the largest angle that could be subtended by this object is d/f.

Note that the limits of f depend on the human visual system: when the object is too near, the eye cannot focus on it — the near limit of f is roughly 0.3 feet (10 cm). An estimate of the required precision for a CAVE follows from the CAVE wall dimension of 8–10 feet (2.4–3.0 m), so f is about 1/24 to 1/30 the screen size.

Next, we have to convert f to the units of our virtual environment. When the current scale of the virtual environment is s, one side of the wall is 10^s units. For example, an 8-foot (2.4 m) CAVE would have f roughly equal to $10^s/24$ in wall units.

Combining the Numbers. We can now put everything together. When the visual angle of an object is smaller than the visual limit, it does not need to be drawn. The condition is that the angle in radians subtended by the object at distance f must be less than the angle subtended by the desired limit of n screen pixels (see Figure 2):

$$\frac{d}{f} = \frac{24d}{10^s} < \frac{n\pi}{2w}$$

Defining $l = \log_{10} d$, we find $24(10^{l-s}) < n\pi/2w$, so that

$$\delta s(n) = l - s < \log_{10}\left(\frac{n\pi}{48w}\right)$$

Note that $n\pi/48w$ is smaller than 1 in general, so δs is normally negative. Plugging in typical values for n and letting $w = 1024$, we find:

$$\delta s(1) = -4.194, \quad \delta s(2) = -3.893, \quad \delta s(3) = -3.717, \quad \delta s(4) = -3.592 .$$

Therefore, if the scale of an object is smaller than the current scale by around 4.2, we can ignore it for rendering purposes without entailing more than a single pixel worth of perceptual error.

4 Implementing Multi-scale Blending

The conceptual goal of keeping all objects drawn at unit scale as the graphical viewing volume sweeps through each object's natural scale is an appealing ideal. However, in practice, very few objects live "alone" in their assigned neighborhood of the order-of-magnitude scale space. From the Earth, at scale 10^7 m, we can see elements of the solar system, at scale 10^{13} m, as well as clouds of stars in our own galaxy, the Milky Way, at scale 10^{21} m. If we wish to show the Milky Way as seen from Earth, we must have *multi-scale* representations of objects at far different scales that can be rationally displayed alongside the Earth without requiring huge scale factors.

In this section, we describe several different issues, problems, and solutions to the mixed scale rendering problem. Our actual implementation involves an elaborate scripting language that supports each of the features described below. In particular, we provide definitions for multiple navigation manifolds, lists of alternate object representations indexed by scale, and animation parameters for hierarchical object motion. Further details will be omitted here for lack of space.

Symbolic interpolators: In many circumstances, certain objects serve as anchors or flags that define a special user context in the visualization. In such circumstances, it is appropriate to render objects at scales that are unconnected to their actual size, but are dictated by their semantic importance. In ordinary cartography, landmarks or objects of legal significance or liability to the map maker are drawn using symbols out of proportion to the conventional size of an object. One example we have implemented keeps the Earth itself visible by using a constant size globe starting above some particular scale threshold.

A related technique (see, e.g., [11]) maintains an entire virtual library of depictions to be used for an object in different scale environments. At one scale, a fully rendered, illuminated 3D object may be appropriate, at another a textured 2D billboard may be correct, while at another one might use fixed-scale text if the object should never completely disappear.

Multi-resolution data structures and octrees: Typical astronomical objects may extend over many orders of magnitude, with a variety of natural representation scales for individual subregions. Such data are normally stored in a hierarchy of resolutions such as an octree. However, to have a fully detailed set of data at all possible points of an enormous astronomical object would be prohibitively expensive. We can sidestep this problem and make the missing detail less obvious to the user by restricting the navigation paths and permitted view directions to permit detailed examination of only certain selected areas. Thus one can in principle design an environment in which the highest levels of detail are typically be limited to only a few regions of the octree representation, permitting substantial economy in data storage. If all the detail is actually required, more sophisticated methods such as those suggested in [16] are needed.

Environment maps: We have already derived a series of formulas determining the level at which a 3D object can be replaced by an asymptotic texture map without requiring an expensive 3D rendering. Such maps are equivalent to so-called "environment maps" that are used to represent distant fixed objects, such as the fixed stars themselves, without rendering them. A spherical texture map, or a series of six orthogonal images with fields of view corresponding to the faces of a cube, will accomplish the desired result. All we need to do is to keep track of the camera motion so we can change the environment map (or switch to 3D detail) if the viewpoint changes significantly.

5   Examples of Scaled Constrained Navigation

We define constrained navigation to be the assignment of a mapping between a controller space and a general field of viewing control parameters [7]. Our typical implementation involves the design of one or more 2D "sidewalks" that define the 3D spatial motion of the user and the view parameters in response to inputs in a limited controller space such as that of a mouse or the CAVE thumb joystick.

Since the response of the user displacement to a unit of controller motion can be completely controlled and customized by the fields stored with the navigation manifold, we can easily adapt the motion response to meet the user requirements of large-scale navigation. In particular, it seems obvious that a "Powers of [Base k]," i.e., logarithmic, scaling of the response is the natural one to use: the farther we get from the Earth's surface, the lower the density of detailed observational data, and the larger the scale of the visible structures that are interesting to depict.

In addition to scaling motion control with the constrained navigation framework, we have studied a number of constraint manifold designs that are well-suited to this work: among these, we describe below the "pond ripple," the "wedge," and the "twist."

Multi-centered Pond Ripple Navigation. In Figure 3, we show a very special disk-shaped navigation manifold. This manifold has not one, but multiple centers of attention, corresponding to the evolving changes in the centers of the rings at each scale level. Tangential motion takes place in the lateral ring direction, centering the viewer's gaze on the current center.

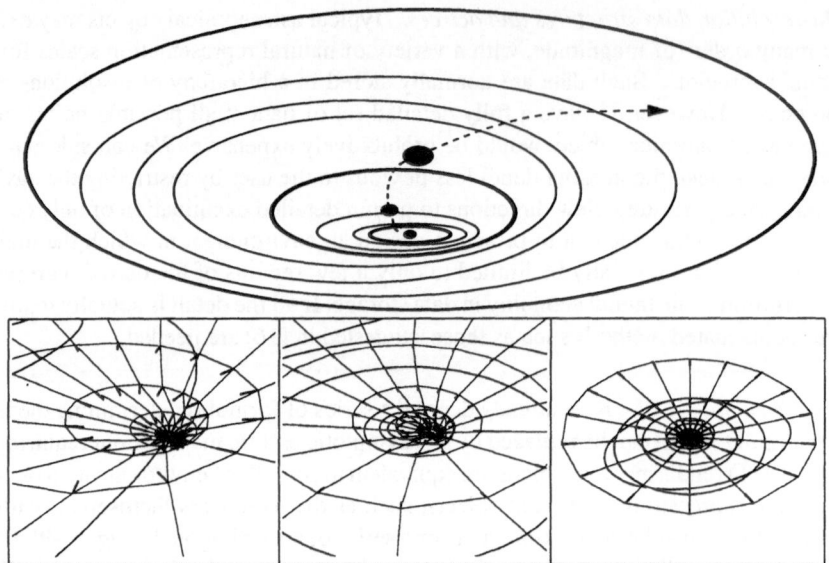

Fig. 3. "Pond ripple" navigation environment with continuous recentering as the viewer orbits around the central object of interest at each scale. The bottom row of diagrams illustrates the fact that the viewpoint directions (short heavy bars) at three different positions focus on three different centers.

Wedge Exponential Journey. In Figure 4, we show a symbolic journey on a wedge-shaped navigation manifold; the width of the wedge as well as the actual distance traveled per unit controller motion expand exponentially with distance from the origin. This allows the viewer to pursue a "Powers of [Base k]" interactive exploration of the space, and to travel between viewpoints in logarithmic time instead of the much less appropriate constant velocity.

Twisting Reorientation. In Figure 5, we show the advantage of the constrained navigation approach for huge scales with smooth evolution between mismatched orientation frames natural to each scale. Beginning with a "wedge" manifold, we twist the frames to get custom orientations. Beginning with an orientation suitable for viewing the Earth, we move out seven orders of magnitude, while twisting so the solar system appears horizontal instead of the Earth's equator; moving out seven more orders of magnitude to the galactic scale, we twist again to orient ourselves to the galactic plane.

6 Conclusions

We have addressed the problem of effective interactive navigation across huge ranges of scales appropriate for spanning the entire physical universe. This was accomplished with a combination of methods, including the following: a systematic treatment of all data at unit scale with (hardware) transformation matrices never exceeding four orders of magnitude in scale range; systematic blending of both iconic (fixed symbolic scale) and very large data groups using multi-scale representations; merging environment maps with mipmaps and octrees to make very distant data visible without violating the unit scale requirement; supporting multi-resolution maps to allow switching to layered unit-scale data representations as we zoom in and out; scaling both spatial navigation control and time-scale of simulations to the scale of the current local data representation; constrained multi-resolution navigation. Future work will focus on additional requirements of time scaling, the multi-resolution problem with very large data sets, techniques for designing navigation constraints to control the required data set sizes, analyzing user responses to the environment, and incorporating the effects of special and general relativity in ways that are intuitive and qualitatively correct without being obtrusive.

Acknowledgments

This research was supported by NASA grant number NAG5-8163. This research was also made possible in part by NSF infrastructure grant CDA 93-03189 and the generosity of the Indiana University Advanced Visualization Laboratory. We thank P.C. Frisch for many useful interactions and contributions.

References

1. Kees Boeke. *Cosmic View: The Universe in Forty Jumps.* John Day, 1957.
2. Arthur N. Cox. *Astrophysical Quantities.* Springer-Verlag New York, Inc., New York, 1999.
3. Charles Eames and Ray Eames. Powers of Ten, 1977. 9 1/2 minute film, made for IBM.
4. M. A. C. Perryman et al., editor. *The Hipparcos and Tycho Catalogues.* ESA Publications Division, The Netherlands, 1997.
5. M. Geller and J. Huchra. Mapping the universe. *Science*, 246:897–910, 1989.
6. Brian Greene. *The Elegant Universe: Superstrings, Hidden Dimensions, and the Quest for the Ultimate Theory.* W.W. Norton, 1999.
7. A. J. Hanson and E. Wernert. Constrained 3D navigation with 2D controllers. In *Proceedings of Visualization '97*, pages 175–182. IEEE Computer Society Press, 1997.
8. Dorrit Hoffleit and Carlos Jaschek. *The Bright Star Catalogue: Fourth Revised Edition.* Yale University Observatory, New Haven, 1982.
9. Michio Kaku. *Introduction to Superstrings and M-Theory.* Springer Verlag, 1999.
10. David Layzer. *Constructing the Universe.* Scientific American Books, 1984.
11. Paulo W. C. Maciel and Peter Shirley. Visual navigation of large environments using textured clusters. *1995 Symposium on Interactive 3D Graphics*, pages 95–102, April 1995.
12. Philip Morrison and Phylis Morrison. *Powers of Ten.* Scientific American Books, 1982.
13. G.M. Nielson, H. Hagen, and H. Mueller. *Scientific Visualization: Overviews, Methodologies, Techniques.* IEEE Computer Society, 1997.
14. The Office of Charles and Ray Eames. Powers of Ten, 1998. CD-ROM educational materials.
15. Bayley Silleck. Cosmic voyage, 1996. 35 minute film, a presentation of the Smithsonian Institution's National Air and Space Museum and the Motorola Foundation.
16. D. Song and M. Norman. Looking in, looking out: Exploring multi-scale data with virtual reality. *Comput. Sci. Eng.*, 1(3):53–64, 1994.
17. E.A. Wernert and A.J. Hanson. A framework for assisted exploration with collaboration. In *Proceedings of Visualization '99*, pages 241–248. IEEE Computer Society Press, 1999.

Editors' Note: see Appendix, p. 288 for colored figures of this paper

Hybrid Model for Vascular Tree Structures

Anna Puig, Dani Tost, and Isabel Navazo

Software Department, Avda. Diagonal, 647, 8 floor, 08018 Barcelona, SPAIN,
anna@lsi.upc.es,
WWW home page: http://www.lsi.upc.es/~anna/

Abstract. This paper proposes a new representation scheme of the cerebral blood vessels. This model provides information on the semantics of the vascular structure: the topological relationships between vessels and the labeling of vascular accidents such as aneurysms and stenoses. In addition, the model keeps information of the inner surface geometry as well as of the vascular map volume properties, i.e. the tissue density, the blood flow velocity and the vessel wall elasticity.

The model can be constructed automatically in a pre-process from a set of segmented MRA images. Its memory requirements are optimized on the basis of the sparseness of the vascular structure. It allows fast queries and efficient traversals and navigations. The visualizations of the vessel surface can be performed at different levels of detail: The direct rendering of the volume is fast because the model provides a natural way to skip over empty data. The paper analyzes the memory requirements of the model along with the costs of the most important operations on it.

1 Introduction

Vascular diseases represent a 10% of the clinical examinations. Current diagnosis methods are based on vessel images obtained by: X-Rays, DSA (Digital Subtraction Angiography), contrasted CT (Computer Tomography) or MRA (Magnetic Resonance Angiography). From these images, physicians must mentally reconstruct the 3D shape of the vessels in order to detect lesions, such as stenoses and aneurysms. This is a difficult task because the brain vascular system has a complex tree-like structure, and the vessels are small, narrow and sparse in comparison to the surrounding volume. The reconstruction and visualization of blood vessels three-dimensional model from a set of slices provides better means of diagnosing and treating vascular pathologies.

Previous papers on this topic address three main approaches:

- Direct visualization of the data, generally performed with the Maximum Intensity Projection (MIP) [1],
- Blood vessels surface extraction using Marching Cubes (MC) [2] and Dividing Cubes (DC) [3] algorithms,
- Construction of a topological description of the vascular system (symbolic model) [4, 5, 6].

The former approach does not require any pre-processing. Even the segmentation, which is required for most operations on MRA data [7] can be avoided with MIP projection. However, MIP lacks depth perception, and therefore it produces ambiguities in the overlapping of vessels, diameters reduction and loss of the smallest vessels. Depth cue can be added to the rendering [8] but, for clinician's use, it is generally necessary to compute in batch several different MIP views and record them as a film. Other visualization strategies require a good segmentation [9] and they are generally too slow for routine diagnosis, although they can be speeded-up by using hardware 3D texture mapping [10].

The main drawback of the second approach is that due to the nature of the vessels, the surface model is composed of a huge amount of tiny faces and the connectivity is difficult to guarantee.

This paper addresses the third approach: the construction of a symbolic model of the vascular tree providing information of the medial line of the vessels and of their diameters. This strategy provides a better understanding of the structure than the two former ones. In addition, it may allow the generation of a simpler and smoother surface model. However, the automatic construction of such a model is complex, because it requires the medial line extraction and diameters computation.

Several symbolic models have been previously proposed. In [4], individual vessels are modeled as a set of cylinders joined onto common circular sections. The model is extracted manually. Its main drawback is that it does not provide a global representation of the topological relationships between vessels. In [5] the vascular structure is represented as a symbolic tree, composed of numbered branches connected at bifurcation nodes, where the diameters of the vessels and the axes orientations are stored. The model computed in [6] is a tree of spline curves marking the center lines of the vessels, plus a sequence of cross-section contours perpendicular to the axis. A smooth surface model composed of tensor product rectangular patches can be fitted on this symbolic model. Both models ([5], [6]) can be constructed automatically. As a drawback, they do not provide a specific representation of aneuryms and stenoses. In addition, they do not give simultaneous information of the surface geometry and the internal property values. The hybrid model presented herein is intended at filling this gap. It is first described in section 2 and next evaluated in section 3 in terms of memory requirements and efficiency of its visualizations.

2 Model representation

The proposed model of the vascular structure provides three layers of information: the global structure of the map (the set of topological relationships between the different components of the map), the vascular surface (the global shape of the internal vascular wall) and the inner volume (the inner properties of the vessels). The volume properties are the density and the blood velocity, as all other relevant physical properties can be defined from them. The blood pressure, for instance, can be derived from the velocity and the vessel diameter and the elasticity can be computed from the density.

2.1 Global structure

Since the physical vascular structure is tree-like, an abstract graph with nodes representing branchings and edges representing vessels presents itself as the obvious choice for the logical definition of the vascular area. The graph is cyclic because the vascular map can present closed paths of nodes, such as the Circle of Willis into the brain, i.e. a particular cerebral arteries network. In addition, the graph is undirected because an explicit relationship of order between nodes does not exist. The flow direction is not represented as the graph direction because in particular regions of the vein system and under specific circumstances, the flow may reverse its direction. Finally, although some vascular pathologies may produce disconnections of the vascular structure, the cerebral map is generally connected. Herein, each connected structure is considered separately as a connected graph.

The edges and the nodes of the graph are composed of a set of labeled segments (see Figure 1). Three different types of vessel segments (features) are considered: normal segments, stenoses and aneurysms. This classification eases the visual detection of accidents, and, as explained later, it reduces the memory requirements of the surface model. All the nodes and the edges segments of the graph are linked with surface information and property values, as described in the next section.

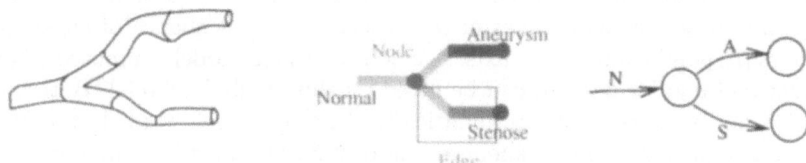

Fig. 1. Representation scheme

As well known, there are two common ways to implement graphs [11]: sequential and linked representations. The simulations performed in this paper are implemented on the basis of the linked representation.

2.2 Surface representation

A natural representation of the inner surface of a blood vessel segment is by sweeping contour cross-sections, generally circles, along a trajectory. The shape of the vessels is such that there are no intersection between contours and its medial line does not self intersect.

Therefore, the surface model is defined as the union of generalized cylinders [12] which represent each segment of the graph. This proposed model fulfills the following properties:

- The skeleton curve $SKLT(t)$, i.e., the sweep trajectory, is the medial axis of the vessel segment. It is a linear approximation of the curve segment which guarantees that every approximated point has a distance less than K to the curve. Different representations of the surface can be obtained varying the value of K. However, to simplify the data structure, in the implementation, K is assumed constant.
- Each section of the generalized cylinder is defined by a Frenet reference Frame which defines a coordinate system based on three orthogonal unity vectors (see Figure 2): a tangent vector $T(t)$, a normal vector $N(t)$ and a bi-normal vector $B(t)$.

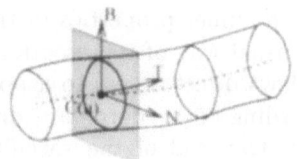

Fig. 2. Frenet Frame illustration.

Thus, the normal plane $L(t)$ at a point t of the skeleton $SKLT(t)$ can be defined with the vectors $N(t)$ and $B(t)$, and the vectors $B(t)$ and $T(t)$ of the normal plane specify the orientation of the curve.

- In each section, the contour curve $C(v)$ of vessel is approximated as a set of points on the normal plane $L(t)$. This approximation allows different levels of precision in the reconstruction from a simple polygon, a circle, an ellipse, to a spline [13]. In regular segments, the contour representation is implicit because the regular sections are circles and thus, it is enough to store their diameters. Cross sections in stenoses, aneuryms and branchings, generally more complex, are represented explicitly.
- It is sufficient to store sections only at points of the skeletal curve where a significant change of curvature or diameter exists and at end-points of the different features. In the regular segments, as the diameter of the cross section varies linearly, only one diameter per cylindrical segment and two diameters per conical segment need to be stored.
- There is a unique correspondence between the points of two consecutive sections. In the branching nodes, this correspondence is based on the criteria defined in [14] and [15].
- There is no blending between two consecutive cross-sections, as vessels do not present twists in their shape. Therefore a linear interpolation between section suffices [12].

The surface model distributed into each element of the graph structure is defined as the clipping region of the generalized cylinders against the node spatial region. Then, the surface data attached to a node is the skeletal curves and the set of end contour curves of the anatomical features which are concatenated in the node.

This surface representation enables the evaluation of the geometry of the boundary model with different levels of detail. The simplest evaluation is a union of generalized cylinders between nodes. A smoother one consists in polygonalizing the surface by tiling between contour curves ([14], [15]). Finally, it can be approximated by tensor product spline surfaces. This property of the model makes it suitable to the variety of operations that must be realized on it. Some operations, such as fluid simulations, require only simple cylindrical approximations while other ones, such as the navigations, need smoother surfaces.

2.3 Volume representation

The volume data represent the inner properties of the vessels, i.e. the density and the velocity. The most usual way of representing volume properties is the voxel model. This model allows direct access to a point, given its coordinates, and ordered traversals according to a coordinate direction. On the contrary, it does not provide a direct traversal of the vascular graph structure. From the occupancy point of view, it has high memory requirements. Even though, considering the sparse structure of the vessels, the voxels corresponding to the vascular map occupy only around a 2% of the model.

An alternative model is a compression of the voxel model using a run-length encoding [16], [17]. It is a suitable model for low occupancy data sets, especially if the original data is full of long constant property values sequences. A first difficulty with this coding is that the direct access to a voxel must be performed as a search on the run-length structure. [16] proposes to use additional indexed tables with pointers to the first voxel of each slice and each row in order to reduce the search to a column length. A second difficulty with a run-length encoded volume is that the volume can only be traversed optimally in the order in which the voxels are encoded. Thus, only a coordinate axis traversal of the model is optimal. Again, [16] solves this problem by precomputing three run-length encoding, one for each of the three coordinate axes. Finally, as the run-length is a spatial enumeration scheme, the volume traversal according to the topological order is, as in the voxel model, computationally expensive. This is a major drawback because topological traversals of the graph are needed in blood flow simulations as well as navigations.

To solve the latter problem the volume representation proposed herein keeps separate volume representation of all the segments of the graph elements (features). This allows a random access to the features, as required in navigations, while taking benefit of the spatial ordering of each separate voxel model. Specifically, the volume representation of each node and segment is a subset of the original voxel model enclosing the bounding box of the feature, such that only the voxels belonging to the feature have a property value (see Figure 3). This

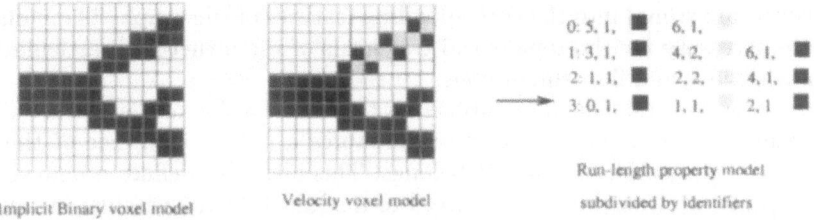

Implicit Binary voxel model   Velocity voxel model   Run-length property model subdivided by identifiers

Fig. 3. Run-length encoding of a property of the original model.

representation involves overlapping of spatially close features of the graph although the property values of vascular voxels are stored only once in the union of all these subsets. The volume models of each feature are coded as run-length according to the ownership of each voxel to the feature. It allows direct accesses to the voxel data of each feature, as well as ordered traversals according to a coordinate direction. Empty voxels of the overlapping regions are no coded several times. It takes advantage of the spatial structure of the voxel model. Thus, it allows a direct access to the property values of any voxel of a segment or a branch.

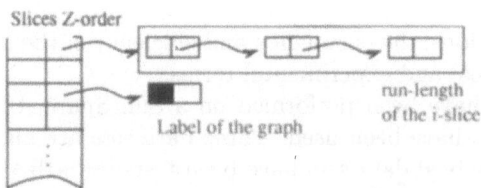

Fig. 4. Run-length encoding of the voxel model in relation to the features of the model.

Keeping the original voxel model in addition to the features volume schemes is redundant and unaffordable in most applications environment because of the memory limitations. Thus, the initial voxel model is segmented and it is compacted by a run-length encoding method on the basis of the vessel ownership property value of each voxel (see Figure 4). This model is necessary to keep the implicit spatial order of the voxel model and to compress the sparseness of the data, without loosing the direct access of a voxel.

2.4 Management of the three information layers

The main characteristic of the three layers model is that it provides simultaneous access to the topology, the surface and the properties. These different queries are managed as follows:

- All the accesses from the graph to the surface (**graph-to-surface**) and to the volume data (**graph-to-volume**) are direct because the surface and the voxels are stored into the corresponding features of the graph. These queries are necessary for the topological traversals of the structure, particularly in navigations and flow simulations.

- In order to establish the relationship between a voxel and a feature of the graph (**voxel-to-graph**), each voxel stores a simple identifier of the segment of the graph to which it belongs. This identifier defines a new volume property of ownership of each voxel to a segment of the vascular structure.
- The access **voxel-to-surface** is solved implicitely. To know if a voxel is traversed by the vessel surface, the feature to which the voxel belongs can be accessed. Next the surface of the feature should be evaluated. This can be done in several ways: at low surface resolution (cylinder test), with more complex test (B-splines test). A different approach consists in simply exploring the set of neighbor voxels and analyzing the gradient value. This latter solution is the classical one used in simple volume models.
- The access **surface-to-voxel** can be performed directly with a given surface point. The voxel to which the point belongs is directly retrieved from the point coordinate values and then, from the voxel, the identifier of the segment of the graph is found (**surface-to-graph**).

3   Evaluation

In the following sections, the occupancy and efficiency of the model are analyzed in terms of theoretical and experimental terms.

All simulations have been performed on a Sun Sparc, Ultra-1 with Solaris 5.5.1. Five data sets have been used. Three data sets are phantom data of different sizes. Two medical data sets have been tested: a MRA data and a SCTA (Spiral Computed Tomography Angiography) data sets. A segmentation preprocess is performed before building the model.

3.1   Occupancy

Let n^3 be the number of voxels of the original data set, and n_b^3 the number of voxels of the bounding box. Let n_v the number of voxels which contains blood vessels. n_v is $0.02 \times n_b^3$. Let n_f be the number of total features of the vascular data with a mean size of n_{ss}^3 voxels.

A classical voxel model of the whole volume requires $28 \times n^3$ bytes (1 integer for the density value and 3 floats for the velocity value at each voxel). A run-length compression of the boundary box representation needs $28 \times 0.02 \times n_b^3 + 6 \times 4 \times n_b^2$ bytes, where the last adder is the occupancy of the three tables ordered according to the three main axis respectively and assuming that the number of features can be coded with one integer. If a double structure of volume data is stored in order to keep both spatial and ordered traversal, the run-length model of the whole data, according to the identifier of the features, takes $0.02 \times n_b^2 \times 8 + 6 \times 4 \times n_b^2$, where the first adder is a coarse approximation of the real cost $n_b^2 \times (log_2 n_f + log_2 n_b)$ bits of a general case of run-length.

In addition to the primary model (i.e. the run-length of labeled voxels) storage requirements, a secondary model for each feature must be coded. If these secondary models are coded with boundary boxes, the mean global occupancy is $28 \times n_f \times n_{ss}^3$ bytes, whereas if these secondary models are run-length code according the ownership of the feature, the memory needed is $28 \times 0.02 \times n_b^3 + 6 \times 4 \times n_{ss}^2$ bytes.

The memory space required to store the graph and the surface is the following. First, the occupancy of the graph is $2 * e$, where e is the number of the edges between branchings. In addition, if each feature has, in mean, three control points and the number of points for each contour is at most 20, the surface representation needs $20 \times 3 \times n_f \times 16$ bytes.

Table 1 shows the memory requeriments of the symbolic model of different input data sets, as well as, the memory requeriments of the classical voxel model and of the mesh of triangles model.

The reduction of the memory requirements ranges from 37.74% to 95, 85%. The maximum efficiency is achieved when the occupancy of the data is low and when the distribution of the data in the volume is more concentrated. This is the case of the model MRA which represents blood vessels. By opposite, model Phantom 3, which represents a skull has also a relatively low occupancy but the data are spread over the whole model. Therefore the compression rate is lower. Finally SCTA produces an overhead of memory requirement because the data, interleaved surface layers, occupy the whole original volume. Summarizing, the model proves to be efficient for sparse and concentrated structures.

Table 1. Comparison of memory requeriments with different set of data.

Data set	Original resolution	Occupancy	Voxels size (Kbytes)	Model size (Kbytes)	Memory reduction
Phantom1	32x32x32	12.72%	32.768 Kb	20.400 Kb	37.74%
Phantom2	64x64x27	8.0%	110.592 Kb	113.484 Kb	102.61%
Phantom3	96x96x69	0.7%	635.904 Kb	129.308 Kb	79.66%
SCTA	128x128x42	21.0%	9633.792 Kb	1597.584 Kb	83.41%
MRA	256x256x124	0.2%	8126.464 Kb	640.080 Kb	92.12%

3.2 Efficiency

The efficiency of the model is next measured in terms of the costs of the visualizations. These costs do not account the pre-process of model construction, which involves a first segmentation step, the medial axis extraction, its processing to automatically detect its features and the construction of the data structures ([18]). The run time of the construction algorithm depends on the model occupancy and its original size. It ranges from seconds for small models to minutes for the SCTA and the MRA data sets.

Figures 5 and 6 show a simplified visualization of topology of the model. Edges segments are approximated by polylines with a circle contour curve at the significant points of the axis. Branching are shown with spheres. This visualization, very fast, is used to manage the user interface of the vascular model and to render the topological relations between different blood vessels.

The visualization of the vascular surface can be performed at different levels of detail such as: simple cylinders at the segments and spheres at the nodes, polygonal approximation of the surface by interpolating contour curves between the curves stored in the model and tiling between curves or by fitting biquadratic splines parameterized with rectangles between successive contours [6].

Figure 7, shows two levels of refinement of the same vessel segment. The first image uses 5 points per contour and no interpolation between the contours stored in the model. The second image uses 20 points per contour and it interpolates 5 contour curves per pair of stored contours. The number of faces is about 100 and 900 respectively. These different levels of detail are useful when the whole model is visualized, and also in navigation where the level of refinement can be computed in relation to the distance of the feature from the observer. Figure 8 shows an image of a walkthrough sequence. The navigation takes advantage of the topological model by clipping nodes and edges that are outside the viewing frustum. Experimentally, the number of edges and nodes to be render simultaneously in a navigation inside the vessels is almost constant, about 4 features. The navigation is therefore much more faster than a complete projection of the model.

The visualization of volume model may be performed by slice-to-slice composition, by MIP projection and by semi-transparent shaded rendering. As an example of the shaded rendering, Figure 9 shows a BTF (Back-To-Front) visualization. Figure 10 shows several boundary boxes of the edges of the graph. Each voxel has a different color depending on the feature to which it belongs. It can be seen that although the boxes may overlap, every vascular labeled voxel belongs only to one box.

Table 2 shows the performance of the volume visualization algorithm, based on the run-length encoding in comparison to full voxel models visualization. The efficiency depends on the level of compactness of the run-length. Results show that the cost decreases a lot in MRA data sets, providing real time rendering. Table 3 compares the cost of a walkthrough driven by the topological model versus a direct rendering of the whole data structure at each frame. The same reasoning as for the surface walkthrough can be done: the approximate number of features visualized per frame is constant and small.

4   Conclusions and future work

The main contribution of this paper is the proposal of a representation scheme for blood vessels. It is a multiresolution hybrid model, which stores symbolic information on the topology and geometry as well as volume and surface data.

Table 2. Performance of the volume BTF-visualization algorithm.

Data set	Original resolution	Voxels size (Kbytes)	Run time voxels	Run-length size (Kbytes)	Run time run-length
Phantom1	32x32x32	32.768 Kb	0.4 sec.	15.992 Kb	0.127 sec.
Phantom2	64x64x27	110.592 Kb	1.48 sec.	108.444 Kb	0.36 sec.
Phantom3	96x96x69	635.904 Kb	2.62 sec.	110.588 Kb	0.17 sec.
SCTA	128x128x42	9633.792 Kb	7.50 sec.	384.676 Kb	2.61 sec.
MRA	256x256x124	8126.464 Kb	28.68 sec.	487.280 Kb	0.38 sec.

Table 3. Performance of the volume walkthrough algorithm.

Data set	Original resolution	Projection whole data	Walkthrough model
Phantom1	32x32x32	0.127 sec.	0.127 sec.
Phantom2	64x64x27	0.36 sec.	0.25 sec.
Phantom3	96x96x69	0.17 sec.	0.045 sec.
SCTA	128x128x42	2.61 sec.	0.304 sec.
MRA	256x256x124	0.38 sec.	0.18 sec.

These characteristics provide flexibility and they enable the realization of all the physicians requirements in a Computed Assisted Neurovascular System.

The model has been implemented and tested with different visualization strategies. Results show that for MRA data a compression rate of 90% can be achieved. Visualizations reduce impressively their cost, especially navigations driven by the topological model.

Work on progress is the integrated visualization of the blood flow velocity with the vascular surface. Another question is how to improve the control of the refinement level by automatically tuning the parameter, which determines the number of control points stored in the coding of the axis model.

This work has been funded by TIC99-1230-C02-02.

References

[1] A. Savopoulos G. Sakas, M. Grimm. Optimized maximum intensity projection (mip). *6th Eurographics Workshop on Rendering*, pages 81–93, June 1995.

[2] G. Gerig, R. Kikinis, and F.A. Jolesz. Image processing of routine spin-echo mr images to enhance vascular structures: Comparison with mr angiography. *3D Imaging in Medicine: Algorithms, Systems and Applications*, pages 121–132, 1990.

[3] H.C. Cline, W.E. Lorensen, R. Kikinis, and F. Jolesz. Three-dimensional segmentation of mr images of the head using probability and connectivity. *Journal of Computer Assisted Tomography*, 14(6):1037–1045, November/December 1990.

[4] C. Barillot, B. Gibaud, J. Scarabin, and J. Coatrieux. 3d reconstruction of cerebral blood vessels. *IEEE Computer Graphics and Applications*, 5(12):13–19, December 1985.

[5] C. Zahlten, H. Jurgens, and H.O. Peitgen. Reconstruction of branching blood vessels from ct-data. *Proceedins of Rostock, Eurographics Workshop on Visualization in Scientific Computing*, 1994.

[6] H.H. Ehricke, K. Donner, W. Koller, and W. Straßer. Visualization of vasculature from volume data. *Computer and Graphics*, 18(3):395–406, 1994.

[7] D. Vandermeulen, R. Verbeeck, L. Berben, D. Delaere, P. Suetens, and G. Marchal. Continuous voxel classification by stochastic relaxation: Theory and application to mr imaging and mr angiography. *Internal Research Report, Medical Imaging Research Lab, ESAT, Belgium*, 1994.

[8] W. Heidrich, M. McCool, and J. Stevens. Interactive maximum projection volume rendering. *Proceedings of the IEEE Visualization 95*, pages 11–18, October 1995.

[9] K.H. Hohne, M. Bomans, A. Pommert, M. Riemer, C. Schierds, U. Tiede, and G. Wiebecke. 3d visualization of tomographic volume data using the generalized voxel-model. *The Visual Computer*, 6(1):28–36, February 1990.

[10] K.J. Zuiderveld. Vr in radiology - first experiences at the university hospital utrecht. *ACM Computer Graphics*, pages 47–48, November 1996.

[11] A.V. Aho, J.E. Hopcroft, and J.D. Ullman. *Data Structures and Algorithms*. Addison-Wesley, 1983.

[12] M/S. Kim, E.J Park, and H.Y. Lee. Modelling and animation of generalized cyllinders with variable radius offset space curves. *The Journal of Visualization and Computer Animation*, 5:189–207, 1994.

[13] T. Maekawa, N.M. Patrikalakis, T. Sakkalis, and G. Yu. Analysis and applications of pipe surfaces. *Computer Aided Geometric Design*, 1997.

[14] J.D. Boissonnat. Surface reconstruction from planar cross section. *Proceedings IEEE Computer Society Conference on Computer Vision and Pattern Recognition*, pages 393–397, 1985.

[15] K. Anjyo, T. Ochi, Y. Usami, and Y. Kawashima. A practical method of constructing surfaces in three-dimensional digitized space. *The Visual Computer*, 3(1):4–12, February 1987.

[16] P. Lacroute. Fast volume rendering using a shear-warp factorization of the viewing transformation. Technical report, Departments of Electrical Engineeering and Computer Science, Stanford University, Setember 1995.

[17] G.T. Herman and H.K. Liu. Three-dimensional display of human organs from computed tomograms. *Computer Graphics and Image Processing*, 9(1):1–21, 1979.

[18] A. Puig. *Contribution to Volume Modeling and to Volume Visualization*. PhD thesis, Software Department, Universitat Politecnica de Catalunya, Barcelona, Spain, October 1998.

Editors' Note: see Appendix, p. 289 for colored figures of this paper

Direct Volume Rendering from Photographic Data

David Ebert[1], Tim McClanahan[2], Penny Rheingans[1], and Terry Yoo[3]

[1] Department of Computer Science and Electrical Engineering
University of Maryland, Baltimore County
[ebert,rheingan]@cs.umbc.edu
[2] Laboratory for Extraterrestrial Physics
NASA Goddard Space Flight Center
xrtpm@lepxgrs.gsfc.nasa.gov
[3] Office of High Performance Computing and Communication
National Library of Medicine
yoo@nlm.nih.gov

Abstract. Direct volume rendering from photographic volume data has the potential to create realistic images of internal volume structure, as well as the structure of boundaries within the volume. While possession of the photographic volume simplifies color calculations in voxel illumination, it complicates opacity calculation. This paper describes a framework for addressing illumination challenges in photographic volume data and presents initial results.

1 Introduction

In recent years, a few photographic volume data sets have become available. The most widely used of these are those of the Visible Human Project (VHP) at the National Library of Medicine[15], but other examples are being created as well. This type of data offers exciting possibilities for realistic volume visualization, since correct color values are known for each voxel. Applications include medical illustration, surgical simulation, and general scientific education. Photographic volume data also offers a challenge to traditional volume rendering techniques.

Traditional direct volume rendering produces an image from a volume of scalar data, using transfer functions from scalar value to color and opacity. For some common data types, such as medical CT volumes, designing an opacity transfer function can be straight-forward. One common approach is to assign high opacity to voxels containing high scalar intensity values, for instance those containing bone. The exact transfer function from density to opacity can be modified to suit the purpose of a particular visualization, but one with monotonically increasing opacity is generally effective. For other data types, such as medical MRI volumes, effective opacity transfer functions are more difficult to design, but the basic problem remains one of mapping one scalar value into another.

The design of the color transfer function, from the scalar value to an RGB triple, is more open-ended. One transfer function might try to map each density value to the color of the material of that density in order to produce a realistic image. If each density uniquely corresponded to a single material type, and each material was a uniform color, such a transfer function would be well-defined. Unfortunately, in medical volumes, both conditions are violated, making realistic color recovery from density values problematic.

Realistically visualizing photographic volume data, where each voxel has a vector RGB value, turns the traditional transfer function design problem inside-out. Now, the color transfer function becomes trivial, since the actual color at each voxel is explicitly known. Design of an effective opacity transfer function is now the major challenge, since neither individual nor average color component values have a natural correspondence with desired opacity.

One fairly successful approach for rendering from photographic volume data has been to compute surfaces or opacity values from an auxiliary volume, for instance a CT volume, and use the photographic data simply for color information. The use of surface techniques results in good representation of the outer boundaries of the object, but reveals little or nothing of the internal structure. With either auxiliary volume approach, the density volume must be registered with the photographic volume, a challenging process which is difficult to automate.

We describe a framework for volume rendering directly from photographic data without the need for a secondary scalar volume to indicate density. Direct volume rendering from the photographic volume eliminates the need to register volumes from different sources and enables the display of internal volume structure along with material boundary information. We describe results of our preliminary experiments using this framework and suggest fruitful directions of continued inquiry.

2 Photographic Volume Data

Photographic color data is becoming increasingly important as a volume information representation. While the acquisition of such data often requires the loss of physical integrity of the sample, tomographic sectioning has been important in anatomy and in pathology. With the growing capability of aggregating multiple photographic cutplane images into 3D volumes, these techniques are growing in importance in other fields. Limitations of MR imaging to generate adequate resolution led the Whole Frog Project at Lawrence Berkeley Laboratory to create an entire frog dataset using cryosection [13]. The use of this dataset is growing as a basis for teaching dissection in biology. The Laboratory for Neurological Imaging at UCLA commonly uses cryosectioning to gain the resolution and contrast required for their intricate analysis of the brain's pathways [16]. Commercial groups now offer mechanical sectioning technology that is used in volume data analysis for geology, medicine, and semiconductor manufacturing [5]. The National Library of Medicine produced one of the most important examples of data acquired through photographic tomographic imaging.

The Visible Human Project was formed to explore the use of digital imaging technology in modern anatomy research and education. A panel study in 1990 recommended as a first project the construction of a digital image library of volumetric data of a normal adult human female and male subject. Data from two subjects, one male and one female, were collected through a variety of methods including the conventional radiological techniques of X-ray CT studies, magnetic resonance imaging MRI, and plain film radiographs. In addition to these conventional clinical studies, the subjects were frozen and sectioned at 1 mm (male subject) and 1/3 mm (female subject) intervals. The exposed surfaces were photographed with 35 mm and 70 mm film and digitized with an electronic camera. Image acquisition was carefully performed and the resulting data is one of the most complete anatomical studies ever performed [15].

Each slice of the digital cryosection data was acquired with a raster resolution of 2048 x 1216 pixels with a horizontal field of view of 25 inches. Voxel dimensions are 0.32 x 0.32 x 1mm in the male dataset. Reduced cost of image storage, networking, and data handling combined with the desire for cubic voxel dimensions prompted more aggressive data acquisition of 0.33 x 0.33 x 0.33 mm voxel resolution in the female dataset. Each subject was encased in frozen gelatin, dyed blue with food additives, to provide physical stability as well as inter-slice opacity. As each slice of the frozen specimen was exposed, the surface was cleaned and sprayed with ethanol to reduce diffuse reflections from frost, masked to eliminate glare from the insulating materials, and a color platen was placed in the field of view for reference. The resulting surface was then photographed with two film and one CCD cameras. The block was dressed with a tray of dry-ice, to refreeze the exposed surface, re-sectioned, and the process repeated. Voids exposed through the sectioning process were filled with blue latex to stabilize the walls, prevent the collection of debris, and to block projection of deeper structures onto the image plane. The blue tint was chosen to ease the subtraction of non-anatomical structures added during the data collection process. Figure 1 shows a sample cryosection cross-section, transecting the eyes and optic nerves of the male subject. The voids of the nasal cavities have been filled with blue latex. Embedded fiducials are visible in the corners. A platen with CYM color dots and an intensity gradient are visible at the bottom of the image for color and grayscale correction. Over 1800 visible light images were acquired for the male, and over 5000 visible light images were collected for the female subject. The resulting datasets are 14 gigabytes and 40 gigabytes respectively.

3   Related Work

Volume rendering as a image generation and reconstruction technique was pioneered in computer graphics by Levoy [9] as well as by Drebin et al. [2]. Both of these early papers included medical data acquired using X-ray CT as examples. X-ray attenuation is a physical property that is particularly amenable to assigning a color transfer function. Levoy in particular explicitly includes color assignment and shading in his rendering pipeline. These initial implementations and much of the derivative work used pseudocolor to improve the visualizations produced through volume rendering. A comprehensive survey of volume rendering is provided by Kaufman [7]. More recent work employing hardware 3D texture graphics capabilities for volume rendering more easily accommodates color data [1]. However, these methods must still generate an opacity transfer function to enable correct compositing of the image planes. Working from segmented data simplifies the problems of opacity assignment, but includes a huge labor investment in hand segmentation. Recent renderings of segmented volume color data of the VHP thorax were published by Zhou [18].

Other techniques for extracting visual information from volumes include isosurface extraction techniques such as the Marching Cubes algorithm [11]. This algorithm and its relatives have been used extensively to display the information within volume data, and have been successfully used on the Visible Human data. However, early work by Lorensen on the Visible Human data either used the X-ray CT data, or separated the red channel of the RGB images to generate isosurfaces from scalar rather than multivalued data [10]. Similar work by the Vesalius group at Columbia University extracts isosurfaces from the color data, and later uses the original color volume as a solid texture to apply color information to the extracted polygonal surfaces [6]. Both of these approaches have involved strictly surface, rather than volume, rendering.

Researchers working with the Visible Human data have analyzed the color gamut of the male dataset in their work on photorealistic volume rendering and virtual dissections [8, 12]. Since reflectance of the light rays is often based on gradients measured in the object volume, careful consideration of the color spaces involved should be part of the work. Sapiro showed that selection of the color space can make dramatic differences when attempting nonlinear image processing [14].

4 Volume Shading

The determination of opacity values and other non-color material properties during the illumination process is a key challenge of direct volume rendering of photographic volumes. Although color information is available in the photographic description of each voxel, the data contains no information on the viewpoint dependent reflective or light transmission properties of the voxels.

The color values of each voxel describe the total reflection from that voxel for a specific set of lighting and viewing parameters. Specifically, the color shows the voxel lit and viewed from directly above. If we treat this total reflection as just diffuse reflection, we can use the same color as the reflection from other viewpoints, since diffuse reflection is independent of view direction. While using a correct bidirectional reflection distribution function for the voxel would yield more accurate results, determining the BRDF from photographic data is still an open research question. Considering the reflection to be simply diffuse results in a reasonable first approximation.

Determining the transmission properties, both opacity and attenuation properties, of each voxel is a larger problem. As with detailed reflection properties, this information is not available in the data. Unlike detailed reflection properties, it cannot simply be ignored. Rendering the volume with constant opacity yields unsatisfying results where little structure of the volume is clearly visible.

While using transmission properties that mimic those of the actual tissue types is attractive, segmentation from color data is particularly difficult. In particular, unlike the density values in CT data, there is no unique relation between color and tissue type.

Our initial approach has been to perform a variant of gradient-based shading on the photographic volume. Gradient-based shading emphasizes boundaries between regions with different properties, in this case different colors. Unlike threshold-based surface rendering approaches, however, gradient-based direct volume rendering still shows regions of lesser gradient. Surface rendering shows only boundaries, while volume rendering can show the entire volume.

5 Implementation

Implementation considerations included the correct color space for gradient calculation and the specific gradient calculations themselves. We also found it useful to use some simple semi-automatic methods to extract segments of the data of particular interest. We chose to apply our methods on the brain section of the data since it represents a comparatively small anatomical volume and contains non-unique color attributes to challenge segmentation processing algorithms. A total of 139 sequential images contained portions of the brain.

5.1 Segment Extraction Preprocessing

In addition to visualizing the entire head, we wanted to experiment with visualizations of just the brain. To this end we used a sequence of semi-automatic filters to strip away unwanted tissue. First, voxels with colors unrepresentative of brain tissue were first removed. Essentially, this step left only voxels colored with a band of shades of yellow and brown. The HSV color model was used to specify desired colors, because of the lack of general intuition about the RGB components of flesh tones. This step stripped away much unwanted tissue, but left outliers in non-target regions that happened to match the specified color bands, as well as gaps within brain tissue in areas where the colors deviated from the specified bands. Next, a 3D flood fill process was used to select only those voxels meeting connectivity requirements with manually selected seed points. After this step, geographically isolated outliers have been eliminated, but internal gaps remain. Finally, an internal dilation filter is used to fill in interior gaps without dilating external image areas.

5.2 Color Space Transformation

RGB image slice values were converted to CIE l*u*v space to obtain a perceptually uniform representation of the color volume. A perceptually uniform color space has the characteristic that equal distances in the color space correspond to equal perceptual differences, at least for reasonably small distances. Using a perceptually uniform color space for gradient calculations allows us to emphasize those features which are noticeable in the photographs, creating a more realistic volume rendering than using a device-derived color space, such as RGB. The CIE l*u*v color space, in particular, also offers the advantage that chromatic and achromatic components of color are described by orthogonal color space dimensions. In the future, this feature will allow us to experiment with biased weighting of the chromatic and achromatic color components, just as the human visual system performs certain scene understanding tasks with segregated achromatic and chromatic color information.

Color space conversions were performed using the methods of Hall [4]. Because precise specifications of the color primaries of the image data were not directly available, we approximated them by the NTSC standard primaries. Although deviations of actual primaries from these are expected to be modest, this approximation does not guarantee the correct absolute CIE l*u*v coordinates for voxels, compromising the device independence of the color space. However, the approximation should preserve the relationships between points in the color space, which is our primary concern since relative judgment, rather than absolute, are the basis for almost all perceptual processes. Specification of the image data color primaries is indirectly available from the color calibration card included in each photograph. Analysis of the RGB values for these physically measurable colors could provide the required calibration information. In fact, analysis of the calibration card colors could address another problem with the Visible Human data, that of inconstant color calibration from slice to slice. This problem presumably results primarily from changes in apparatus illumination over the course of the data collection process, and is apparent as dark and light stripes across coronal or sagital slices. Since we know that the card did not undergo substantial changes in color over the course of data collection, color differences from card image to card image must represent changes in illumination and camera parameters. Ideally, a single process could be used to calibrate the image data to the color card, as well as to the CIE color primaries. We have not yet undertaken this process, but as better calibration data becomes available, the approximations made in the color model transformation could simply be corrected.

5.3   Rendering Process

We performed rendering experiments on both the unsegmented and the segmented brain photographic (rgb) data (139x600x500), which were converted to CIE l*u*v space. The volume gradient was approximated using a standard central difference approximation with a change to account for the distance of the colors in the CIE l*u*v space instead of using the voxel density differences:

$$gradient.x = CIE\_distance(voxel[x-1][y][z], voxel[x+1][y][z])$$
$$gradient.y = CIE\_distance(voxel[x][y-1][z], voxel[x][y+1][z])$$
$$gradient.z = CIE\_distance(voxel[x][y][z-1], voxel[x][y][z+1])$$

where

$$CIE\_distance(voxel1, voxel2) =$$
$$\sqrt{(voxel1.l - voxel2.l)^2 + (voxel1.u - voxel2.u)^2 + (voxel1.v - voxel2.v)^2}.$$

For the density of each voxel, the magnitude of the gradient was used. The volume renderer used is a modified volume ray tracer that uses atmospheric accumulation, attenuation, illumination, and shadowing [17,3]. The opacity transfer function that was used is the following:

$$opacity = (gradient\_magnitude * scalar)^{exponent}$$

6 Results

The results demonstrating the effectiveness of using the CIE l*u*v color space distance as the basis for volume rendering from photographic data can be seen in Figures 2a and 2b. Figure 2a shows the brain rendered with a lower opacity scalar, 0.9, and an exponent of 1. Figure 2b shows a more opaque volume created by a higher opacity scalar, 1.0, and using an exponent less than 1 (specifically, 0.8) to increase the importance of small gradient changes within the volume. Both images highlight the tissue boundaries between the grey matter and the Corpus Collosum and also reveal a lateral ventricle within the brain.

Figure 3a, 3b, 3c, and 3d show volume rendering from the unsegmented photographic data. Figure 3a shows a side-view image generated by a higher exponent value (exponent=1.2) making the interior tissue structures more transparent. Figure 3b shows the same view image generated with a lower exponent (exponent=0.9) to increase the importance of gradients within tissue types. Both of these images show tissue boundaries between grey matter and the Corpus Collosum, a lateral ventricle, portions of the skull, and the sinus cavity. Figures 3c and 3d are top view images of the unsegmented photographic data showing the range of images that can be generated by varying the exponent to show all internal tissues (Figure 3c, exponent=0.5) or to show primarily tissue boundaries (Figure 3d, exponent=1.0).

7 Summary and Future Directions

We have proposed a basic framework for addressing the rendering challenges presented by photographic volume data. Preliminary results show that color gradient in a perceptually uniform color space can be used to perform gradient-based volume rendering from photographic data.

Many open issues remain. These include color calibration and internal registration methods, the optimal weighting of color space dimensions in gradient calculations, more realistic reflectance models, and determination of segment-based densities directly from the photographic volume.

8 Acknowledgments

This work supported in part by the National Science Foundation under Grants No. ACIR 9996043 and ACIR 9978032.

References

1. B. Cabral, N. Cam, and J. Foran. Accelerated volume rendering and tomographic reconstruction using texture mapping hardware. In *1994 Symposium on Volume Visualization*, pages 91–98, 1994.
2. R.A. Drebin, L. Carpenter, and P. Hanrahan. Volume rendering. In *Computer Graphics (SIGGRAPH '88 Proceedings)*, number 22, pages 65–74, 1988.
3. David S. Ebert and Richard E. Parent. Rendering and animation of gaseous phenomena by combining fast volume and scanline A-buffer techniques. In Forest Baskett, editor, *Computer Graphics (SIGGRAPH '90 Proceedings)*, volume 24, pages 357–366, August 1990.
4. Roy Hall. *Illumination and Color in Computer Generated Imagery*. Springer-Verlag, 1989.
5. T. Hazeldine. Annular slicing: Physics, materials, geology, sedimentology and medicine make use of precision annular cutting. so does the semiconductor industry. *Semiconductor*, Oct. 1997.
6. C. Imielinska. Technical challenges of 3d visualization of large color data sets. In R. A. Banvard and P. Cerveri, editors, *Proceedings of the Second Visible Human Project Conference*. US Department of Health and Human Services, Public Health Service, National Institutes of Health, October 7-8 1998.
7. Arie Kaufman. Volume visualization. *ACM Computing Surveys*, 28(1):165–167, 1996.
8. J. Kerr. Photorealistic volume-rendered anatomical atlases and interactive virtual dissections of the dissectible human. In R. A. Banvard, editor, *Proceedings of the Visible Human Project Conference*. US Department of Health and Human Services, Public Health Service, National Institutes of Health, October 7-8 1996.
9. Marc Levoy. Display of surfaces from volume data. *IEEE Computer Graphics and Applications*, 8(3):29–37, 1988.
10. William E. Lorensen. Marching through the visible man. In *Proceedings of IEEE Visualization '95*, pages 368–373. IEEE Press, October 1995.
11. William E. Lorensen and H. E. Cline. Marching cubes: A high resolution 3d surface construction algorithm. In *Computer Graphics (SIGGRAPH '87 Proceedings)*, volume 21, pages 163–169, 1987.
12. J. Marquez. Radiometric inhomogeneities in the color cryosection images of the vhp. In R. A. Banvard, editor, *Proceedings of the Visible Human Project Conference*. US Department of Health and Human Services, Public Health Service, National Institutes of Health, October 1-2 1996.

13. W. Nip and C. Logan. Whole frog technical report. Technical Report LBL-35331, University of California, Lawrence Berkeley Laboratory, 1991.
14. G. Sapiro and D. Ringach. Anisotropic diffusion of multivalued images with applications to color filtering. *IEEE Transactions of Image Processing*, 5:1582–1586, 1996.
15. S. Spitzer, M. J. Ackerman, A. L. Scherzinger, and D. Whitlock. The visible human male: A technical report. *Journal of the Americal Medical Informatics Association*, 3(2):118–130, 1996.
16. A. W. Toga and J. C¿ Mazziotta. *Brain Mapping: The Methods*. Academic Press, 1996.
17. Roni Yagel, David S. Ebert, J. N. Scott, and Y. Kurzion. Grouping volume renderers for enhanded visualization in computational fluid dynamics. *IEEE Transactions on Visualization and Computer Graphics*, 1(2):117–132, June 1995. ISSN 1077-2626.
18. Ruixa Zhou and Earl Henderson. Visualization of the visible human anatomical images. In R. A. Banvard and P. Cerveri, editors, *Proceedings of the Second Visible Human Project Conference*. US Department of Health and Human Services, Public Health Service, National Institutes of Health, October 1-2 1998.

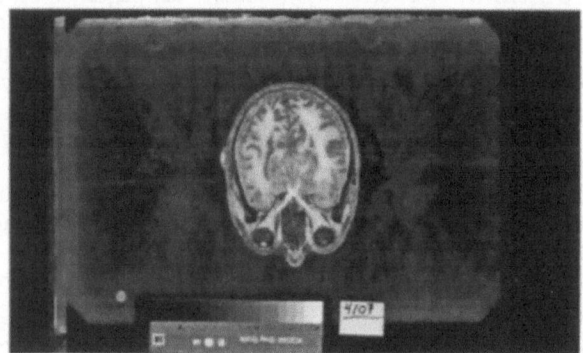

Fig. 1. Sample slice of the digital photographic data transecting the eyes and optic nerves of the Visible Human Male Dataset.

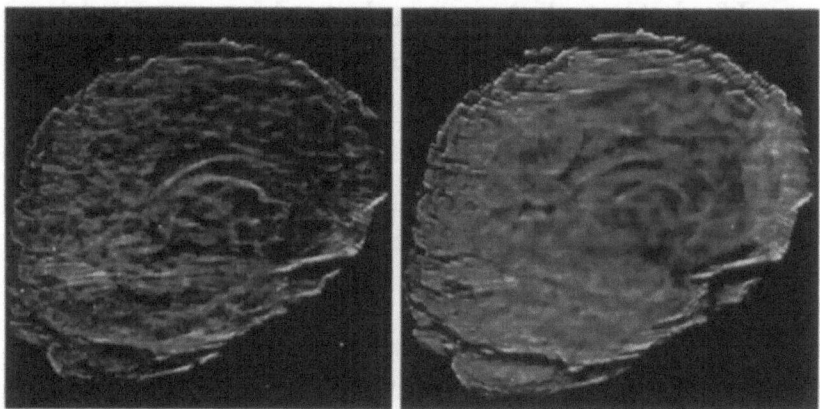

Fig. 2. Volume rendering of segmented photographic Visible Male brain, without and with enhancement of low gradient areas.

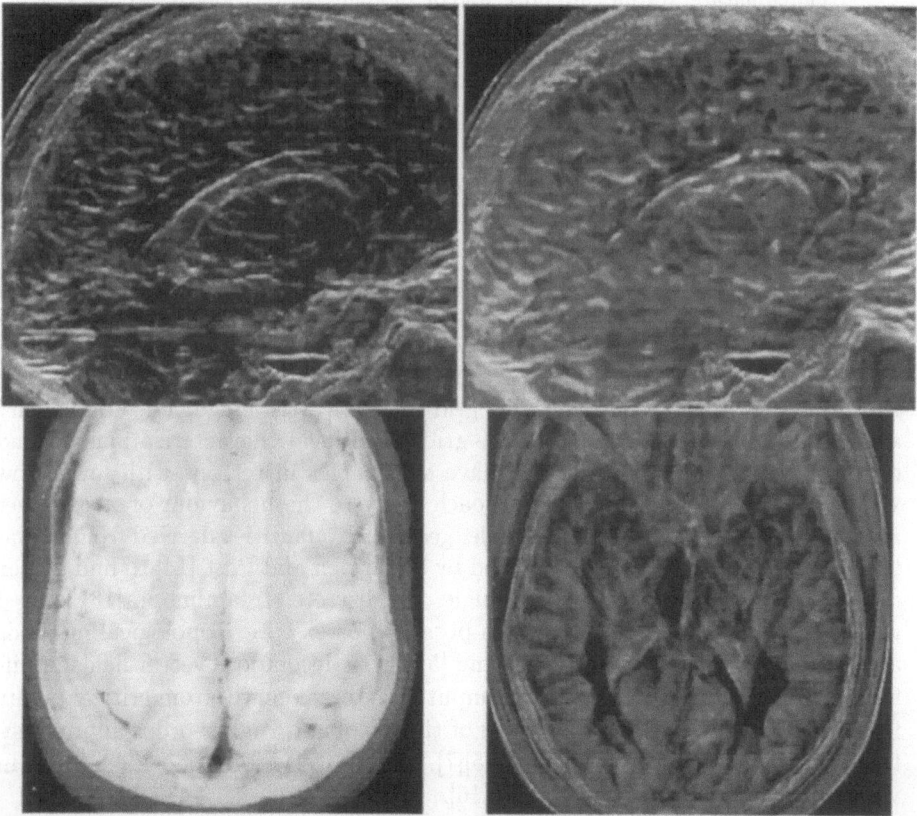

Fig. 3. Volume rendering of photographic Visible Male head, without and with enhancement of low gradient areas from side and top views.

Variational Approach to Vector Field Decomposition

Konrad Polthier and Eike Preuß

Technische Universität Berlin
{polthier, preuss}@math.tu-berlin.de

Abstract. For the feature analysis of vector fields we decompose a given vector field into three components: a divergence-free, a rotation-free, and a harmonic vector field. This Hodge-type decomposition splits a vector field using a variational approach, and allows to locate sources, sinks, and vortices as extremal points of the potentials of the components. Our method applies to discrete tangential vector fields on surfaces, and is of global nature. Results are presented of applying the method to test cases and a CFD flow.

1 Introduction

Features of vector fields strongly affect the characteristics of flows and their physical behavior. Among the most important features are vortices and pairs of sources and sinks. In many applications, vortices must be avoided to avoid energy losses, or sources and sinks must be located, for example to understand atmospheric behaviors. Although feature analysis is an important area only few technical tools are available for their detection and visualization.

A number of heuristic criteria are currently used for vortex detection [11][6]. Physical quantities of the underlying grid such as vortex magnitude and helicity are located at isolated vertices of the grid. Such local characterizations depend on the chosen neighbourhood and have deficiencies in regions with lower flow velocity. A slightly more global approach analyses the behaviour of streamlines and other integral curves. For example, geometric quantities derived from curvature properties of streamlines are used by Sadarjoen and Post [14] to find vortex cores, and the polyhedral winding angle of a discrete streamline in [13] is used to detect closed streamlines around a possible vortex core. Topological methods as introduced by Helman and Hesselink [5] try to decompose vector fields in different global regions of interest by computing integral curves from critical points found by local linear approximations of the Jacobian. Higher-order approximations yield different decomposition [15][7]. For an overview of known methods in vector field visualization see Teitzel [16].

In this paper we present a variational approach for the decomposition of a given vector field in different components, a divergence-free, a rotation-free, and a harmonic remainder. Instead of trying to define a discrete version of the Jacobian $\nabla \xi$ of the discrete vector field and its splitting in a stretching tensor

S and a vorticity matrix Ω, we derive a Hodge type decomposition by minimizing certain energies. This more global point of view reduces the dependency on local measurement inaccuracies and discretization artifacts. Our approach comes along the lines with a definition of discrete differential operators of higher differentiability order on piecewise linear functions and vector fields [9].

The application of our decomposition is two-fold. First, the derived vector field components have distinguished properties which are mixed in the original vector field. Second, two components are the gradient respectively the co-gradient of potential functions. The existence of potential functions allows to identify features of the original vector field as local extrema of the associated potentials which are easily detectable.

2   Related Work

Methods for direct vortex detection are often based on the assumption to have regions with high amounts of rotation or of pressure extrema. See for example Banks and Singer [1] for an overview of possible quantities to investigate. The deficiencies of first-order approximations have been widely recognized, and, for example, tried to overcome with higher-order methods [12][15].

The Jacobian $\nabla \xi$ of a differentiable vector field ξ in $\mathbb{R}^2$ and $\mathbb{R}^3$ can be decomposed in a streching tensor S and a vorticity matrix Ω

$$\nabla \xi = \frac{1}{2}(\nabla \xi + \nabla \xi^t) + \frac{1}{2}(\nabla \xi - \nabla \xi^t) =: S + \Omega.$$

The eigenvalues of the diagonal matrix S correspond to the compressibility of the flow, and the off-diagonal entries of the anti-symmetric matrix Ω are the components of the rotation vector.

In the present paper we choose a different approach by defining higher order differential operators. On discrete curved surfaces it is rather non-trivial to define higher order derivatives point-wise and in local coordinates. Therefore we follow a different approach and define discrete differential operators as total quantities being integrated over local regions. For example, the discrete divergence of a vector field is defined at an edge midpoint as an integrated quantity over the adjacent triangles. Similar concepts are, for example, used for the (total) Gauß curvature at vertices of polyhedral surfaces [10].

On the other hand, our approach has contact with weak derivatives used in finite element theory where the formal application of partial integration is used to shift the differentiation operation to differentiable test functions. In fact, the integrands of our discrete differential operators div_h and rot_h have been obtained from $\nabla \xi$ by formal partial integration with test functions.

We avoid the derivation of the more technical details of the vector field decomposition and prefer to cite the important theorems in section 3. A discussion of the experimental results is given in section 4 where we discuss the decomposition of different test cases.

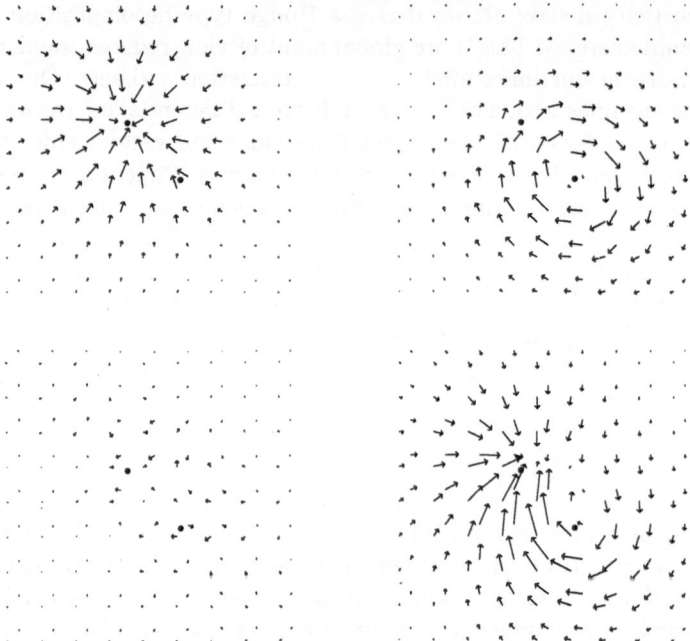

Fig. 1. Decomposition of test vector field (bottom right) in rotation-free (upper left), divergence-free (upper right) and harmonic component (bottom left). The two dots in each image indicate the centers of the original potential. Notice, that the dots in the combined vector field do not seem to lie in the centers indicated by the vector field, although they do. But the components clearly recover the original centers. Compare figure 3 for the associated potentials.

3   Decomposition of Discrete Vector Fields

In this section we are heading for a decomposition of tangential vector fields in Λ^1 into simpler components with special properties. A vector field in Λ^1 is integrable if it is the gradient of a function in S_h. To define other types of vector fields let us recall some differential operators and define discrete equivalents. For details of this section we refer to [9].

Let $\mathfrak{T}$ be the triangulation of a curved surface immersed in $\mathbb{R}^3$ (or more general, $\mathbb{R}^n$) which is assumed to be simply connected, for simplicity. On $\mathfrak{T}$ we have the function space

$$S_h := \left\{ u \;\middle|\; \begin{array}{c} u_{|T} \text{ is a piecewise linear function on each triangle } T \in \mathfrak{T}, \\ \text{and } u \text{ is continuous on } \mathfrak{T} \end{array} \right\}.$$

of piecewise linear functions. They are uniquely determined by their values at vertices. We consider the space of piecewise constant vector fields

$$\Lambda^1 := \{\xi \mid \xi_{|T} \text{ is a constant, tangential vector on each triangle } T \in \mathfrak{T}\}.$$

The divergence of a differentiable vector field $v = (v_1, v_2)$ in the euclidean plane with respect to local coordinates (x, y) is defined as $\operatorname{div} v = v_{1|x} + v_{2|y}$, and the rotation is a vector of length $|\operatorname{rot} v| = v_{2|x} - v_{1|y}$ normal to the plane. Recalling that for a vertex $p \in \mathfrak{T}$, the set $\operatorname{star} p$ consists of all triangles having p as vertex, and for an edge midpoint m, $\operatorname{star} m$ consists of the two triangles having m in common, we define discrete versions of both differential operators, the discrete divergence and the discrete rotation:

Definition 1. *Let ξ be a piecewise constant vector field on $\mathfrak{T}$. Then the discrete divergence respectively discrete rotation of ξ are defined at each edge midpoint m by*

$$\operatorname{div}_h \xi(m) := \int_{\partial \operatorname{star} m} \langle \xi, \nu \rangle$$

$$\operatorname{rot}_h \xi(m) := \int_{\partial \operatorname{star} m} \langle J\xi, \nu \rangle$$

where J denotes the rotation of a vector by 90° degrees in the plane of a triangle, and ν the exterior normal along the oriented boundary of $\operatorname{star} m$.

Our main application of the discrete differential operators is the characterization of integrable vector fields. We recall that a smooth vector field ξ is gradient of a differentiable function if and only if its rotation vanishes, and, correspondingly, $J\xi$ is gradient if the divergence of ξ vanishes. In the discrete setting we have a similar result:

Theorem 1. *Let $\xi \in \Lambda^1$ be a piecewise constant vector field on a simply connected triangulation $\mathfrak{T}$. Then we have*
1. *$\xi = \nabla u$ with a function $u \in S_h$ if and only if $\operatorname{rot}_h \xi(p) = 0$*
2. *$\xi = J\nabla w$ with a function $w \in S_h$ if and only if $\operatorname{div}_h \xi(p) = 0$.*

This result leads to the possibility of decomposing a vector field on triangulated surfaces in a rotation-free, a divergence-free, and a harmonic vector field.

Theorem 2. *Let $\mathfrak{T}$ be a triangulation of a compact surface. Any vector field $\xi \in \Lambda^1(\mathfrak{T})$ has a unique decomposition*

$$\xi = \nabla u + J\nabla w + v$$

with $u, w \in S_h$, $v \in \Lambda^1$ and normalization

$$\int_\Omega u = 0, \quad \int_\Omega w = 0 \text{ and } \operatorname{div}_h v = 0, \ \operatorname{rot}_h v = 0.$$

The function u respectively w is obtained by minimizing the energy determined by the value $\langle \nabla u - 2\xi, \nabla u \rangle$ respectively $\langle \nabla w + 2J\xi, \nabla w \rangle$ integrated over $\mathfrak{T}$. The harmonic vector field component v is defined as the remainder $v := \xi - \nabla u - J\nabla w$. The normalization of u and w only fixes a specific offset, and it has no influence on the gradients.

In our numerical experiments the energy minimization is performed with a conjugate gradient method leading to minimizers u and w in S_h of the respective energies. These functions directly determine the three vector field components ∇u, $J\nabla w$, and v. Please note, for efficiency, one does not need to store the three components explicitly since they are determined by the scalar-valued potential functions u and w. Further, if one is interested only, say, in identifying the vortices of a vector field ξ, then it suffices to calculate w and identify its local extrema, i.e. without actually performing the decomposition.

4   Results

The first test case consists of an artificial vector field generated as the sum of a gradient vector field corresponding to a point potential and three rotation vector fields corresponding to three potentials whose gradient has been rotated by 90° degrees, see figure 2. Application of the minimization algorithms leads to two functions u and w. The upper images in figure 2 display ∇u and $J\nabla w$. The algorithm has clearly separated the gradient field from the three vortices.

The visualization is made with the line integral convolution method [2]. We used a filter kernel with variable length to emphasize regions with vectors of small length.

Another interesting fact is visible in figure 1 where the original vector field is the sum of a single potential and vortex. The two dots in each image indicate the centers of the original potentials. Notice, that the dots in the combined vector field seem to deviate from the centers indicated by the vector field, although they do are in the correct place. Obviously, the summation of both vector fields leads to a misleading transition of the visible vortex center and potential. It is remarkable, that the reconstructed components clearly recover the original centers.

We have applied the algorithm to a flow from a CFD simulation, see figure 4. The rotation-free component of the incompressible flow around a cylinder vanishes as expected, except at inlet and outlet. The divergence-free component has additional artifacts at the boundary which are currently not well-understood.

A slight drawback of the present method is that it does not allow interaction of the user to steer the decomposition. Compare de Leeuw and Post [3] for interactive vortex detection.

Leeuw and van Liere [4] proposed a hierarchical ordering of flow features to reduce the complexitiy and suppress high-frequency patterns. It would be an interesting approach to combine their ideas with the decomposed fields presented here. An approach might be to first smooth the obtained principal functions to reduce high-frequencies.

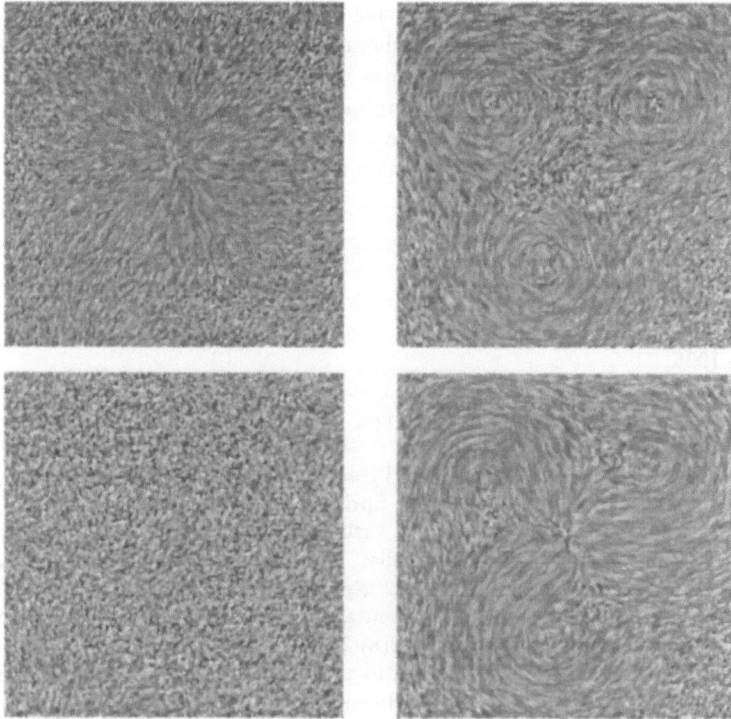

Fig. 2. Test vector field consisting of three vortices and a potential. The algorithm clearly separates the components. The harmonic component (lower left) has only a minor size.

The vector field components derived from gradients of the potentials are directed, and, therefore, emphasizing the vector directions in the LIC images will enroll additional information. Here tools for displaying the orientation of vectors using oriented textures developed in [8] would clearly add information to the LIC images.

5   Conclusions and Future Work

We have presented a variational approach for the decomposition of piecewise linear and piecewise constant vector field on triangulated surfaces in a divergence-free and a rotation-free component with a harmonic remainder. In many sample cases we obtained a nearly perfect separation of different features. We notice that in test cases with incompressible flows the divergence-free component vanishes as expected, but the harmonic remainder has a significant size. Here future study must concentrate on the currently unclear physical properties and features associated with the harmonic component.

The existence of different components of the vector field enables to separate features of the vector field, and study them for each component separately. Beside an extension to three-dimensional flows in volumes, there is also a connection to other feature extraction concepts required. Especially useful should be that fact, that the decomposition allows to study the potentials associated with the components rather than the vector fields as indicated in figure 3.

References

1. D. Banks and B. Singer. A predictor-corrector technique for visualizing unsteady flow. *IEEE Transactions on Visualization and Computer Graphics*, 1(2):151–163, June 1995.
2. B. Cabral and L. C. Leedom. Imaging vector fields using line integral convolution. *SIGGRAPH 93 Conference Proceedings*, 27:263–272, 1993.
3. W. C. de Leeuv and F. H. Post. A statistic view on vector fields. In M. Göbel, H. Müller, and B. Urban, editors, *Visualization in Scientific Computing, Eurographics*, pages 53 – 62, Wien, 1995. Springer Verlag.
4. W. C. de Leeuv and R. van Liere. Visualization of global flow structures using multiple levels of topology. In E. Gröller, H. Löffelmann, and W. Ribarsky, editors, *Data Visualization '99*, pages 45 – 52. Springer Verlag, 1999.
5. J. Helman and L. Hesselink. Representation and display of vector field topology in fluid flow data sets. *Computer*, 22(8):27–36, August 1989.
6. D. Kenwright and R. Haimes. Vortex identification - applications in aerodynamics: A case study. In R. Yagel and H. Hagen, editors, *Proc. Visualization '97*, pages 413 – 416, 1997.
7. D. N. Kenwright, C. Henze, and C. Levit. Feature extraction of separation and attachment lines. *IEEE Trans. on Visualization and CG*, 5(2):135–144, 1999.
8. L. Khouas, C. Odet, and D. Friboulet. 2d vector field visualization using furlike texture. In E. Gröller, H. Löffelmann, and W. Ribarsky, editors, *Data Visualization '99*, pages 35 – 44. Springer Verlag, 1999.
9. K. Polthier. Hodge decomposition of vectorfields. *in preparation*, 1999.
10. K. Polthier and M. Schmies. Straightest geodesics on polyhedral surfaces. In H.-C. Hege and K. Polthier, editors, *Mathematical Visualization*, pages 135–150. Springer Verlag, Heidelberg, 1998.
11. M. Roth and R. Peikert. Flow visualization for turbomachinery design. In R. Yagel and G. Nielson, editors, *Proc. Visualization '96*, pages 381 – 384. IEEE Computer Society Press, 1996.
12. M. Roth and R. Peikert. A higher-order method for finding vortex core lines. In D. Ebert, H. Hagen, and H. Rushmeier, editors, *Proc. Visualization '98*, pages 143 – 150. IEEE Computer Society Press, 1998.
13. I. Sadarjoen, F. Post, B. Ma, D. Banks, and H. Pagendarm. Selective visualization of vortices in hydrodynamic flows. In D. Ebert, H. Hagen, and H. Rushmeier, editors, *Proc. Visualization '98*, pages 419 – 423. IEEE Computer Society Press, 1998.
14. I. Sadarjoen and F. H. Post. Geometric methods for vortex extraction. In E. Gröller, H. Löffelmann, and W. Ribarsky, editors, *Data Visualization '99*, pages 53 – 62. Springer Verlag, 1999.
15. G. Scheuermann, H. Hagen, and H. Krüger. Clifford algebra in vector field visualization. In H.-C. Hege and K. Polthier, editors, *Mathematical Visualization*, pages 343–351. Springer-Verlag, Heidelberg, 1998.

154

16. C. Teitzel. Adaptive methods and hierarchical data structures for interactive three-dimensional flow visualization. Dissertation, Friedrich-Alexander-Universität Erlangen-Nürnberg, IMMD 32/09, Computer Graphics Group, September 1999.

6 Figure Appendix

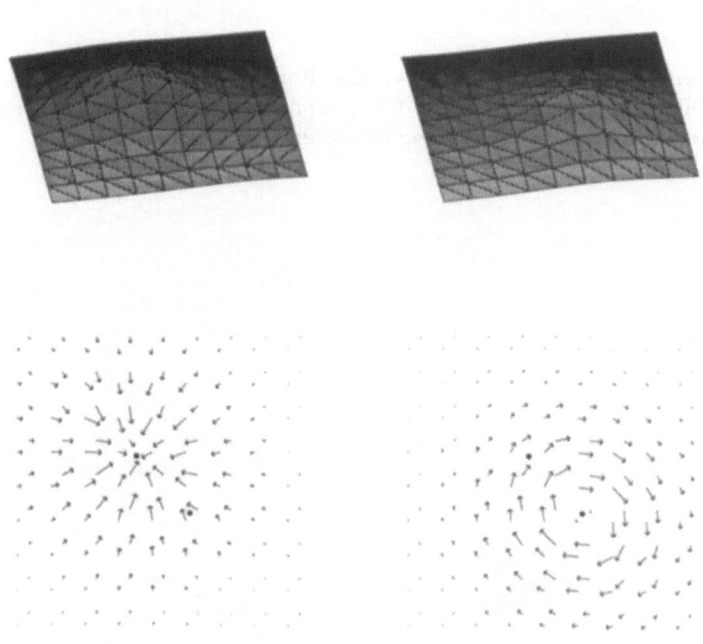

Fig. 3. Rotation-free (left column) and divergence-free (right column) components of the vector field decomposition shown in figure 1 with associated potentials. Features of the original vector field are easily identified as local extrema of the potentials.

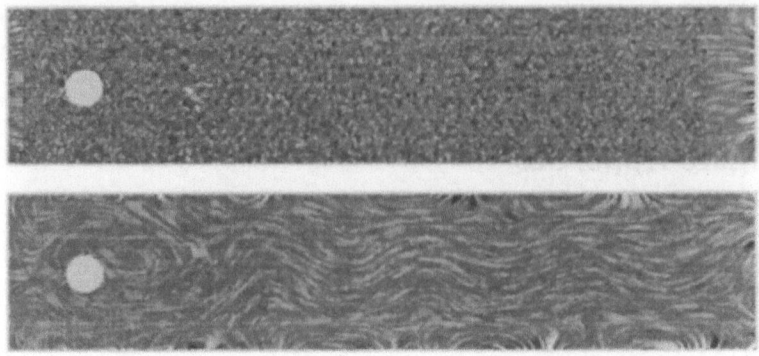

Fig. 4. Incompressible flow around a cylinder. Rotation-free component (top) vanishes except at inlet and outlet. Divergence-free component (bottom) has additional artifact at boundary. Courtesy Michael Hinze, TU-Berlin.

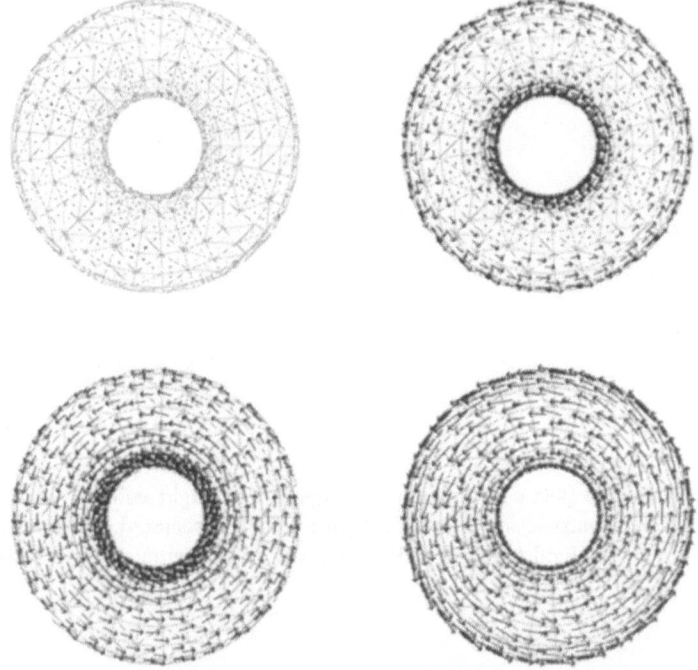

Fig. 5. Decomposition of the vector field generated by rotating the torus around the vertical axis (bottom right). The rotation-free component (upper left) vanishes because the field belongs to an isometry of the ambient space. Since the torus is not-simply connected we have a typical harmonic component (bottom left) belonging to the incompressible, rotation-free component of the underlying flow. The divergence free component (upper right) shows the expected behavior where the vortex maxima are a one-parameter family along the upper and lower latitudes.

Integrated Multiresolution Geometry and Texture Models for Terrain Visualization

Konstantin Baumann, Jürgen Döllner, Klaus Hinrichs

Department of Computer Science, University of Münster
{kostab, dollner, khh}@uni-muenster.de

Abstract. In this paper, an approach for integrating multiresolution representations of terrain geometry and terrain texture data is presented. A terrain is modeled by a regular grid, which can be partially refined by local TINs in order to represent morphologically complex terrain parts. The multiresolution models for terrain texture data and geometry data are closely related: The rendering algorithm selects geometry and texture patches based on screen-space error criteria. Multiple texture hierarchies, which may represent different thematic information layers, can be bound to one terrain model. Multiple textures lead to a drastic improvement of visual quality: Topographic textures can be used to provide pixel-precise shading, alpha textures can be used to restrict or to highlight thematic textures. Multiple textures facilitate the development of visual interaction tools such as *magic lenses,* and texture animations. Multitexturing permits an efficient implementation of these concepts.

1 Introduction

In many kinds of virtual environments digital terrain models play a central role as fundamental tools to present and communicate spatial information. Various hierarchical data structures are suited for representing terrains, e.g., hierarchical TINs [6], R-trees [12], restricted quadtree triangulations [16], and progressive meshes [10]. Most multiresolution modeling schemes support a specific type of input data such as arbitrarily distributed data points for triangulated irregular networks (TINs), or regularly distributed data points for grids. In general, real-world terrain data sets are composed of data of different types. For example, a cartographic terrain model can include grid data describing the digital elevation model (DEM) and microstructures describing structures at a finer resolution such as topographically complex or interesting terrain parts (e.g., riverbeds illustrated in Fig. 1).

Texture data represent another important category of terrain data. Multiresolution modeling can be extended to texture data as well. In particular, for real-world terrain models, large, high-resolution textures need to be processed which usually do not fit into graphics texture memory or even into main memory. Moreover, texturing can be employed to implement visual tools such as magic lenses (see Section 5.1). Therefore, multiresolution modeling for digital terrain models should consider both geometry and texture data.

2   Related Work

Hierarchical triangulations based on TINs have been applied to generate multiresolution models which can be used by level-of-detail (LOD) algorithms (de Floriani et al. [6], Gross et al. [9], Voigtmann et al. [17], and Xia et al. [18]). Regular grids have been used for multiresolution modeling (Falby et al. [8]) and for real-time, continuous LOD rendering (Duchaineau et al. [7], Lindstrom et al. [14], and Pajarola [16]). Hoppe [10] introduced the *view-dependent progressive mesh* which has been further optimized for real-time terrain rendering, and the *geomorph*, a technique to minimize *popping* effects in the terrain representation during changes of the level of detail. Chen et al. [3] discussed a method for combining LOD techniques with *image-based modeling and rendering* techniques to take advantage of the frame-to-frame coherence in screen-space. In many applications the visual quality with respect to topographic terrain features or thematic terrain data is as important as rendering performance. Recent developments (e.g., Hoppe [11], and Xia et al. [18]) take into account the visual quality by considering surface normals, but do not provide an explicit control of the terrain shading as proposed in this paper. Most LOD techniques are limited with respect to the management of large-scale texture data: In contrast to the LOD mechanism for geometry data no similar mechanism is provided for texture data. Lindstrom et al. [15] proposed a method which handles a single large-scale texture related to a LOD terrain geometry. However, an efficient treatment of multiple logical texture layers is required for the interactive exploration and manipulation of terrain data (e.g., terrain visualization used for landscape analysis and planing).

3   Data Structures for Integrated Multiresolution Modeling

We construct a *hybrid terrain model* by a regular grid, called the *reference grid*, and from a collection of TINs, called *microstructures*, which are associated to grid cells and refine the terrain representation within the cell domain. Each of the refining microstructures must adapt itself continuously to the neighboring grid cells and microstructures (see Fig. 1). Grid cells are refined where complex morphology has to be represented (e.g. riverbeds, streets, or ridges). This hybrid terrain model combines the advantages of grids and TINs: it leads to a memory-efficient and morphologically precise terrain representation. Handling the details as precisely as possible is important because we perceive a terrain model mainly by the terrain shading and terrain silhouette, and both depend on geometric details.

3.1   Generic Multiresolution Data Structure for Terrain Geometry

This section defines a multiresolution model for geometry data, the *approximation tree*, which is generic with respect to the type of terrain data as required by hybrid terrain models. Let $P = \{p_1,...,p_n\}$ be a set of n data points in the xy-plane. Let $D(P)$ be the minimal axis-parallel bounding box of set P, and let $G = (P,h_G)$ be a terrain model defined by the point set P and an elevation function $h_G : D(P) \rightarrow R$, which calculates elevation values for points of D(P) by interpolating the height values for data points of P. The domain of G is defined as D(G) = D(P).

An *approximation tree* $A_{s,d}(G)$ for a terrain model G is represented by a tree; its nodes are called *geometry patches*. Each geometry patch N represents a rectangular region $D(N) \subseteq D(G)$ and approximates the terrain surface G in that region by an approximating terrain surface $G(N) = (P(N), h_{G(N)})$. The set P(N) consists of at most s data points: $|P(N)| \leq s$. Furthermore, the four corner points of D(N) must be contained in P(N). The way the node calculates the data points P(N) depends on the *approximation strategy* adopted by the node. For example, a grid-based node will select evenly spaced points from a grid data set, whereas a TIN-based node will select points from an arbitrary data set based on an error criterion.

The *geometric approximation error* ε(N) of a geometry patch N is defined by the maximal vertical distance between the terrain models G(N) and G:
$\varepsilon(N) = \max_{p \in D(N)} |h_{G(N)}(p) - h_G(p)|$.

Each terrain patch N can have at most d child nodes. The child nodes are constructed as follows: If the geometric approximation error ε(N) exceeds a certain threshold $\varepsilon \geq 0$, the domain D(N) is decomposed into a set of at most d rectangular, disjoint subdomains $D(N_i)$. The strategy for decomposing a patch depends on the type of the node: a grid-based node applies a quadtree-like subdivision, whereas a TIN-based node is subdivided by a line parallel to the x- or y-axis. For each subdomain $D(N_i)$, a child node N_i of N is constructed which approximates the terrain surface in that subdomain. The domain of the root node of the approximation tree $A_{s,d}(G)$ is D(G) covering the whole domain of the terrain G.

3.2 Multiresolution Data Structure for Terrain Textures

Multiresolution modeling for texture data in the context of multiresolution terrain models is motivated by the following observations:

- Visualization applications are likely to use texture data up to several hundreds of megabytes, for example, in cartographic applications. However, graphics hardware imposes constraints on the size of textures. For example, common OpenGL implementations can process textures up to 1024 x 1024 pixels and constrain the actual size to a power of 2, i.e., 2^m x 2^n pixels.
- The selection of a level-of-detail texture depends on the texture approximation

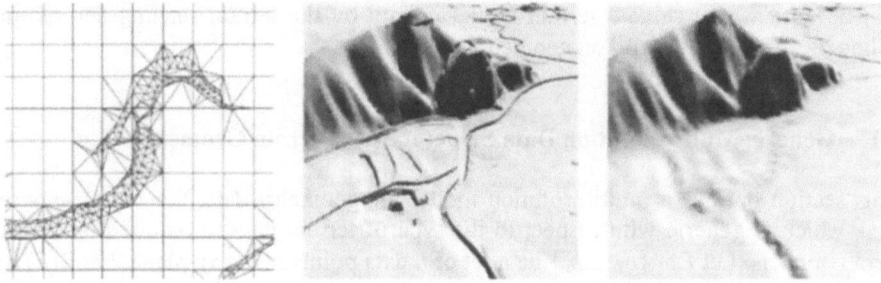

Fig. 1. Hybrid terrain model specified by a reference grid and partially refined by TINs (*left*), view of the terrain model with TINs (*center*) and without TINs (*right*)

error and the patch geometry.

Let T be a geo-referenced 2D texture used for a terrain model G with $D(T) \supseteq D(G)$. The texture pyramid $\Delta(T)$ of a terrain texture T consists of a sequence of textures T_i with decreasing resolution. Conceptually, each texture is created by scaling down the predecessor texture by a factor of ½. The first texture of the sequence is the original terrain texture, the last texture consists of 1 x 1 pixels.

In analogy to the approximation tree for geometric data, we define a similar tree for multiresolution textures, the *approximation tree for terrain textures* $A_{s,d}(G,T)$. The nodes of a tree $A_{s,d}(G,T)$ are called *texture patches*. A texture patch M has the following properties:

- M is associated with exactly one geometry patch $N_M \in A_{s,d}(G)$.
- The domain of M covers the domain of N_M: $D(M) \supseteq D(N_M)$.
- M references that image part S_M with the highest resolution in $\Delta(T)$ which completely covers the domain $D(N_M)$ and fulfills the constraints of the rendering system.
- If S_M is not an image part of the first texture in the texture pyramid $\Delta(T)$ (i.e., it is not part of the original terrain texture T) and the geometry patch N_M has child nodes, then the texture patch M has child nodes, too.
- If the texture patch M has child nodes, then M and N_M have the same number of child nodes, and the domain of a child of M is equal to the domain of the corresponding child of N_M.

The texture resolution of M is considered optimal for the geometry patch N_M. But the geometry patch can also be rendered with any parent texture patch M' of M because its domain covers the domain of M and therefore of N_M, too. In such a case, the texture resolution is non-optimal for the geometry patch, but if the geometry patch is far away from the viewer this reduction of resolution is not visible and can speed up rendering significantly, since less texture data have to be processed.

The rendering algorithm traverses an approximation tree recursively, calculates visual approximation errors, and selects geometry patches and texture patches based on user-defined quality criteria.

3.3 Visual Approximation Errors

Let $A_{s,d}(G)$ and $A_{s,d}(G,T)$ be a geometry approximation tree and an associated texture approximation tree. Let $N \in A_{s,d}(G)$ be a geometry patch with an approximating terrain surface $G(N)$, and let $M \in A_{s,d}(G,T)$ be a texture patch, $D(M) \supseteq D(N)$. Let $B(N)$ be the minimal 3D axis-parallel bounding box of $G(N)$.

The visual approximation errors are defined if the bounding box $B(N)$ intersects the current view volume. If the bounding box intersects, then the point p of the bounding box closest to the camera and inside the view volume can be determined. The visual approximation errors are calculated as follows:

- *Visual geometry approximation error* $\alpha(N)$: Construct a line segment centered at p, parallel to the z-axis (the direction of elevation) having length $\varepsilon(N)$ where $\varepsilon(N)$ denotes the geometric approximation error. Projecting that segment onto the view plane, the visual geometry approximation error $\alpha(N)$ is the length of the projected line segment measured in pixels (see Fig. 2).

- *Visual texture approximation error* $\gamma(M,N)$: Determine the width w and height h (in the terrain coordinate system) of a texel of the texture patch M. Construct two line-segments centered at p: one is of length w parallel to the x-axis, the other is of length h parallel to the y-axis. Projecting both onto the view plane, the visual texture approximation error $\gamma(M,N)$ is the maximum length of both projected segments measured in pixels.

The visual approximation errors $\alpha(N)$ and $\gamma(M,N)$ are a measure for the visual quality of a geometry patch N and a texture patch M. If we render the approximating terrain surface $G(N)$, it is ensured that each of the pixels of $G(N)$ differs by at most $\alpha(N)$ pixels compared to the original terrain surface G. In analogy, for the texels of the approximating terrain texture M, we can expect that the texels are sufficiently dense for the actual camera settings.

3.4   Recursive Rendering of Approximation Trees

A hybrid terrain model G represented by a tree $A_{s,d}(G)$ together with a terrain texture T represented by a tree $A_{s,d}(G,T)$ is rendered recursively, starting with the pair of root nodes (N_0,M_0). For a pair of nodes (N,M), the algorithm works as follows:

1. If the bounding box $B(N)$ of the current geometry patch does not intersect the current view volume, the recursive rendering stops at this point (*hierarchical view-volume culling*).
2. Otherwise, the visual approximation errors $\alpha(N)$ and $\gamma(M,N)$ are calculated. If $\alpha(N)$ is larger than a user-defined geometric threshold or if $\gamma(M,N)$ is larger than a user-defined texture threshold, then the rendering calls itself recursively for the child patch pairs (N_i,M_i) (*recursive refinement*). If M has no child texture patches M_i, M is used instead.
3. If both visual errors are ok, assign to M the parent texture patch M' of M until the visual texture approximation error $\gamma(M',N)$ exceeds the user-defined texture threshold. The resolution of the parent's texture is lower but this way we guarantee that the resolution comes close to the user-defined texture threshold (*reduction of texture data*).
4. Render the approximating terrain surface $G(N)$ together with the texture determined in the previous step, and terminates the recursion (*rendering*).

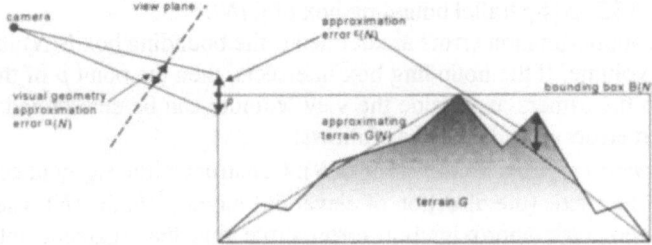

Fig. 2. Visual geometry approximation error $\alpha(N)$ and its calculation

The thresholds for both errors allow the user to prioritize either interactivity or visual quality. Low thresholds lead to higher visual quality, higher thresholds accelerate rendering. Step 3 of the algorithm ensures that the selected texture patch matches the resolution needed for projecting it onto the screen without a visual distortion, but with as few texture data to be processed as possible.

Gaps between two adjacent geometry patches are possible. The gaps are at most as big as the user-defined geometry approximation error. To close the gaps, walls are inserted between geometry patches of different levels of detail. The walls have the same texture coordinates as the edge between the two geometry patches, so they get colored the same way and are visually not recognizable for a reasonable visual geometry approximation error threshold (commonly smaller than 4 pixels).

Only a part of the texture pyramid is actually required to render a single frame. Therefore, texture patches load their texture data on demand, spanning a separate thread. While the texture data is being loaded, the texture data of the parent texture patch can be used. In this case, texture resolution is not optimal, but interactivity is ensured because the application is not blocked and can use at least a reasonable approximation of the required texture. Furthermore, we make use of a mechanism called *file-to-memory mapping* provided by modern operating systems. Each image of the texture pyramid is kept on secondary storage and is only mapped to memory.

4 Shading by Topographic Textures

In many terrain visualization systems the visual quality depends directly on the geometric resolution of the approximating terrain due to the underlying *Gouraud shading*. The vertex normals of a triangle determine the shading of the whole triangle, leading to shading artifacts if triangles become large or thin. For a level-of-detail terrain model this implies that the lower the resolution, the more topographic details get lost, and that shading changes if the LOD-dependent geometry changes. In our approach, the terrain shading relies on *topographic textures*.

A *topographic texture* is precalculated and applied as regular terrain texture to the terrain model, reintroducing topographic details that might be removed during the geometric simplification process. A similar approach has been made in the context of *appearance-preserving simplification strategies* (e.g., Cignoni et al. [4], and Cohen et al. [5]). Terrain models shaded by topographic textures have the following properties:

1. The visual quality of a terrain model depends on the quality of the topographic textures because they encode visually the terrain's morphology.
2. The geometric complexity needed to achieve high-quality images of terrain models is considerably less compared to Gouraud shading because topographic textures are applied pixel-precisely (see Fig. 3).
3. The visual effects of LOD changes can be minimized by topographic textures because the constant shading *hides* discontinuities in the geometric representation during a change in the LOD.

A topographic texture consists of luminance values and depends on the high-resolution terrain model, the terrain surface properties, the lighting conditions, and on special design rules. In cartography, for example, design rules have been developed

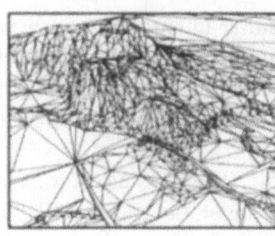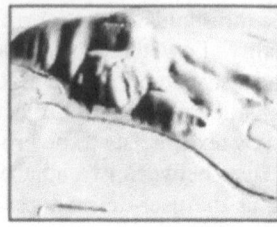

Fig. 3. Wire-frame representation of an approximating terrain model (*left*), Gouraud-shaded model (*center*), shaded by a topographic texture (*right*)

which improves the perception of morphologically important terrain parts such as peaks, pits, valleys and ridges. Note that the ambient and diffuse light, which are the most important ones for terrain shading, do not depend on the camera settings which justifies the precalculation of a topographic texture.

Furthermore, we are not limited to the lighting and shading models provided by the underlying 3D rendering system. Fig. 4 (*left*) shows an automatically generated topographic texture which takes into account self-shadowing of a terrain model. Fig. 4 (*right*) shows a topographic texture calculated for the Himalaya Mountains. The shades and lighting conditions are chosen in such a way that the morphology can be perceived easily.

5 Applications of Multiple Textures

In many applications it is necessary to map two or more thematic textures onto a terrain surface (e.g., topographic texture, road map, and land use information). The approximation tree has been extended to handle more than one texture tree, i.e., a geometry patch can be associated with texture patches of several different texture trees. As a consequence, information layers can recalculate their texture data without affecting the textures of the other layers.

It is important that each texture remains independent: textures of several information layers cannot be merged into one final 2D texture due to their different domains

Fig. 4. Topographic texture with self-shadowing and combined with a cartographic texture (*left*), topographic texture based on cartographic shading rules applied to a terrain model of the Himalaya (*right*)

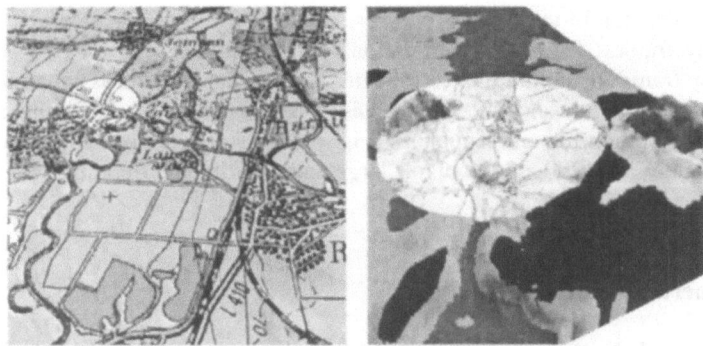

Fig. 5. Applications of multiple textures in terrain visualization: Interactive highlight lens (*left*), thematic lens adding information to a local area (*right*)

and resolutions. In addition, it is not feasible to merge large-scale textures with several hundreds of megabytes in real-time. Furthermore, multiple textures facilitate the implementation of dynamic textures and texture-based animations.

5.1 Texture Lenses and Texture Animations

In Fig. 5 (*left*) an additional luminance texture is used to implement a highlight lens which is combined with a cartographic and a topographic texture. The lens can be used, for example, to control the visual focus of an observer during a presentation. Due to the use of multiple textures the luminance texture needs not to be merged with the high-resolution cartographic and topographic textures and therefore can be moved across the terrain in real-time.

In Fig. 5 (*right*) a thematic lens exhibits a cartographic texture inside a circular region surrounded by a high-contrast height texture which visualizes discrete height regions, everything combined with a topographic shading texture. The visibility of the cartographic texture is restricted by a visibility-restricting texture. Note that visibility-restricting textures normally have low memory requirements because low resolutions (e.g., 128 x 128 pixels) are sufficient.

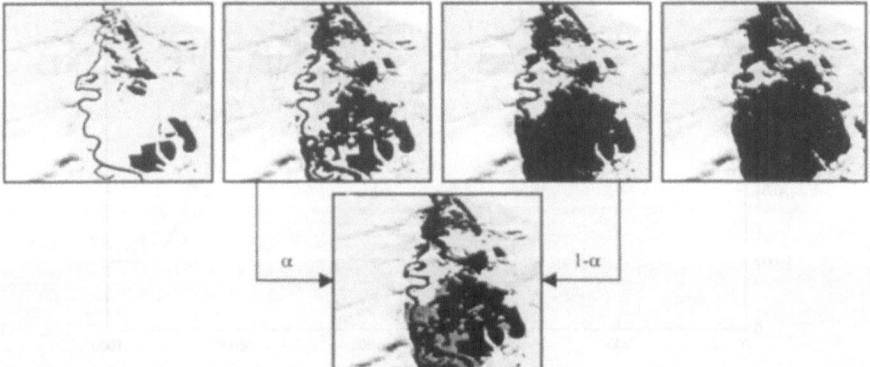

Fig. 6. Key texture frames of a flooding animation (*top*) and an interpolated texture (*bottom*)

To visualize spatio-temporal processes, one can use multiple textures to specify texture key frames. During the animation, we interpolate between two consecutive texture key frames using multitexturing with appropriate weights assigned to both key frame textures. No intermediate texture has to be created which allows us to animate even high-resolution texture sequences. Fig. 6 shows a flooding animation. The texture key frames are 1200 x 2400 pixels large and describe the flooding state in a landscape at concrete time stamps.

5.2 Experiments and Results

All screen shots presented in this paper have been taken from the *LandExplorer* [13], a prototype implementing the described concepts [1]. All time measurements were performed on a standard PC equipped with a 350 MHz Pentium II processor, 128 MB RAM, Riva TNT graphics card with 16 MB graphics memory, and running Windows NT 4.0 (SP6). The window was 640 x 480 pixels large in true-color. We used a terrain data set consisting of about 500.000 triangles (a 640x320 reference grid plus additional fine-structures introducing about 130.000 triangles). The original data set needs more than 5 seconds to render without level-of-detail techniques. The size of the topographic texture is 2500 x 5000 pixels (13 MB, gray-scale) and the thematic texture is 3200 x 6400 pixels (62 MB, RGB) large. The method using Gouraud shading is the slowest method, even rendering with two textures is almost always faster.

6   Conclusions

The hybrid terrain model improves the visual quality of terrain models because microstructures have a great impact on the perception of a terrain model. The rendering

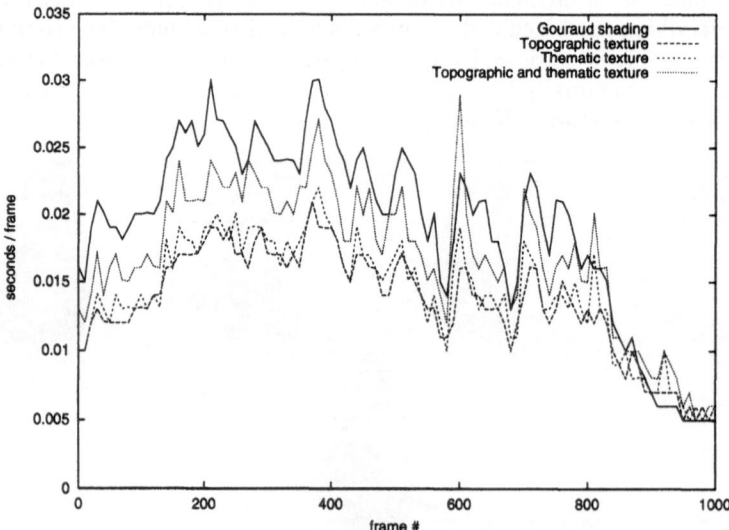

Fig. 7. Timings for different visualization techniques

process considers both screen-space geometric errors and texture errors which control visual quality and geometric correctness. As a direct application of multiple textures, shading can be implemented by topographic textures improving the visual quality dramatically because they outwit human perception by providing detailed and LOD-independent shading information for a coarse geometry. Using multiple textures proves to be feasible because multitexturing for its implementation is available on modern graphics hardware. Currently, a cartographic visualization system [2] is being developed based on the presented multiresolution model.

References

[1] K. Baumann, J. Döllner, K. Hinrichs, O. Kersting: *A Hybrid, Hierarchical Data Structure for Real-Time Terrain Visualization*. Proc. Computer Graphics Intern. '99, 85-92, 1999.

[2] G. Buziek, J. Döllner: *Concept and Implementation of an Interactive, Cartographic Virtual Reality System*. Proceedings of the International Cartographic Conference ICC '99, Ottawa, 637-648, 1999.

[3] B. Chen, J. Swan, E. Kuo, A. Kaufman: *LOD-Sprite Technique for Accelerated Terrain Rendering*. Proceedings IEEE Visualization '99, 1999.

[4] P. Cignoni, C. Montani, C. Rocchini, R. Scopigno: *A general method for preserving attribute values on simplified meshes*. Proceedings IEEE Visualization '98, 59-66, 1998.

[5] J. Cohen, M. Olano, D. Manocha: *Appearance-Preserving Simplification*. Proceedings of SIGGRAPH '98, 115-122, 1998.

[6] L. De Floriani, P. Magillo, E. Puppo: *Efficient Implementation of Multi-Triangulations*. Proceedings IEEE Visualization '98, 1998.

[7] M. Duchaineau, M. Wolinsky, D. Sigeti, M. Miller, C. Aldrich, M. Mineev-Weinstein: *ROAMing Terrain: Real-time Optimally Adapting Meshes*. Proceedings IEEE Visualization '97, 81-88, 1997.

[8] J. Falby, M. Zyda, D. Pratt, R. Mackey: *NPSNET: Hierarchical Data Structures for Real-Time Three-Dimensional Visual Simulation*. Computers & Graphics, 17(1):65-69, 1993.

[9] M. Gross, R. Gatti, O. Staadt: *Fast Multiresolution Surface Meshing*. Proceedings IEEE Visualization '95, 135-142, 1995.

[10] H. Hoppe: *Smooth View-Dependent Level-of-Detail Control an its Application to Terrain Rendering*. Proceedings IEEE Visualization '98, 35-42, 1998.

[11] H. Hoppe: *New quadric metric for simplifying meshes with appearance attributes*. Proceedings IEEE Visualization '99, 59-66, 1999.

[12] M. Kofler, M. Gervautz, M. Gruber: *The Styria Flyover - LOD management for huge textured terrain models*. Proc. Computer Graphics International '98, 444-454, 1998.

[13] Land*Explorer*. WWW-Site: http://www.mamvrs.de/geovisualiz.htm, 1999.

[14] P. Lindstrom, D. Koller, W. Ribarsky, L. Hodges, N. Faust, G. Turner: *Real-Time, Continuous Level of Detail Rendering of Height Fields*. Proc. SIGGRAPH '96, 109-118, 1996.

[15] P. Lindstrom, D. Koller, L. Hodges, W. Ribarsky, N. Faust, G. Turner: *Level-of-detail Management for Real-Time Rendering of Phototextured Terrain*. GVU TR 95-06, 1995.

[16] R. Pajarola: *Large Scale Terrain Visualization using the Restricted Quadtree Triangulation*. Proceedings IEEE Visualization '98, 19-26, 1998.

[17] A. Voigtmann, L. Becker, K. Hinrichs: *A Hierarchical Model for Multiresolution Surface Reconstruction*. Graphical Models and Image Processing, 59:333-348, 1997.

[18] J. Xia, J. El-Sana, A. Varshney: *Adaptive Real-Time Level-of-detail-based Rendering for Polygonal Models*. IEEE Transactions on Visualization and Computer Graphics '97, 3(2):171-183, 1997.

A Framework for Interactive Hardware Accelerated Remote 3D-Visualization

Klaus Engel, Ove Sommer, and Thomas Ertl

University of Stuttgart, IfI, Visualization and Interactive Systems Group
{engel, sommer, ertl}@informatik.uni-stuttgart.de
http://wwwvis.informatik.uni-stuttgart.de

Abstract. In this paper we present a framework that provides remote control to Open Inventor or Cosmo3D based visualization applications. A visualization server distributes a visualization session to Java based clients by transmitting compressed images from the server frame buffer. Visualization parameters and GUI events from the clients are applied to the server application by sending CORBA (Common Object Request Broker Architecture) requests.

The framework provides transparent access to remote visualization capabilities and allows sharing of expensive resources. Additionally the framework opens new possibilities for collaborative work and distance education. We present a teleradiology system and an automotive development application which make use of the proposed techniques.

1 Introduction

The rapid evolution of todays digital communication networks enables access to a huge amount of scientific data and remote computation capabilities, like shared memory multiprocessor machines or special high-end graphics hardware. Concerning this development we believe that techniques have to be developed to enable the visualization of scientific data using remotely available high-end visualization architectures from any Internet-connected desktop computer.

In the development of todays desktop computers two contrary trends can be observed. On the one hand the computation and rendering capabilities of modern low-cost PCs are quickly rising. One the other hand the network computer (NC) is a very simple and inexpensive device that acts as a thin client to more powerful server machines.

Up to now hardware accelerated rendering has required local rendering. For example, in the X-Windows system a remotely started 3D application uses local 3D acceleration hardware for rendering. Local rendering enables high interaction rates. However, there are certain conditions under which local rendering is impossible or undesirable. For example, typical data sets from scientific simulations and measurements neither can quickly transferred nor can be handled on modern desktop computers because of their immense size. On the other side there is a class of high-end servers, supercomputers and workstations with special 3D graphics acceleration hardware, numerical computation power and high-performance IO bandwidth that provide the necessary means to handle large scientific data sets. Furthermore local rendering often requires the transfer of sensitive data from servers to clients. This may be undesirable because of security

168

Server   Client

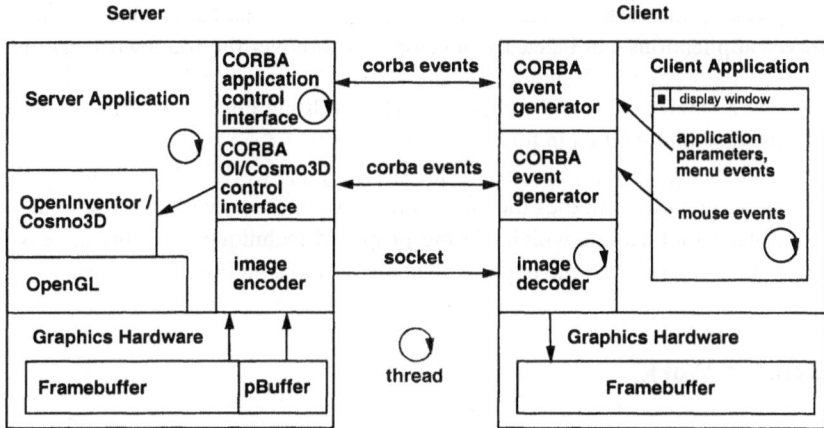

Fig. 1. The client-server scenario used in this framework. One or more clients remotely observe and control an application on a server by using two CORBA interfaces and a socket connection.

reasons, e.g. for patient data in medical applications its more secure to transfer (encrypted) image data instead of original data.

We developed a framework for the remote visualization of large scientific data sets. It is able to use 3D acceleration hardware of the server system and allows to use these features from any Internet-connected PC interactively. The general scenario used for the framework consists of an high-end visualization server for rendering images and one or more clients to provide a user interface, display the rendered images, and control the visualization (Fig. 1). There are no special requirements for the choice of the clients. Even a PDA with TCP/IP network connection and adequate display resolution would be sufficient.

On the server side an Open Inventor or Cosmo3D based visualization application is rendering either images on-screen into the frame buffer or off-screen into the pbuffer. Thereby two different scenarios can be derived: A master user controls the session on the server machine and uses the visualization application locally. Images rendered on-screen into the frame buffer are encoded and transferred to all clients using a socket connection. The second scenario uses a display-less server that runs the server job in the background and renders into the pbuffer. The pbuffer is a special protected graphics memory block which allows hardware accelerated off-screen rendering. Images are read from this buffer into main memory, encoded and transferred to the attached clients using a socket connection.

The Java client enables transparent remote access to the visualization capabilities of the high-end server from any window system supporting platform. The received image data is decoded, stored as a Java2D buffered image and drawn into the frame buffer. Mouse and keyboard events generated on the client are transported to the server machine using CORBA method calls. The server provides an Open Inventor or Cosmo3D interface for these events that passes the parameters to objects and functions that handle the events on the server side. Thereby, when using Open Inventor, a manipulator widget can be remotely picked and dragged. Application parameters are transferred to the server using a second CORBA event interface.

Using our framework, which consists of several C++ and Java classes, the client and server applications can be easily developed. Moreover the framework enables the conversion of any existing visualization application into a remotely controllable one by adding only a small amount of additional code. Multiple clients are able to share the view on the data and interact in turn.

In the following section we will describe some related work in the field of web based visualization. Section 3 outlines the architecture of our framework. Section 4 explains two exemplary applications which use the proposed techniques. Results are given in section 5. We will conclude the paper with some remarks on future activities.

2   Related Work

In the past years several approaches for scientific visualization on digital networks were investigated. One of the first progressive applications for volume visualization was presented by Lippert et al. [7].

Hendin introduced a VRML based volume visualization tool [5], which uses three stacks of perpendicular slices. We introduced techniques for fast volume clipping, collaborative work, and data size reduction [2] as an extension to Hendin's approach.

An opposite approach was proposed by Ma et al. The web based volume visualization system called DiVision allows users to explore remote volumetric data sets using a web-browser [10]. The system computes images on a visualization server, which are transferred to the client and inserted into a graph. However, the application does not support interactive manipulation.

In the virtual network computing (VNC) system by Richardson et al. [11], server machines supply applications, data, and desktop environments that can be accessed from any Internet-connected machine using a wide variety of machine architectures. It is a remote display system which allows the user to view a computing desktop environment from anywhere on the Internet. VNC does not support to use remote 3D graphics acceleration hardware.

Just recently SGI announced OpenGL Vizserver [8]. From the limited amount of information which is currently available it is understood that OpenGL Vizserver will enable a single Onyx2 workstation to distribute visualization sessions to multiple UNIX operating system desktop workstations by transmitting compressed images from the Onyx2 frame buffer. While the system only seems to work inside organization networks and only on SGI machines, our framework allows transparent access to any high-end server from any Internet-connected desktop PC.

3   The Framework

The framework consists of several C++ classes for the server side and Java modules that are used to build a client application. Because of the modular structure of the framework it can be applied to any visualization application and new codecs can be added.

3.1 Server modules

A stand-alone visualization application can easily be converted into a remotely accessible one by adding a CORBA interface and slightly modifying the scene graph. Currently the framework was adapted for Open Inventor and Cosmo3D.

Open Inventor: Open Inventor traverses the scene graph in a fixed order from top to bottom and left to right. Because of this behavior we can add a new scene graph node named PBufferNode that switches to the pbuffer rendering context in front (to the left) of the contributing nodes, generally at the leftmost position. The SocketNode reads the frame buffer content into main memory and performs the encoding and transmission of the image data. It is inserted behind (to the right) of the contributing nodes, generally at the rightmost position in the scene graph.

As soon as the scene graph has changed or new mapping parameters (e.g. modified transfer functions) have been received a render action is applied to the root node of the scene graph. Then the following sequence of steps is performed:

1. The render action traverses the scene graph from top to bottom and left to right.
2. As soon as the PBufferNode is reached, the render method switches to the pbuffer rendering context. If there was no PBufferNode node added to the scene graph the rendering context remains on-screen.
3. The render action continues to traverse through the scene graph and the shape nodes draw geometry into the current rendering context.
4. As soon as the SocketNode is visited, the content of the frame buffer is read into main memory. It compresses the image data using one of the available compression methods and provides the encoded data to all connected clients via a given socket port.
5. The render action continues to traverse the scene graph.

Open Inventor provides convenient mechanisms to convert 2D events received from a client to 3D events. The steps that are performed can be summarized as follows:

1. The client application registers interest in particular events with its window system.
2. The client application receives an event from its window system.
3. The client calls the appropriate server method using CORBA and delivers the necessary parameters (e.g. mouse position, mouse button) to the server.
4. The server application translates the event into a SoEvent (for mouse events SoLocation2Event).
5. The SoEvent is inserted into an instance of the SoHandleEventAction class.
6. The handle event action is applied to the top node of the scene graph. This action traverses the scene graph. Each node implements its own action behavior. When a node (typically a manipulator) is found, it handles the event.
7. If necessary a render action is applied to the scene graph.

The CORBA main loop and the Open Inventor main loop are running in two separate threads. As Open Inventor is not thread-save the received client events can not be applied to the scene graph immediately. The Ticker class is an Open Inventor engine (derived from SoEngine) that stores SoHandleEventActions created by client event receiving CORBA methods. The Ticker class is called from the Open Inventor main loop in fixed time steps and triggers a SoHandleEventAction.

Cosmo3D: The scene graph structure provided by Cosmo3D differs from that of Open Inventor. No information is inherited horizontally in the Cosmo3D scene graph, which is traversed downwards from top to bottom in each branch. OpenGL Optimizer offers different kinds of scene graph traversal actions: On the one hand there are depth-first traversal actions which do their work in the same order as Open Inventor actions do. On the other hand breadth-first traversal actions can be applied for parallelization using multiple processors. Thus, a new Cosmo3D scene graph node, which implements the corresponding functionality of the described `PBufferNode` and `SocketNode`, is derived from the Cosmo3D class `csGroup`. Its method `drawVisit()` contains a pre- and a post-traversal section. The former switches the rendering context to pbuffer while the latter section starts the encoding and transmission of the image data.

In contrast to the Open Inventor scenario the new Cosmo3D scene graph node is used as a root node which enfolds the original scene graph as its subgraph.

Alternatively, the scene graph can be left untouched and just one function call has to be added at the beginning of the method `opXmViewer::swapBuffers()`. This function determines the current drawing buffer, calls `glReadPixels()` and initiates the image encoding and transmission.

Incoming mouse or key events are interpreted and handled after a pre-processing step by the corresponding methods in the `opXmViewer` class as if they had been appeared on the local site. For example, mouse events are sent to a method which converts the incoming data and emulates the method `processPendingEvents()` of `opXmViewer`.

3.2 Client modules

The client modules are implemented using the JAVA2 platform. We provide the following classes:

RenderArea: The basic render area `vis.inventor.RenderArea` is a drawing area for frame buffer content, that was received from the visualization server. For this purpose the Java2D `java.awt.image.BufferedImage` class is used. The drawing area also relays mouse events that are sent to the server. It is derived from the `java.awt.Panel` class and can be added into any Java container. The `vis.inventor.FullViewer` class, derived from this class, provides the look-and-feel and functionality of Open Inventors `SoXtFullViewer`, which includes a decoration around the render area. This decoration is make up out of thumb wheels, sliders, and push buttons. It also supports a context menu that allows to change the Open Inventor rendering type in several ways.

Decoders: `vis.imagedecoder.Decoder` is the abstract base class for all codecs we have implemented. New decoders can be integrated into the framework by deriving new classes from this base class. We provide the decoders `vis.imagedecoder.ZLIBDecoder`, `vis.imagedecoder.LZODecoder`, `vis.imagedecoder.RLEDecoder` and `vis.imagedecoder.RAWDecoder`.

172

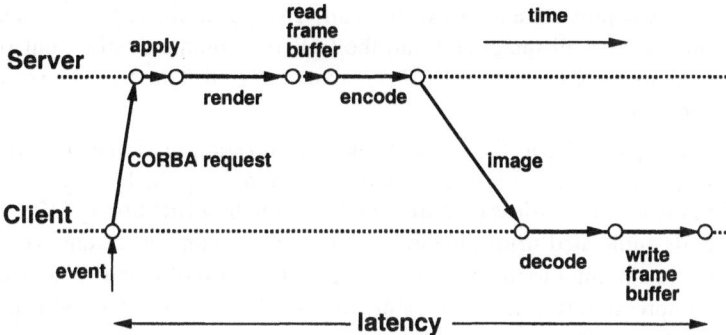

Fig. 2. The latency in between manipulation and image update consists of request, event translation, rendering, frame buffer read, encoding, transfer, decoding and display time.

SimpleViewer: We also provide a class `vis.viewers.SimpleViewer` which can be used as a client to observe any remote visualization session of an application that was adapted using our framework. For this purpose only some lines of code have to be added to a stand-alone visualization application.

3.3 Network Communication

The transmission of images and the remote control of the server application are strictly decoupled. The image data is streamed through a TCP socket connection and the application, mapping and rendering parameters are applied to the server application by using CORBA method calls. Once new mapping or rendering parameters were received by the server they are applied to the visualization. Then the rendering is performed, the frame buffer is copied into main memory, the data is encoded, transferred to the clients, decoded and finally copied into the frame buffer of the client (Fig. 2). These steps determine the overall latency of the application between manipulation and image update.

CORBA interfaces: The CORBA interface to the server is divided up into two parts:

- The events for the render area of the server application are sent to the server using the interface `RenderArea`. For a server that uses Open Inventor the interface `FullViewer` extends the `RenderArea` interface with functionality of the Open Inventor `SoXtFullViewer` class.
- Application specific parameters are accessed via the `Application` interface. We provide a base interface which can be extended to make the functionality of a server application accessible. For example, if the application allows to add clipping planes we would have to add a method `addClippingPlane` to the interface and implement the appropriate server method.

Image compression and transfer: The image transmission is decoupled from the CORBA interfaces because we wanted to be able to quickly replace the image transfer codec with more sophisticated ones without any changes in the remaining networking code. It could be as well integrated into the CORBA communication, but right now we transfer the image data using a TCP socket connection. The user can select one of following encoding types:

- **RAW:** The codec returns the original data, thus no compression is performed. This method can be used in a high bandwidth network to keep the latency low.
- **ZLIB:** Performs a loss-less compression based on the ZLIB library [12]. The algorithm finds duplicated strings in the input data. The second occurrence of a string is replaced by a pointer to the previous string, in the form of a pair (distance,length). ZLIB compression is a standard feature of Java since version 1.1 and is performed via fast native code.
- **LZO:** Performs a loss-less compression based on the LZO library [9]. LZO is a block compression algorithm - it compresses and decompresses a block of data. Block size must be the same for compression and decompression. LZO compresses a block of data into matches (a sliding dictionary) and runs of non-matching literals. LZO favors speed over compression ratio. As the LZO codec is not a standard method of Java, the LZO decompression is accessed via a Java Native Interface (JNI) call.
- **RLE:** Performs a loss-less compression based on run-length encoding.

Despite of the image compression the amount of data that has to be transmitted to the client is still too large for low-bandwidth network connections like an ISDN connection. In order to allow high interaction rates on such connections we additionally apply an image size reduction while interacting with the data. For example, when a manipulator is picked and dragged we transfer the images with half or quarter resolution and scale the images to full size on the client. As soon as the user stops dragging the manipulator a full frame is transmitted. The combination of compression and image size reduction provides sufficient frame rates even on 56k modem network connections.

4 Collaborative visualization environments

In areas where specialists are separated by distance the work-flow efficiency can be improved by collaborative applications. For example, such applications allow users to discuss the visualized data sharing the same view. Furthermore, expensive experts can be consulted and distance education or advanced training can be held. Additionally, our approach provides simultaneous access to a server application for multiple users. Thus the capabilities of expensive hardware can be sharded by low-cost client systems.

In this section we will present two applications that were extended by our framework to enable collaborative work. The first one is an application that uses 3D texture mapping hardware of high-end graphics workstations to visualize medical volume data. The second one is employed in the car development process for visualization of huge time-dependent finite element models.

4.1 Teleradiology

The use of 3D texture mapping hardware has become a powerful visualization option for direct volume rendering [1, 15]. Unfortunately, up to now 3D texture mapping hardware is still restricted to high-end graphics workstations. Now one can make the hardware capabilities accessible to almost any client system by using our framework.

For this purpose an interactive stand-alone texture based rendering application for medical volume data has been extended (Fig. 3, left). It has been integrated into the Open Inventor framework in order to obtain the whole flexibility and functionality offered by this graphics API. By introducing a new class, the volume renderer has been represented as a separate object within the hierarchical structure of the scene graph. This allows convenient use of built-in manipulators, sensors, editors and other predefined classes, methods and features (light sources, anti-aliasing, perspective/orthogonal projection, fly, walk, trackball) [4, 13].

First a few lines of code were added in order to extend the scene graph using the SocketNode we previously introduced. With this small modification it is possible to join a visualization session passively using a web-browser and observe the visualization. To allow remote control of the application we had to extend the base CORBA application interface with additional functionality (e.g. update of transfer function, adding of clipping planes, etc.). On the client side a viewer application was developed that provides the same look-and-feel as the stand-alone application (Fig. 3, right). We used the vis.inventor.FullViewer class as the display area in the main window, added some buttons to the decoration, and reimplemented the menu bar and the dialogs of the stand-alone application in Java (e.g. the transfer-function dialog).

Recapitulating, our approach allows working groups to discuss medical volume data sets collaboratively. 3D texture supporting graphics workstations, which were too expensive for many hospitals, can now be used remotely from any desktop PC. No patient data is transferred through the network and the security of the image stream can be ensured by using SSH socket tunneling.

4.2 Visualization of crash-worthiness simulations

Another example where the presented technique is very useful is the visualization in the car development process. In cooperation with the BMW Group we developed a Cosmo3D / OpenGL Optimizer based application which is in productive use in the pre- and post-processing of crash-worthiness simulations.

The car bodies are represented by about 500.000 mainly four-sided finite elements. During simulation the first 120 ms are computed and the coordinates of the deforming mesh are stored in 60 time steps together with the tracked parameters into a result file. Those result files often contain more than 1.5 GB of data.

Our application *crashViewer* builds a Cosmo3D scene graph that describes the car body for each time step. We developed an OpenGL Optimizer based viewer which allows the engineer to visualize and animate the crash. Furthermore, the crash performance can be analyzed by directly mapping the tracked scalar parameters as colors onto the geometry or by visualizing the force flux by force tubes [6].

To represent the large scale Gouraud-shaded time-dependent geometry with constant topology for each time step in a Cosmo3D scene graph, as proposed in [14], approximately 700 MB of main memory is required. Since only high-end graphics workstations are equipped with such a lot of main memory a technique was sought to make those expensive resources accessible to low-end systems. Hence, we extended the stand-alone visualization application by our approach which offers a solution by transferring image data from any OpenGL supporting workstation to arbitrary window supporting client systems.

If a meeting of the analyzing engineers is too time-consuming because they are, for example, located at distant sites, the image transfer allows for a collaborative discussion on the crash-worthiness of the current model variant. One engineer starts the visualization application which is able to provide the rendered images in encoded data stream form as previously outlined.

There are two scenarios: first, the other engineer starts a Java application that offers a minimal set of functionality of the original C++-based visualization application. 2D mouse events and keystrokes triggered on the client side are transmitted back to the server application and interpreted there like described in section 3.1. In the second scenario where one engineer will communicate some results to one or more engineers who do not have to interact with the model the former one advises an URL to other participants. They start any HTML browser, download a HTML page from the given URL which includes a Java applet (see Fig. 4). This applet encapsulates the Simple-Viewer described in section 3.2.

Summarizing, our approach allows collaborative working groups to discuss simulation results in distributed heterogeneous environments. There are low requirements to participating client systems. Additionally, in regard to security aspects, for example, if third party engineers of subcontractors are involved, the pure data will stay in-house; instead just image data is transferred. We expect, that this technique will facilitate the collaboration between accordant working groups of BMW and Rover, where it will be tested in the next few months.

5 Results

In this section we show results for the proposed techniques. On the server side all tests were run on a SGI Onyx2 workstation equipped with two 195 MHz R10000 processors and 512 MB of main memory. A SGI O2 workstation with the same processor and 128 MB of main memory and a 333 MHz Celeron PC equipped with 64 MB of main memory were serving as the client systems. The O2 was linked via a 100 MBit Ethernet network connection and the PC was linked using a 64 kBit ISDN Internet connection. We used the medical volume visualization environment with a 512x512x106 CT data set. A typical image sequence with frames of 704 pixels width, 576 pixels height and 24 bits depth was used.

First we analyze the frame rates that can be achieved over the local network and the Internet connection (Table 1). While interacting with the volume the images were rendered and transferred at quarter resolution (176x144 pixels), after interaction a full frame was transmitted. This leads to faster rendering, encoding, transfer, decoding and display times. When using the stand-alone non-networked visualization application we achieved an average frame rate of 4.1 for the full frames and 25.4 frames per second for quarter frames. Using the ISDN connection no interactive rates were achieved using full frames. This is negligible during interaction because then we are transferring

method	data size full	data size quarter	LOCAL full	LOCAL quarter	LAN full	LAN quarter	ISDN full	ISDN quarter
RAW	1.2 mb	76 kb			1.9	19	-	-
ZLIB	106 kb	6 kb	4.1	25	2.2	14	0.09	1.5
LZO	150 kb	9 kb			2.4	16	0.06	1.3
RLE	160 kb	10 kb			2.5	16	0.06	1.2

Table 1. The listed frame rates were achieved when visualizing a CT data set (512x512x106) locally, remotely using ISDN and remotely in a LAN network. Quarter resolution frames (176x144 pixels) were sent while interacting with the volume and full resolution frames (704x576 pixels) were sent after interaction. The average amount of transferred data for a full and quarter sized frame is denoted in the left part of the table.

quarter sized images only. Thus interactive refresh rates could be achieved. Only if the user stops to interact (e.g. releases a manipulator) the user has to wait some seconds for the full size frame. Best frame rates were achieved using ZLIB compression because the network bandwidth is the limiting factor for transmission via ISDN. We will see later that ZLIB compression has the best compression ratio. For a local network connection best frame rates were achieved using RLE compression because the encoding and decoding times are the critical values.

Secondly, we compare the encoding time and the compression ratios when using ZLIB, RLE and LZO compression. The average encoding time was 180 milliseconds for ZLIB, 50 milliseconds for LZO and 30 milliseconds for RLE. The average decoding time on the O2 was 93 milliseconds for ZLIB, 29 milliseconds for LZO and 13 milliseconds for RLE. Obviously, concerning the compression ratio ZLIB compression is ahead of the other compression methods (Fig. 5). RLE performs nearly as well as LZO compression. RLE is the simplest and thus the fastest compression method we investigated. The compression ratios of all methods of course depend on the covering of the screen space by the visualization. That is why a magnification leads to lower compression factors. Because of these results we currently favour ZLIB compression on low bandwidth channels and RLE compression on high bandwidth channels.

6   Conclusions and Future Work

We have presented a framework which allows remote high-end visualization from any Internet-connected, Java-enabled desktop PC. The introduced techniques were demonstrated by two applications. A volume renderer for 3D medical data, that uses special 3D texture hardware, can now be used remotely from any Internet-connected PC. We will effect an application study with the Department of Neurosurgery of the University of Erlangen-Nuremberg. Furthermore, a visualization application for large scale data sets of crash-worthiness simulations was extended by the presented framework to enable collaboration in the car development process.

As a main result of our work it is now possible to remotely explore huge scientific data sets on specialized server hardware using low-cost clients. We showed that this is even possible using a low bandwidth channel like an ISDN connection. The transfer of GUI events and application parameters requires a much lower bandwidth than the download of rendered images. Exactly this scenario is given while using low-cost broadband Internet connections like cable modems or satellite connections.

An area of future work involves the development of specialized image-streaming codecs for computer generated image sequences. First results using video streaming codecs were presented in [3]. However, currently available codecs are based on the needs of Internet video streams. Those streams have different characteristics than rendered ones (variable frame rate, partial changes in consecutive frames, ...). One possible approach would be to use lossy compression while interacting with the data and loss-less compression when having still images. Also image encryption techniques are necessary for transferring sensitive data over the Internet.

References

1. B. Cabral, N. Cam, and J. Foran. Accelerated Volume Rendering and Tomographic Reconstruction Using Texture Mapping Hardware. In A. Kaufman and W. Krüger, editors, *1994 Symposium on Volume Visualization*, pages 91–98. ACM SIGGRAPH, 1994.
2. K. Engel and T. Ertl. Texture-based Volume Visualization for Multiple Users on the World Wide Web. In *5th Eurographics Workshop on Virtual Environments*, 1999.
3. Klaus Engel, Ove Sommer, Christian Ernst, and Thomas Ertl. Remote Visualization using Image-Streaming Codecs. In *Proceedings of Symposium on Intelligent Multimedia and Distance Education*, Baden-Baden, Germany, August 1999.
4. P. Hastreiter, H.K. Çakmak, and Th. Ertl. Intuitive and Interactive Manipulation of 3D Data Sets by Integrating Texture Mapping Based Volume Rendering into the OpenInventor Class Hierarchy. In K. Spitzer Th. Lehman, I. Scholl, editor, *Bildverarbeitung fuer die Medizin: Algorithmen, Systeme, Anwendungen*, pages 149–154. Inst. f. Med. Inf. u. Biom. d. RWTH, Aachen, Verl. d. Augustinus Buchhandlung, 1996.
5. Ofer Hendin, Nigel John, and Ofer Shochet. Medical Volume Rendering Over the WWW using VRML and JAVA. In *Proceedings of MMVR*, 1997.
6. S. Kuschfeldt, O. Sommer, T. Ertl, and M. Holzner. Efficient Visualization of Crash-Worthiness Simulations. *IEEE Computer Graphics and Applications*, 18(4):60–65, 1998.
7. L. Lippert, M.H. Gross, and C. Kurmann. Compression domain volume rendering for distributed environments. In *Proceedings Eurographics '97*, pages C95–C107, 1997.
8. SGI Newsroom. SGI Brings Advanced Visualization to the Desktop with OpenGL Vizserver. http://www.sgi.com/software/vizserver/.
9. M.F.X.J. Oberhumer. Lzo. http://wildsau.idv.uni-linz.ac.at/mfx/lzo.html.
10. J. Patten and K.-L. Ma. A Graph Based Approach for Visualizing Volume Rendering Results. In *Proc. of GI'98 Conference on Computer Graphics and Interactive Techniques*, 1998.
11. T. Richardson, Q. Stafford-Fraser, K. R. Wood, and A. Hopper. Virtual Network Computing. *IEEE Internet Computing*, 2(1), January 1998.
12. G. Roelofs. Zlib. http://www.cdrom.com/pub/infozip/zlib/.
13. O. Sommer, A. Dietz, R. Westermann, and T. Ertl. An Interactive Visualization and Navigation Tool for Medical Volume Data. In V. Skala, editor, *Proc. 6th International Conference in Central Europe on Computer Graphics and Visualization '98*, pages 362–371, 1998.
14. Ove Sommer and Thomas Ertl. Geometry and rendering optimizations for the interactive visualization of crash-worthiness simultations. In *Proceedings of SPIE, Visual Data Exploration and Analysis VII*, volume 3960, January 2000. to appear.
15. Rüdiger Westermann and Thomas Ertl. Efficiently Using Graphics Hardware in Volume Rendering Applications. In *Computer Graphics Proceedings SIGGRAPH '98*, Annual Conference Series, pages 169–177. ACM SIGGRAPH, July 1998.

Editors' Note: see Appendix, p. 291 for colored figures of this paper

Appearance-Based Virtual-View Generation for Fly Through in a Real Dynamic Scene

Shigeyuki Baba[1,2], Hideo Saito[1,3], Sundar Vedula[1], Kong Man Cheung[1], and Takeo Kanade[1]

[1] Carnegie Mellon University, Pittsburgh PA 15213, USA
[2] Sony Corporation, Tokyo 141-0001, Japan
[3] Department of Information and Computer Science, Keio University, Yokohama 223-8522, Japan

Abstract. We present appearance-based virtual view generation which allows viewers to fly through a real dynamic scene. The scene is captured by synchronized multiple cameras. Arbitrary views are generated by interpolating two original camera-view images near the given viewpoint. The quality of the generated synthetic view is determined by the precision, consistency and density of correspondences between the two images. All or most of previous work that uses interpolation extracts the correspondences from these two images. However, not only is it difficult to do so reliably (the task requires a good stereo algorithm), but also the two images alone sometimes do not have enough information, due to problems such as occlusion. Instead, we take advantage of the fact that we have many views, from which we can extract much more reliable and comprehensive 3D geometry of the scene as a 3D model. The dense and precise correspondences between the two images, to be used for interpolation, are derived from this constructed 3D model. Our method of 3D modeling from multiple images uses the Multiple Baseline Stereo method and Shape from Silhouette method.

1   Introduction

Recently, synthesizing virtual images from multiple real images have found many applications including virtual reality, tele-presence and stereoscopic displays. Some applications, such as off-line generation of appealing graphics, can use a manual procedure [2], but many others require automation. This requirement for automation is especially strong in dealing with dynamic scenes. This paper aims at a completely automated method for synthesizing virtual images.

Model Based Rendering is one technique for image synthesis. First, a 3D model is reconstructed from the multiple images. Then, the colors on the real images are used to form the texture of the 3D model. Using conventional rendering techniques, virtual images are generated from the color textured 3D model. Wheeler et al. [16] proposed a method for accurate 3D model reconstruction from multiple view range images and demonstrated that the generated 3D shape and reflectance model can synthesize high-quality virtual view images. Debevec et al. [2] created precise synthetic images of a static scene whose model is constructed by an interactive 3D modeling system. Faugeras et al. [17] developed a system which can generate 3D models of a static environment semi-automatically. Our group [7] demonstrated automated creation of 4D (3D + time) models for time-varying scenes, together with texture mapping and rendering of new views. These methods have the advantage of handling the occlusion problem as they make use of the 3D models. However, texture mapping onto the constructed 3D model with errors may cause blur of synthesized virtual images.

Image Based Rendering method does not require any 3D models for synthesizing virtual images. Plenoptic methods that represents the radiance as a function of the position and the directions is one of the popular methods for Image Based Rendering [5,9,10]. The computation cost of these methods is less than that of Model Based Rendering. However, the creation of higher quality synthetic images requires a large number of original images. Another approach is generating synthetic images using correspondences between the original images such as View Interpolation [1] and View Morphing [13]. In those methods, correspondences between the original images must be specified for warping the original images to generate intermediate views. The correspondences are generally given manually [1,13], by the use of optical-flow [3] or by the use of dense stereo matching [12,15].

In this paper, we present a view interpolation approach which we call *Appearance-Based Virtual-View Generation*. First, a 3D model, which has enough geometrical information of a scene, is reconstructed from multiple images by using "Multiple Baseline Stereo" (MBS) [11] and "Shape from Silhouette" (SS) [4,6]. Taking advantage of the fact that we have 3D models of the scene, geometrically accurate correspondences are derived from the 3D models. The precise and dense correspondences generate virtual views at arbitrary viewpoints without losing pixels even in occlusion regions. In the following sections, we describe the details of each process.

2 Three Dimensional Model Reconstruction

3D models are reconstructed from multiple images captured in a facility called "The 3D Room" [8] by using either MBS method or SS method depending on the complexity of an objects in any given scenes. The former method is used for complex objects and the latter method is used for simpler objects.

Before the execution of these methods, calibration of all cameras is required (we currently have 49 cameras in a room). We use Tsai's camera calibration algorithm [14]. Tsai's camera model has 11 parameters, consisting of five intrinsic parameters and six extrinsic parameters. To implement this algorithm, we built a calibration device using 64 LEDs on single plane. The 8x8 LEDs are placed at an interval of 300mm uniformly. Camera calibration images are taken at five different positions by changing the height of this device. Once we know the relationship between the image coordinates and the world coordinates by this measurement, all the camera parameters are computed with this algorithm.

In execution of MBS, some neighboring cameras are chosen for each of the 49 cameras. Depth images are generated for each camera and all of the depth images are merged into a single 3D model, which is represented by triangle meshes, by using volumetric merging at each time frame. The region of interest is specified during the execution of volumetric merging process in order to obtain the desired 3D objects in the dynamic scene.

As for Shape from Silhouette, foreground (silhouette) images are generated for each camera before the computation of 3D model. Background subtraction is performed for the input images from each of the 49 cameras and dilation and erosion processing is performed to improve the quality of foreground images. After generating foreground images of all cameras, all of the images are back-projected into 3D space. Each camera viewpoint and its foreground image define a bounding volume. The 3D model can be reconstructed from intersecting volumes of multiple bounding volumes defined by these foreground images.

3   Deriving Pairwise Correspondence from 3D Model

The 3D model reconstructed from multiple camera-view images is used to derive correspondences between any neighboring camera image pairs. Figure 1 shows how to derive the correspondences from the 3D model. View 1 and view 2 are the views of a pair of neighboring cameras. First, the intersection of the ray from the point a in view 1 with the surface of the 3D model is computed by using camera calibration data. Then, the intersecting point A on the surface of the 3D model is projected onto view 2 and the projected point a' is computed. That is, the point a' in view 2 is the corresponding point of the point a in view 1. If there are points whose pixel rays have no intersecting point on the surface of the 3D model like the points as shown figure 1, those points have no corresponding points. This procedure is performed for all the projected objects in view 1 and the correspondences from view 1 to view 2 are derived. We also need the correspondences from view 2 to view 1 to overcome occlusion problems.

After the derivation of the correspondences in the entire view, disparity vectors are defined for all of the corresponding points like the vector d_a and d'_a shown in figure 1. These vectors are used for estimating the pixel position in the virtual views.

4   Virtual View Generation

We extend the View Interpolation method, so that the correspondences, derived from a 3D model, can generate virtual views at arbitrary viewpoints without losing pixels even in occlusion regions.

4.1   Review of View Interpolation

Once the correspondences between two neighboring views are derived, synthesized views at arbitrary viewpoints between those views can be generated. We

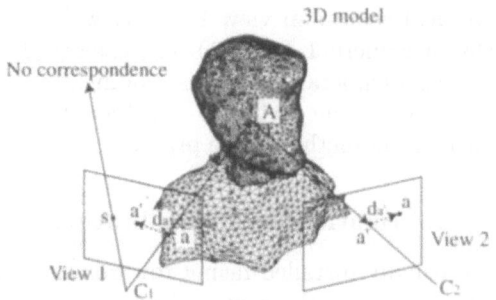

Fig. 1. The scheme for correspondence derivation between a pair of neighboring views, using a 3D model.

use an interpolation algorithm which is based on the related concepts of "View Interpolation" [1] and "View Morphing" [13] to generate the synthesized views. This interpolation algorithm involves the computation of the position and the color of the pixels using the correspondences between two images, described by the following equations.

$$P_i = w_1 P + w_2 P' , \tag{1}$$

$$I_i(P_i) = w_1 I(P) + w_2 I'(P') . \tag{2}$$

where

$$w_1 + w_2 = 1$$

P and P' are the position of the corresponding points in the view 1 and the view 2, respectively. $I(P)$ and $I'(P')$ are the color of the corresponding points in view 1 and view 2 as well. P_i is the interpolated position and $I_i(P_i)$ is the interpolated color and w_1 and w_2 $(w_1 + w_2 = 1)$ are weighting factors.

4.2 Zooming

We have to account for the fact that focal lengths of various cameras may be different, when dealing with multiple cameras for capturing real views. In this situation, if two neighboring camera-views with different focal lengths are chosen, the object size in the virtual views change during the movement of viewpoints. To avoid this problem, it is necessary to add a zooming image feature to the view interpolation. We modify the view interpolation equation as follows;

$$P_i = w_1 \left\{ (P - C) \frac{f_v}{f} + C \right\} + w_2 \left\{ (P' - C') \frac{f_v}{f'} + C' \right\} , \tag{3}$$

$$I_i(P_i) = w_1 I(P) + w_2 I'(P') . \tag{4}$$

where

$$w_1 + w_2 = 1$$

C and C' are the optical centers in view 1 and view 2, respectively. f and f' are the focal lengths of camera 1 (view 1) and camera 2 (view 2). f_v is the focal length of the virtual camera. With this modification, the virtual camera can zoom in and out in accordance with the focal length f_v. This modification makes the view interpolation method more practical.

4.3 Viewport Transformation Using Calibration Data

Multiple cameras are usually installed facing towards the center of the object. However, it is difficult to adjust the center of the objects to the exact optical center of each camera-view, even for static objects. If there is an offset between the center of the objects and the optical center in the view, the objects in the virtual view may move out of the field of view during zooming as described in the previous section. To avoid this problem, we transfer the viewport so that the objects can be placed at the center of the virtual view.

Since the calibration data for each camera is computed, we can define the projection matrices using the intrinsic and extrinsic parameters for each camera. Then, if the center of the objects in the world coordinates is defined, it can be projected onto each view using those matrices. Comparing the position of this projected point and the optical center in the views, the transformation value for re-centering objects in views can be computed. Using these transformation values, the center of the objects can be shifted to the optical center in the virtual view.

4.4 Pseudo Correspondences for Handling Occlusion

It is not unusual that the camera views have occluded regions in the scene. For instance, if we have two cameras and a L shaped object in a scene, as shown in figure 2, a part of the surface may be in an occlusion region for those views. If we compute the correspondence point of q in view 1 by using the scheme described in the section 3, it is back-projected onto the occluded surface of the 3D model and there is no consistent corresponding point in the view 2. For such an occluded area, we can not generate interpolated views by the method described by equations (1) and (2).

We solve this problem by introducing the concept of *Pseudo Correspondences*. The pseudo correspondences can be derived from the 3D model and then applied to the view interpolation, described by the equations (1) and (2). In figure 2, the point q in view 1 has no corresponding point in view 2. The back-projected 3D point Q of the 2D point q on the occluded surface can be virtually re-projected onto the point q' in view 2, even though it can not be seen from view 2. We name such correspondences *"Pseudo Correspondences"* In addition, the point r is back-projected onto the point R on the surface of the 3D model and re-projected onto the point r' in view 2, the same point as q'. Hence, q' ($= r'$) has both the pseudo corresponding point q and the real corresponding point r in the view 1. Applying these pseudo correspondences to equation (1), the position of the pixels in the virtual view can be interpolated. However, the color of the pixel

184

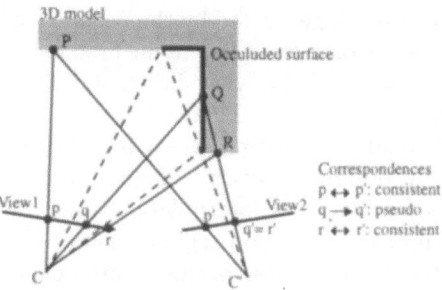

Fig. 2. Consistent and pseudo correspondences.

cannot be interpolated by equation (2) because the occluded surface cannot be seen from the view 2. In this case, the color should be simply chosen from the colors of the point in the neighboring two views.

To apply pseudo correspondences to view interpolation, we generate the two interpolated images using two directed correspondences, from view 1 to view 2 and from view 2 to view 1, separately. This means that two interpolated images are generated at one virtual view point. Then, the two interpolated images are blended into a single image. This implementation is described by the following equations:

$$I_w\left(u + w_1 d_u\left(u, v\right), v + w_1 d_v\left(u, v\right)\right) = I\left(u, v\right), \tag{5}$$

$$I'_w\left(u + w_2 d'_u\left(u, v\right), v + w_2 d'_v\left(u, v\right)\right) = I'\left(u, v\right) \ .$$

where

$$d_u\left(u, v\right) = \left\{\left(u' - u'_c\right)\frac{f_v}{f'} + u'_c\right\} - \left\{\left(u - u_c\right)\frac{f_v}{f} + u_c\right\},$$

$$d_v\left(u, v\right) = \left\{\left(v' - v'_c\right)\frac{f_v}{f'} + v'_c\right\} - \left\{\left(v - v_c\right)\frac{f_v}{f} + v_c\right\},$$

$$d_u\left(u, v\right) = -d'_u\left(u, v\right), d_v\left(u, v\right) = -d'_v\left(u, v\right) \ .$$

Both $I_w(u, v)$ and $I'_w(u, v)$ are the warped images generated by using the correspondences from view 1 to view 2 and from view 2 to view 1, respectively. $I(u, v)$ and $I'(u, v)$ are the original images at the two neighboring viewpoints. d and d' are the disparity vectors which are computed along with the derivation of the correspondences, described in the section 3. (u_c, v_c) and (u'_c, v'_c) are the optical centers in the view 1 and the view 2, respectively. These equations include zooming with focal length, f, f', f_v as described in the section 4.2.

After the generation of the two warped images, these are blended into a single interpolated image by using the following equation:

$$I_i\left(u, v\right) = \begin{cases} w_1 I\left(u, v\right) & \text{if } I\left(u, v\right) \neq 0 \text{ and } I'\left(u, v\right) = 0, \\ w_2 I'\left(u, v\right) & \text{if } I\left(u, v\right) = 0 \text{ and } I'\left(u, v\right) \neq 0, \\ w_1 I\left(u, v\right) + w_2 I'\left(u, v\right) & \text{otherwise} \ . \end{cases} \tag{6}$$

The color of the warped pixel, generated by the pseudo correspondence, is simply chosen from either view 1 or view 2 like as in first two cases of this equation. Figure 3 shows the process of this view interpolation algorithm. View 1 and view 2 are the original views taken by two cameras. Two warped images are generated using the weighting factors, $w_1 = 0.6$ and $w_2 = 0.4$. Each warped image is generated by different correspondence data as described above. By blending these warped images, an interpolated view is generated. The circled areas in this figure show the occlusion regions in the views. Using conventional view interpolation, the color in these occluded area cannot be recovered. With pseudo correspondences, an interpolated view can be generated without losing pixels in these regions as shown in figure 3.

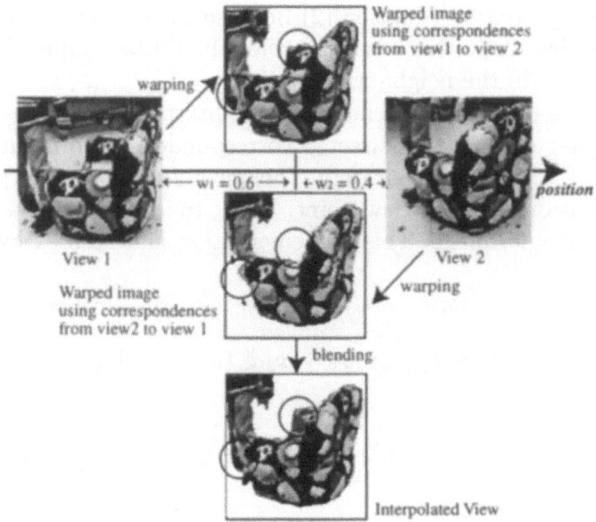

Fig. 3. View interpolation with pseudo correspondences.

5   Experimental Results

5.1   The 3D Room

In order to reconstruct 3D models of dynamic events efficiently and automatically, we have developed a facility called "3D Room" [8], in which dynamic events can be captured by synchronized cameras and be digitized as 3D representations with time frame. Forty nine cameras are installed at various points in the 3D room: 10 cameras on each of the four walls and 9 cameras on the ceiling. A PC cluster system (currently 17 PCs) can digitize all the NTSC video signals from the cameras simultaneously in real time as uncompressed and lossless color images (YUV422) at full video rate (30 fps). Each PC is used for the digitization of the video signals as well as image processing.

5.2 Virtual View Generation with Various Weighting Factors

Figure 4 shows example results of the appearance-based virtual view interpolation. In this example, twelve interpolated images whose weighting factors are 0.0, 0.2, 0.8 and 1.0 are generated from the original images of the two cameras (cam #29 and cam #30) at three different time frames (0, 50 and 100). If the weighting factor is either 0.0 or 1.0, the quality of the interpolated images is the same as the original images of the camera #29 or the camera #30. As a result, when we view the scene from the same viewpoint as an original camera, we obtain the full quality image. This is one of the advantages of our view interpolation algorithm. In most of the model based rendering methods, the texture rendered onto the model surface is blurred by the error of the recovered 3D model, that results in blurred virtual view images even if the virtual view point is the same as the real camera position. Moreover, the occlusion area can be successfully interpolated in each virtual view because of the pseudo correspondences. We have developed a GUI-based viewer application for viewing virtual views, synthesized by using our methods. With this viewer, users can easily specify the virtual camera position by using a mouse and fly through a real dynamic scene.

5.3 Virtual View Generation from Camera-Views with Different Focal Length

Figure 5 shows another example of the appearance-based view interpolation. In this example, the original views are taken from two cameras with different focal length. In order to avoid changing the image size among the virtual view points, we use same focal length f_v of the virtual camera. Using the algorithm described in sections 4.2 and 4.3, the interpolated views are generated with same focal length and the objects in those views are successfully centered using the camera calibration data.

6 Conclusion

Our method, which we call the Appearance-Based Virtual-View Generation of dynamic events, uses a 3D model to derive accurate correspondences between the original views. We have defined *Pseudo Correspondences* in order to avoid the occlusion problems. Since our correspondences contain geometric information, virtual views are generated at arbitrary viewpoints without losing pixels even in occlusion regions. Virtual view generation based on Image Based Rendering can be implemented using simple and fast 2D image processing techniques. That is, once the correspondences are derived from the 3D model, processing time of the virtual view generation does not depend on complexity of the 3D objects like the other image based rendering methods. Zooming and centering features are

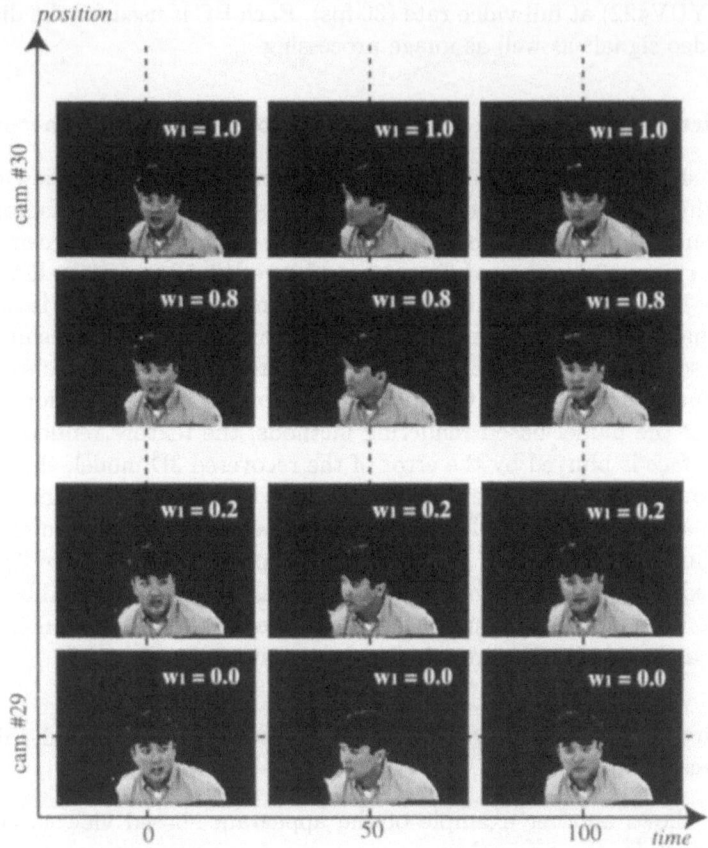

Fig. 4. Example results of Appearance based view generation.

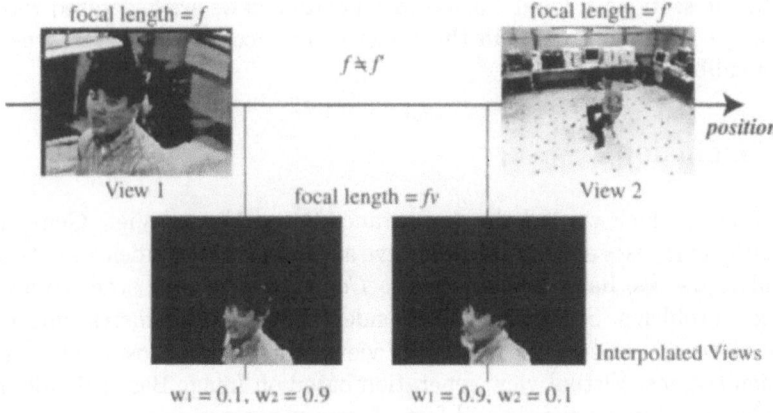

Fig. 5. Example results of Appearance based view generation, using two cameras with different focal length.

also implemented by using the transformation of the disparity vectors and the viewport. Thus the Appearance-Based Virtual-View Generation combines both accuracy and flexibility in the creation of virtual worlds from real views.

References

1. Chen, S., Williams, L.: View Interpolation for Image Synthesis. Proc. of SIG-GRAPH'93. (1982) 279–288
2. Debevec, P., Taylor, C., Malik, J.: Modeling and Rendering Architecture from Photographs: A Hybrid Geometry and Image-Based Approach. Proc. of SIG-GRAPH'96. (1996)
3. Avidan S., Shashua, A.: Novel View Synthesis by Cascading Trilinear Tensors. IEEE TVCG. Vol.4. No.4 (1998) 293–306
4. Potmesil, M.: Generating Octree Models of 3D Objects from Their Silhouettes in a Sequence of Images. Computer Vision, Graphics and Image Processing. **40** (1987) 277–283
5. Gortler, S.J., Grzeszczuk, R., Szeliski, R., Cohen, M.F.: The Lumigraph. Proc. of SIGGRAPH'96. (1996)
6. Chein, C.H., Aggarawal, J.K.: Identification of 3D Objects from Multiple Silhouettes using Quadtrees / Octrees. Computer Vision, Graphics and Image Processing. **36** (1986) 100–113
7. Kanade, T., Rander, P.W., Narayanan, P.J.: Virtualized Reality: Constructing Virtual Worlds from Real Scenes. IEEE Multimedia. Vol.4. No.1 (1997)
8. Kanade, T., Saito, H., Vedula, S.: The 3D Room: Digitizing Time-Varying 3D Events by Synchronized Multiple Video Streams. CMU-RI-TR-98-34 (1998)
9. Katayama, A., Tanaka, K., Oshino, T., Tamura, H.: A Viewpoint Dependent Stereoscopic Display Using Interpolation of Multi-Viewpoint Images. SPIE Proc. Vol.2409. Stereoscopic Displays and Virtual Reality Systems II (1995) 11–20
10. Levoy, M., Hanrahan, P.: Light Field Rendering. Proc. of SIGGRAPH'96 (1996)
11. Okutomi, M., Kanade, T.: A Multiple-Baseline Stereo. IEEE Trans. on PAMI. Vol.15. No.4 (1993) 353–363
12. Narayanan, P.J., Rander, P.W., Kanade, T.: Constructing Virtual Worlds Using Dense Stereo. Proc. ICCV'98 (1998)
13. Seitz, S.M., Dyler, C.R.: View Morphing. Proc. of SIGGRAPH'96 (1996) 21–30
14. Tsai, R.: A Versatile Camera Calibration Technique for High-Accuracy 3D Machine Vision Metrology Using Off-The-Shelf TV Cameras and Lenses. IEEE J.of Robotics and Automation RA-3. 4. (1987) 323–344
15. Vedula, S., Rander, P.W., Saito, H., Kanade, T.: Modeling, Combining and Rendering Dynamic Real-World Events from Image Sequences. Proc. 4th Conf. Virtual Systems and Multimedia. Vol.1 (1998) 326–332
16. Wheeler, M.D., Sato,Y., Ikeuchi, K.: Consensus Surfaces for Modeling 3D Objects from Multiple Range Images. DARPA Image Understanding Workshop (1997)
17. Faugeras, O., Laveau, S., Robert, L., Csurka, G., Zeller, C.: 3-D Reconstruction of Urban Scenes from Sequences of Images. INRIA Technical Report. No.2572 (1995)

SMARTLINK: An Agent for Supporting Dataflow Application Construction

Alexandru Telea, Jarke J. van Wijk

Eindhoven University of Technology,
Den Dolech 2, Eindhoven 5600 MB, The Netherlands,
alext@win.tue.nl, http://www.win.tue.nl/math/an/alext
vanwijk@win.tue.nl, http://www.win.tue.nl/cs/tt/vanwijk

Abstract. Visual programmable dataflow systems are an effective way to build a large class of visualization applications from existing software modules. However, the appeal of dataflow systems is often decreased as their users have to get familiar with libraries containing hundreds of different modules. Classical documentation systems such as hypertext or example suites are not always effective, as they lack the context of the user's questions and problems. We present a new visual dataflow programming assistant that is simple to use, offers context-sensitive help derived from the user's own behavior, and smoothly integrates in the effective point-and-click visual programming metaphor. We illustrate our approach with real-life usage examples.

1 Introduction

A recurring problem for scientific computing and visualization practitioners is the need to easily build new experimental applications for 3D data processing and visualization, imaging, or feature extraction. In many cases users need to build an entire suite of applications, e.g. when experimenting with different combinations of algorithms to explore a given dataset. Writing a new program for every trial application is prohibitively expensive in terms of time and convenience, especially for non experts in programming. Visual programming dataflow systems are a well known solution to this problem. Application building is done in such systems by picking visual representations of code components (or modules) in a module browser GUI and connecting them interactively to form a dataflow network. Users with little programming expertise can thus quickly create a suite of applications by reusing existing algorithms and data structures in a simple, intuitive way.

Although praised, visual dataflow programming systems are often less easy to use in practice than expected. Such systems usually come with large libraries offering hundreds of different modules spanning application areas as diverse as imaging, 3D rendering, CFD visualizations, charting, and volume visualization. The expected effectiveness of visual dataflow programming is noticeably diminished by the difficulty to find the 'right' modules to build the desired application among this large set of available choices. This difficulty is increased by the fact that the provided module set can freely grow with user programmed modules.

Often users have had to rewrite existing modules simply since they were unable to locate them in the already existing large libraries. Schemes that address this documentation problem such as online manuals and help agents have proven to be of a limited use in case of a large number of modules and users with limited expertise.

We have addressed this problem by devising the SMARTLINK help agent. SMARTLINK provides a new manner of assisting users visually in building dataflow applications by effectively exploiting their domain-specific knowledge and learning their preferences and interests. The concept presented here is a visual supplement to classical documentation browsers and can be used in a larger context than visual dataflow programming solely. We first discuss the limitations of the classical visual dataflow systems and of their help agents in Section 2. In section 3, we present the SMARTLINK concept. Section 4 extends the concept to object-oriented dataflow systems. Section 5 concludes the paper with a discussion of the method and ideas for future development.

2   Background

The dataflow programming paradigm is widely used by many systems, whether visual environments such as AVS or Express[4], Iris Explorer, or Khoros[5], or programming frameworks such as VTK[6], Open Inventor[2], or Java3D[3]. Application construction in such systems is an iterative task of finding the right modules and assembling them by connecting their inputs and outputs to create the desired dataflow network.

Whether visually programmable or not, such systems offer two main application building mechanisms. Firstly, a typing mechanism is provided for the modules' ports which enforces that only compatible ports can be connected. Object-oriented typed systems [6, 2, 8] take this a step further as OO typing automates some of the module input-to-output data conversions. End users are thus partially relieved of the burden of converting data explicitly by inserting conversion modules, which simplifies the network construction. Secondly, online help and documentation tools assist users in finding out information about a given module, such as the types and meaning of its ports, its operational semantics, an example of use, and hyperlinks to related modules.

Most of the improvement related to dataflow systems since their advent more than ten years ago has concentrated on areas such as external code integration and multi-language support [4, 8], object-oriented (OO) architectures [2, 6, 7], web-based integration and multiprocessing[4], and providing more modules and user interaction tools. However, the basic problem of assisting users in building dataflow applications has not received much attention. In practice, this task is mostly approached by a combination of the following three methods:

- modify an existing 'sample' dataflow network
- browse the online documentation to find out which module fits a given problem context
- ask human assistance (e.g. a more experienced colleague)

However effective, the above methods do not scale well in the context of a general-purpose dataflow system with hundreds of modules for various application domains, used by an unexperienced, possibly isolated user. Ideally, a visual dataflow programming help agent should posess several qualities, among which four are presented next (see also Fig. 1 a).

Qualities \ Methods	reuse existing examples	documentation and FAQs	experienced colleagues
simplicity	medium	low	high
specificity	medium	can be anything	high
learning capability	no	no	yes
availability	medium/high	medium/high	usually low

a)

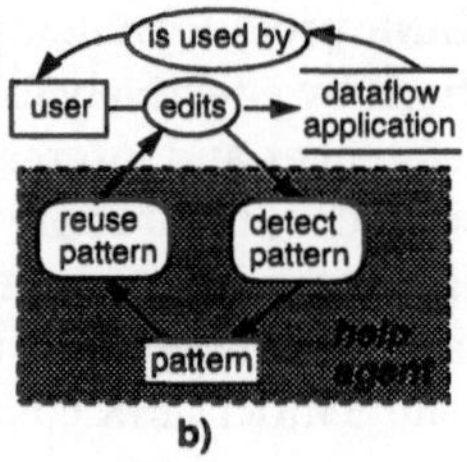

b)

Fig. 1. Visual programming help agents and qualities (a). The help agent in the network construction process (b)

Simplicity: Dataflow application construction should be simple to use and learn for novice users as well. Reusing an existing dataflow network by editing it is sometimes a viable solution. However, when no suitable example is available one is left with the tedious task of browsing huge amounts of module documentation. Using such documentation systems can be difficult, as these are usually built to reflect the module libraries' *structures*, and not their *use*. A known example for this are the OO component documentation browsers [3, 6]. To use such browsers, one must be familiar with OO notions such as inheritance and subtyping along which the documentation is structured. Understanding such documentation is often an all or none process, which is clearly undesirable for users that need only specific answers. Moreover, going back and forth between documentation browsing and network visual editing is clearly a disruptive process which one would like to avoid.

Specificity: The most ubiquitous questions of dataflow network building are not 'what does the module `ImageWarpFilter` do ?' but 'which module should I use to connect *this* data reader to *that* filter?' or 'what should I use to view *this* filter's data output?'. These questions have a clearly localized, context dependent nature. Consequently, generic documentation systems such as manual pages are not appropriate here, as these do not capture the question's context dependency. A documentation mechanism to capture more context is the frequently asked questions (FAQs) list. However, FAQs may also grow large, become less simple to use, and provide less specific answers. Generally speaking, the more specific and context-dependent the question is, the less probable it is to find the answer in general-purpose documentation.

Learning capabilities: As users build dataflow applications, they detect certain patterns such as the necessity or preference to use a certain filter in combination with some reader. Examples and documentation are however by definition static and thus can not reflect the user's accumulating knowledge unless they are rewritten as new information is learnt. Often users have no other alternative but manually create their own custom annotations and reuse them time and again.

Availability: In all the above, using the knowledge of another human is clearly the most effective way to learn to design new dataflow applications. However, experienced users from whom to learn are often not available. Availability is usually not a problem for automatic help agents such as documentation and examples.

Summarizing the above, we would like to have a simple to use, context-specific, intelligent automated agent to assist the construction of dataflow networks. Such an agent (the shaded area of the dataflow-like diagram in Fig. 1 b) should detect, capture, and exploit the user's behavioral patterns during network editing to assist the editing process. So far, these are performed explicitly by the human user during the learning process. Next, we present the construction and use of an automated network construction help agent.

3 The SMARTLINK Agent

The proposed solution is based on the observation that dataflow application building is mostly learnt by examples. Users find out that, for example, an Actor's output should be connected to a Viewer's input to view the actor, or sometimes to a Writer to write the actor to file. Next, when an Actor is present when building a new network, one infers a Viewer or a Writer might be needed, so a Viewer is instantiated and connected to the Actor.

We have exploited the above by constructing SMARTLINK, an agent that assists dataflow network building. Currently SMARTLINK is implemented atop of the VISION general-purpose object-oriented visualization system [7]. However, the presented method can be easily added to any dataflow system. To describe the SMARTLINK agent, we first introduce a few terms. A *dataflow graph* consists of nodes, which represent module ports, and *links*, which represent data flows between these ports. A *path* is a sequence of links in the graph from one port to another. The *input* of a path is its first link's input port and its *output* is its last link's output port.

The SMARTLINK agent maintains a database able to store a set of links for all module ports in the dataflow system. These links represent the 'most used connections' into which that port is involved. For example, the 'output' port of an Actor might store two links, one to Viewer's 'input' port, and one to Writer's 'input' port. For every link an integer weight is stored that represents how many times that connection has been done. The database contains thus accumulated knowledge on system typical usage patterns. The database construction and the usage of the information it stores are described next.

3.1 Database Construction

The SMARTLINK database is initially empty. As the user employs the dataflow system to edit networks, SMARTLINK updates the database by looking at the port connect operations executed. When two ports p_1 and p_2 are connected, the two links $p_1 - p_2$ and $p_2 - p_1$ are looked up in the database. If found, the link's weight is incremented by 1, else a new link connecting the two ports and having weight 1 is inserted in the database. Figure 2 shows this for the example discussed above by depicting the database after an `Actor` was connected to a `Viewer` (a), then to a `Writer` (b), and finally again to a `Viewer` (c). Conceptually, these links form a separate *preferences graph* which indicates which module ports were connected throughout the system's utilization.

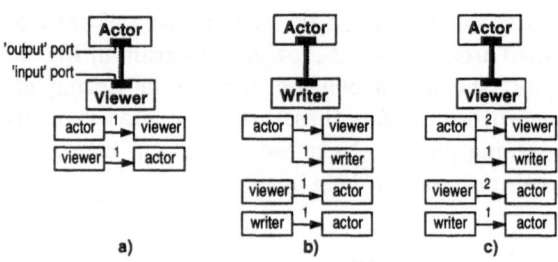

Fig. 2. SMARTLINK database for three network construction stages

3.2 Database Use

The information stored in the SMARTLINK database is used to assist network construction process by an interactive help agent. The purpose of the help agent is to answer the two types of questions discussed in Section 2, i.e. 'how can one connect two given modules?' and 'what can be connected to a given module?'. SMARTLINK addresses these two questions by its two operation modes called 'join' and 'cascading'. To illustrate these operation modes, we shall use a visualization example involving modules from the VTK library [6]. As described elsewhere[7], we have integrated VTK in VISSION such that its over 300 C++ components can be manipulated as visual icons in VISSION's network editor.

Join Mode: In this mode, SMARTLINK provides suggestions for the connection of two given module ports. We shall exemplify this by a VTK-based dataflow application for visualizing isosurfaces of a quadric function (Fig. 3). The pipeline consists of a scalar quadric function definition `VTKQuadric` sampled onto a regular grid by `VTKSampleFunction`. Isosurfaces are extracted from the sampled dataset by `VTKContourFilter`, mapped to geometric primitives by `VTKDataSetMapper`, then to drawable objects by `VTKActor`, and finally viewed in a `VTKViewer`. Additionaly, `VTKLookupTable`, `VTKProperty`, and `VTKDataSetInspector` control the colormap used, the actor's material properties, and the mapping of scalars to colors respectively.

Suppose now that all the user knows is that he wants to *visualize* the *isosurfaces* of a *quadric* function. He may start the network construction by creating a `VTKViewer`, a `VTKContourFilter`, and a `VTKQuadric` respectively, but he doesn't

194

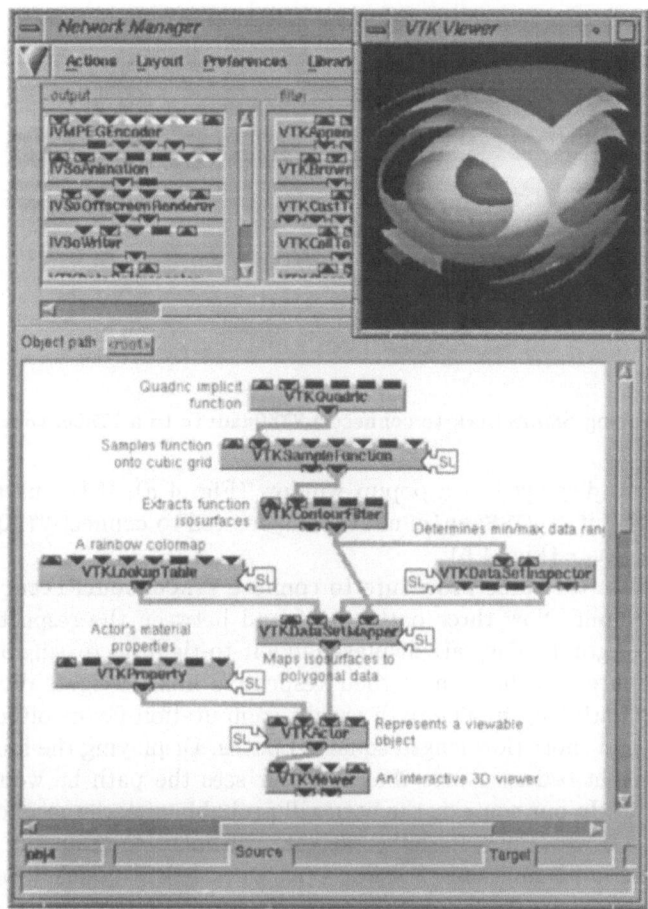

Fig. 3. VTK visualization example of quadric function isosurfaces

know how to continue. In a classic setup, he would have to browse the documentation or existing network examples to see what modules should be added to complete the network. In our case, however, all he needs to do is to connect visually **VTKQuadric**'s output with **VTKContourFilter**'s data input port. As the ports are not directly compatible, the system waits a short while (e.g. one second) to ensure that the connection attempt was not accidental, and then initiates a breadth-first search in the preferences graph from **VTKQuadric**'s output to **VTKContourFilter**'s data input port. The search is driven via a best-first heuristic by exploring links in decreasing weight order. Next, the system performs the same search in the opposite direction, i.e. from **VTKContourFilter**'s input to **VTKQuadric**'s output. More paths connecting the two ports may be found in this way, as the preferences graph is not symmetric with respect to the links' directions. This is so since SMARTLINK may forget links from the preferences graph, as explained in Section 5. In our case, a path passing via **VTKSampleFunction** is found, as the system 'remembers' that this connection was done previously.

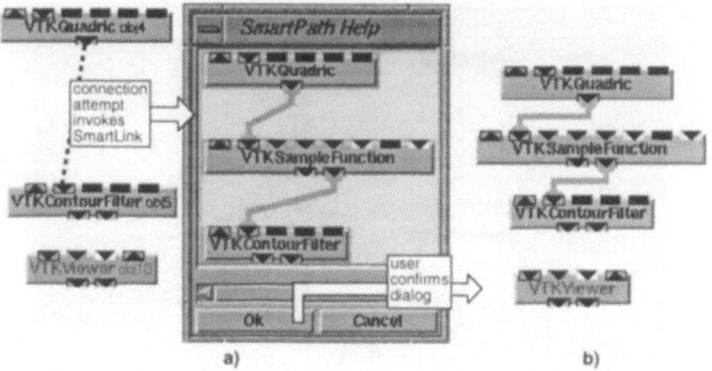

Fig. 4. Using SMARTLINK to connect a `VTKQuadric` to a `VTKContourFilter`

The path is displayed in a popup window (Fig. 4 a). If the user selects the path by clicking it, a `VTKSampleFunction` is created to connect `VTKQuadric` and `VTKContourFilter` (Fig. 4 b).

Next, we use the same procedure to connect `VTKContourFilter`'s output to `VTKViewer`'s input. Now three paths are found between the respective ports in the preferences graph. They are displayed in left-to-right decreasing order of their weights, computed as the sum of their respective links' weights divided by the square of the path length. The path weight computation favors often used (high weight sum) and short (low length squared) paths. Displaying the found paths in decreasing weight order ensures that the user sees the path he would probably best prefer first. In our context, the user will probably pick the second one. If this path is picked a few times, its links' weights will increase gradually until it will be found more important than the first one, when it will be displayed first. The

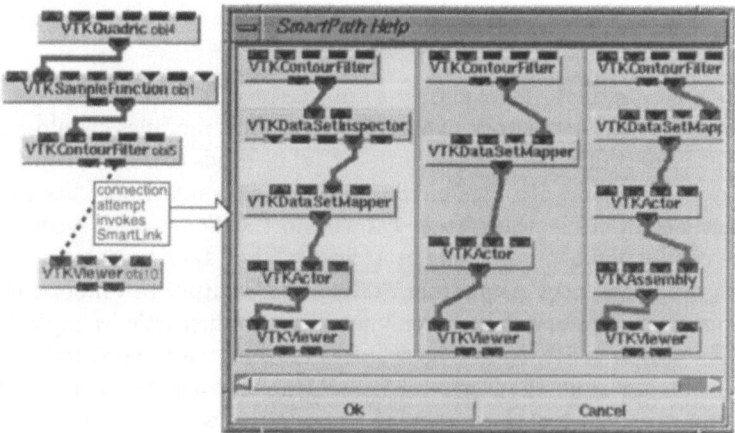

Fig. 5. SMARTLINK display of multiple paths

system has thus learnt from the user's behavior.

196

Cascading Mode: In this mode, SMARTLINK offers suggestions for modules that can follow from a given port. So far we have created the main part of the pipeline in Fig. 3 with only two mouse clicks. SMARTLINK offers also a second way to assist the network construction. In this so-called 'cascading mode', the user can simply click on any existing module port to find out which other modules can be connected there. For example, if the quadric visualization user didn't know how to view the isosurfaces output by VTKContourFilter, he could click on VTKContourFilter's output to ask 'what could be connected here?'. A menu would pop up listing all the modules that VTKContourFilter's output prefers to connect to, in decreasing order of the found links' weights. For example, to change the default colormap and material properties, two clicks on VTKDataSetMapper's 'colormap' and VTKActor's 'properties' ports, respectively, are sufficient (Fig. 6 a-c).

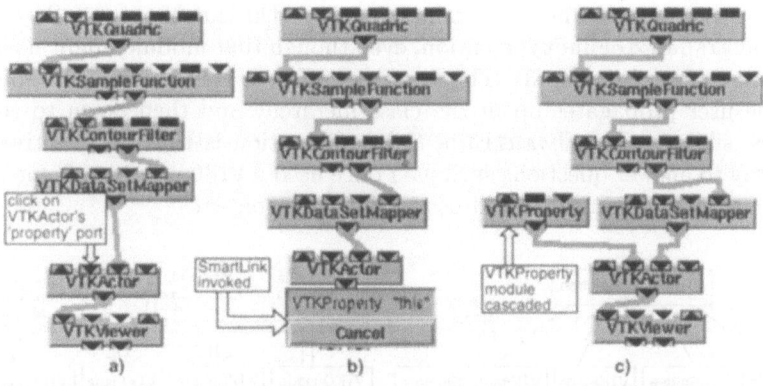

Fig. 6. SMARTLINK cascading operation. a) Port 'properties' of VTKActor selected. b) Popup menu with cascading option. c) Module VTKProperty cascaded

Overall, four mouse clicks are needed to construct the quadric visualization network from the three key initial modules. The six modules inserted automatically by SMARTLINK are shown marked with 'SL' in Fig. 3. Without SMARTLINK, this would have meant searching, instantiating, and connecting these modules in the right places. This would have taken longer even for an experienced user who knew exactly what and where to insert.

4 SMARTLINK and Object Orientation

VISSION is built on an object-oriented basis. VISSION modules are compiled C++ classes with inputs and outputs typed by user defined C++ types. Interpreted C++ is used for run-time scripting and type checking. Modules are organized as OO class libraries where derived classes inherit similar semantics and purpose, and specialize some feature [1]. For example, data reader modules inherit features such as input and output ports from a base class declaring the data reading interface and specialize the update operation. VISSION's OO foundations offer also an advantage regarding the SMARTLINK mechanism, as the next example shows.

VTK provides a class `vtkImplicitFunction` to represent all scalar functions $f(x, y, z) = 0$. This class declares an interface to evaluate the function, while subclasses such as `vtkQuadric`, `vtkSphere`, `vtkPlane`, and `vtkImplicitBoolean` define concrete quadric, spherical, planar, and boolean combinations of implicit functions respectively (Fig. 7). In dataflow terms, it is `vtkImplicitFunction` that declares the data output port via which `vtkQuadric` connected to the `vtkSampleFunction` in our visualization example. When this connection is made, SMARTLINK adds its information to `vtkQuadric`'s output and also to the `vtkImplicitFunction` baseclass output (Fig. 7 b, steps 1-2). When SMARTLINK is subsequently used to assist connecting `VTKSampleFunction`'s input, four possibilities are offered, namely `vtkQuadric`, `vtkSphere`,`vtkPlane`, and `vtkImplicitBoolean`. Similarly, when the user requests help on connecting e.g. `vtkPlane`'s output, SMARTLINK offers the `VTKSampleFunction` option, even though that module might have never used before (Fig. 7 b, step 3). The perceived effect is that the information learnt from the user propagates up in the class hierarchy and then down to the used modules' siblings. The SMARTLINK-OO combination is thus an effective, automatic way to answer questions such as 'I once used a `VTKQuadric` in some context - show me other modules I could use in the same context'.

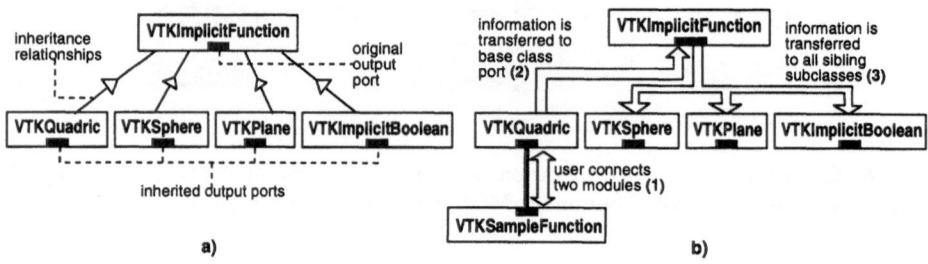

Fig. 7. Effect of object-orientation on SMARTLINK

5 Discussion and Future Work

We have presented SMARTLINK, an agent that makes visual dataflow programming easier and more attractive for non experienced users by exploiting the context information they inherently provide. The method does not replace, but augments, classical help media such as hypertext and examples, as follows. First, its use is simple (a single extra mouse click per usage) and does not disrupt the visual point-and-click network editing. Second, assistance is given automatically in a context specific manner, rather than in a generic, situation independent way. Third, the system learns from the user, so different users will be offered assistance that depends on their previous behavior. For example, an imaging researcher's database will contain information mainly on imaging modules, while that of a CFD engineer will contain mainly CFD-related links. To prevent database pollution with spurious information or its unbounded growth, we limit the number of links stored per module port to a small amount (e.g. 10) and manage them in a most recently used, highest weight first fashion. The system can thus 'forget' sparsely used links or links done by mistake. Finally, the method's combination with the object-oriented module libraries' structure is a simple but powerful way for generalizing from the learnt information. In contrast to most OO help systems, our method is simple to use as it hides the OO aspect completely from non OO-experienced end users.

Users have found SMARTLINK very effective. Experienced users employed the cascading building method from a few 'key' modules to speed up network construction considerably. Novice users would copy the database of an expert in the field and get domain-specific online assistance for free, but block the agent's learning mode to prevent it learning from their mistakes. This emulates well the help of a specialized user, as the database mirrors the human's system usage pattern. Database importing would interactively supplement documentation, tutorials, and examples.

SMARTLINK can be expanded in several ways. Different learning algorithms (i.e. weight increasing and path weight computation functions) may sample the user's behavior more effectively. A promising direction is to directly visualize the user behavior derived information. This would allow the identification of dataflow systems' usage patterns, and provide a compact, visual way to understand and work with large (dataflow) libraries. Finally, SMARTLINK could assist other programming tools that assemble typed components, such as C++ or Java visual compilers.

References

1. B. STROUSTRUP, *The C++ Programming Manual*, Addison-Wesley,1993.
2. J. WERNECKE, *The Inventor Mentor*, Addison-Wesley, 1993.
3. H. SOWIZRAL, K. RUSHFORTH, M. DEERING, *The Java3D API Specification*, Addison-Wesley, 1998.
4. C. UPSON, T. FAULHABER, D. KAMINS, D. LAIDLAW, D. SCHLEGEL, J. VROOM, R. GURWITZ, AND A. VAN DAM, *The Application Visualization System: A Computational Environment for Scientific Visualization.*, IEEE Computer Graphics and Applications, July 1989, 30–42. See also http://www.avs.com.
5. KUBICA, RASURE *The Khoros Application Development Environment*, Experimental Environments for Computer Vision and Image Processing, H.I Christensen and J.L Crowley eds, World Scientific, 1994.
6. W. SCHROEDER, K. MARTIN, B. LORENSEN, *The Visualization Toolkit: An Object-Oriented Approach to 3D Graphics*, Prentice Hall, 1995
7. A. C. TELEA, J. J. VAN WIJK, VISSION: *An Object Oriented Dataflow System for Simulation and Visualization*, Proceedings of the IEEE VisSym'99 Visualization Symposium, Springer, 1999
8. C. GUNN, A. ORTMANN, U. PINKALL, K. POLTHIER, U. SCHWARZ, *Oorange: A Virtual Laboratory for Experimental Mathematics*, Technical University Berlin. http://www-sfb288.math.tu-berlin.de/oorange/

Editors' Note: see Appendix, p. 292 for colored figure of this paper

Design of Visualizations for Urban Modeling

L. Denise Pinnel, Matthew Dockrey, A.J. Bernheim Brush, and Alan Borning

Department of Computer Science and Engineering
University of Washington
Seattle, WA 98195-2350 USA
{denisep,mrd,ajb,borning}@cs.washington.edu

Abstract. Urban planning experts often use computer models to help
evaluate alternative land use policies, particularly as they interact with
transportation and environmental decisions. The greatly increased data
volume provided by new land use models makes their effective use difficult without suitable visualization tools. We present UrbanView, a visualization system for urban modeling, and describe a user study to
determine appropriate visualizations for the urban modeling domain.

1   Introduction

Patterns of land use and available transportation systems play a critical role in
determining the economic vitality, livability, and sustainability of urban areas.
Transportation interacts strongly with land use: different kinds of transportation
systems induce different patterns of land use, while at the same time, different
kinds of land use induce demands for different kinds of transportation systems.
Both land use and transportation have strong environmental effects, in particular
on emissions and resource consumption.

Urban planning experts use computer models to simulate and evaluate different land use polices. UrbanSim, an urban modeling system being written by
our collaborators at the University of Washington, is several orders of magnitude more complex than existing models. This resulting increase in data volume,
coupled with the diverse types of analyses experts must perform makes the use
of new visualization techniques essential.

While there has been a large amount of work on geographical information
systems (GIS), very little if any rigorous research has been done evaluating the
usefulness of other types of visualizations for this domain. Our goal is to examine
various visualization types to find the appropriate visual representations for
urban modeling tasks.

We first describe our system architecture which includes UrbanSim, our urban modeling system, and UrbanView, our visualization system specialized for
urban modeling. Then we discuss a user study we conducted to determine what
types of visualizations are useful for urban modeling tasks.

2   Urban Development Project Architecture

Figure 1 shows the system architecture. UrbanSim and UrbanView are described
in the following sections.

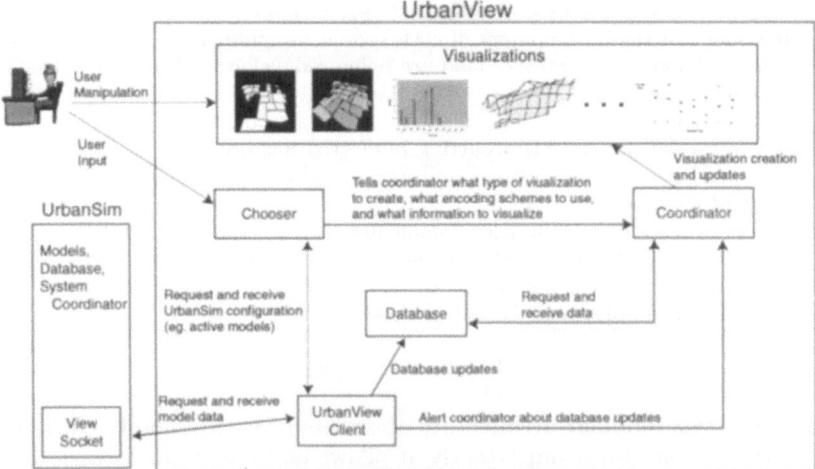

Fig. 1. Urban Development Project Architecture

2.1 UrbanSim

UrbanSim [1, 8] is a land use modeling system designed to model the development of urban areas. In particular, it models the interactions between land use and transportation infrastructure, along with the resulting environmental effects.

In current practice, land use modeling, if done at all, employs a very simplified, aggregate land use model. The unfortunate consequence is that the model is not sensitive to important policy alternatives such as changing zoning, urban growth boundaries, or taxes and incentives. UrbanSim is intended to provide a much more detailed model of land use, which explicitly allows different policy alternatives to be modeled and compared. UrbanSim is composed of an object store and numerous component models that simulate various actors in the urban development process. For example, there are component models that simulate business creation and closure, household and business movement and location choices, and developer decisions such as the character, density, and scale of property development.

A user interacts with UrbanSim to create scenarios that specify alternative packages of policies, economic and demographic forecasts, and other exogenous inputs. The system is then executed for a specified number of years (typically 20), and the results are analyzed.

2.2 UrbanView

UrbanView is the visualization generation system we are building to interface with UrbanSim. With UrbanView a user can create visualizations that represent model execution decisions, resulting land use phenomena and model data output in useful informational form.

UrbanView is composed of an object store, a visualization coordinator, a visualization chooser, and individual visualization components. As illustrated

1. Where is the development of different types of land occurring?
2. How much of the development of each type is Greenfield vs. redevelopment?
3. How much and what types of land are being redeveloped? From what use into what use? Where?
4. What is the distribution of density of new development?
5. How much development is occurring in desired locations?
6. How many acres of agricultural and forest land are being consumed by development? How fast?
7. Why is the model building/developing the parcels/buildings it is?
8. Where are households of each type generally locating?
9. Where are the biggest population gains and losses?
10. How much is employment decentralizing?

Fig. 2. User Tasks

in Figure 1, this modular architecture allows us to easily modify each component in the system. Most importantly, it allows us to experiment with different visualization chooser modules during user testing.

There are a wide range of possibilities for user interaction with the chooser module, from complete user control over the creation of visualizations to an autonomous visualization generator based on user task input. The results from this user study and future user studies will determine where on this continuum of user interaction the chooser should be placed to optimally support users.

The coordinator creates the individual visualizations from visualization specifications, keeps track of each of the visual displays, and notifies the visualizations of data updates. The individual visualization components then request their data directly from the object store.

UrbanView will dynamically interact with the UrbanSim component models as they are executing, as well as create static visualizations of model results. We are currently in the process of linking UrbanView and UrbanSim which will allow us to present users with interactive visualizations in addition to the existing static ones.

3 User Study

It seems clear that certain visualizations are extremely useful and much more effective than textual descriptions for specific contexts. However, there is not one perfect visualizations for an urban modeling data set. Different visualizations are often necessary for each different user task. We present an analysis of visualization types that provides some initial results for understanding what makes visualizations useful and how to create them for the urban planning domain.

For the study, all of the two- and three-dimensional map visualizations were generated using UrbanView. At the time of the tests we were still implementing the other visualization types, which we have now completed.

3.1 Study Method

Five participants, two graduate students and three professors, all from the Department of Urban Design and Planning at the University of Washington, took

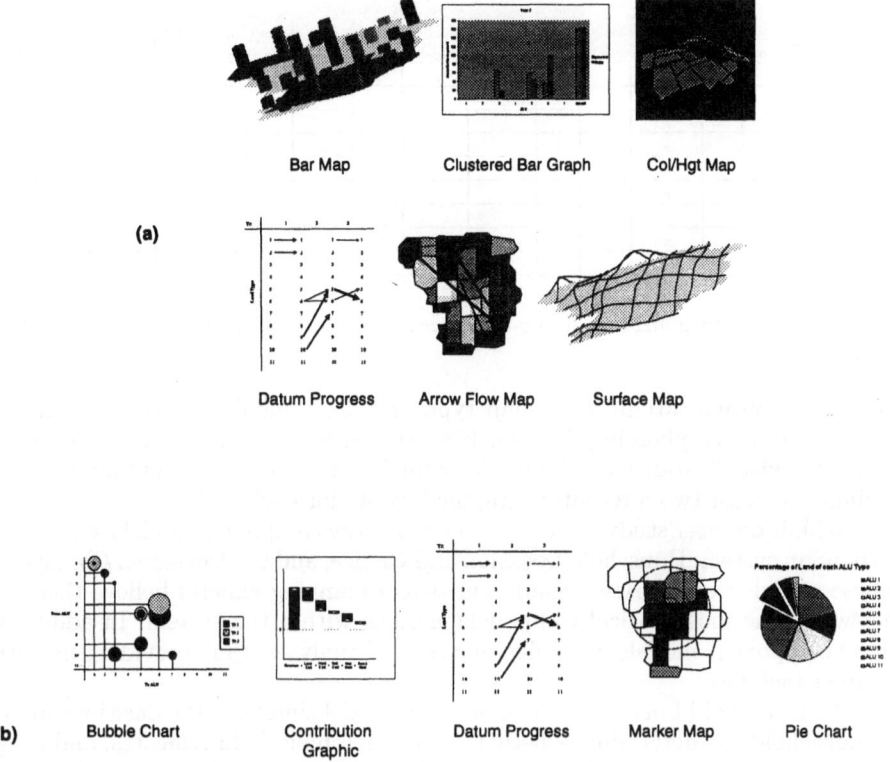

Fig. 3. Sample visualization layouts for (a) comparison processing activities and (b) quantitative processing activities

part in the user study. Each participant was given a series of tasks, enumerated in Figure 2, that could be answered using the information provided by UrbanSim model output. These tasks were selected by an urban land use expert and intended to be representative of the questions an analyst asks while using a system such as UrbanSim.

The entire user test consisted of ten questions, each having between three and seven corresponding visualizations, for a total of 55 different visualizations. Each visualization contained all the information necessary to complete the task. The participants were asked to evaluate, for each task, the usefulness of each of the visualizations presented on a scale of one (most useful) to five (least useful), and then to rank the visualizations in order of preference.

Our visualization designs focused on the understanding and explanation of land use as simulated by the UrbanSim system. Visualizations such as the ones presented to participants in this study would be used by analysts while they are studying the system's output and behavior. To choose the sample visualizations for each task, we examined the current GIS visualizations prevalent in the urban planning domain and designed visualizations based on previous research.

Figure 3 shows some of the different visualization types presented for Comparison and Quantitative processing activities as described in Section 4.1. Many

Types	Color Intensity	Distinct Colors	Color Scale	Bars	Stacked Bars	Cluster Bars	Lines	Area Height	Terrain/ Surface	Arrows	Num./ Values	Marker Size	Marker Shape
Graphs		x		x	x	x	x						
Pie Chart		x											
2D Map	x	x	x							x	x	x	
3D Map	x	x	x	x	x			x	x	x			
Symbol Chart													x
Bubble Chart		x										x	
Progression										x	x	x	
Contribution		x		x					x	x			
Table		x								x			

Fig. 4. Types of visualizations cross-referenced by the encodings they can utilize effectively.

of our test visualizations were map type visualizations due to the spatial nature of the urban planning domain. However, in order to verify our belief that map type visualizations are the most useful for urban planning and analysis we included at least two alternative graphic layouts for each task.

To limit our user study we focused on two of the component models within the UrbanSim system: Household Location and Choice, and the Developer/Redeveloper. These models were chosen because our urban planning experts believe these to be two of the most critical component models within the system. In addition, these component models use different types of analyses, thus providing us with a larger task base.

The Household Location and Choice sub-model simulates the decisions made by households. It determines whether or not each household relocates, and if relocating, determines the relocation site. The Developer/Redeveloper sub-model simulates the decisions made by builders as to whether or not to develop or redevelop property, and if so, where, what, and how much to build.

4 Data Analysis

We analyzed the data collected during our user tests to determine favored visualization types and encoding methods. We believe the most promising approach to generating visualizations for analysis and understanding of the urban modeling domain is to design them based on user tasks and their corresponding cognitive processing requirements.

Observation of visualizations show that the makeup of a visualization is more complex than the simple classification scheme used by Lohse et al. [5]. In addition to the overall schematic layout and base look of visualizations, visualizations represent data using information encoding schemes. Thus, rather than using a single type classification of visualizations, we break visualizations down into base type (map, bar chart, line graph) and encoding method(color, marker, arrows, height). Classifying the test visualizations by type and encoding methods resulted in thirty-two distinct visualization type variations as shown in Figure 4. Color examples of these visualization type variations can be found in our technical report [7].

Our analyses focused on matching visualization base types and visualization encoding methods to both specific cognitive processing activities and complete tasks. Given the spatial nature of the urban planning domain, we expect that

	Spatial	Comparison	Alternatives	Options	Trends	Relations	Aggregation	Qualitative	Quantitative	Description
Task 1	x	x			x			x		
Task 2		x							x	
Task 3	x				x	x			x	
Task 4					x	x	x	x		
Task 5	x		x		x				x	
Task 6							x		x	
Task 7			x			x		x		x
Task 8	x	x			x			x		
Task 9	x	x		x			x	x		
Task 10					x	x	x		x	

Fig. 5. Breakdown of tasks into human cognitive processing activity requirements. We describe the breakdown of Task 3 into the required cognitive processing activities in the text.

map type visualizations will be the most widely preferred type of visualization. For those activities and tasks where map visualizations were favored, we explored the relationship between map encoding schemes and processing activities and user tasks. We also looked at processing activities and user tasks where map based visualizations were not preferred even though the domain is spatially oriented.

4.1 Cognitive Processing Based Analysis

Previous research [2–4, 6] has shown that effective visualizations allow users to substitute quick perceptual inferences for more difficult logic inferences. Thus the display of information is dependent upon the cognitive processing required by a user's task.

To analyze the data collected during our user study we classified the ten tasks based on the cognitive processing activities required for each task. Consider for example Task 3 from Figure 2, one of the more complex tasks in the study. This multi-step task poses the questions: How much and what types of land are being redeveloped? From what use into what use? Where? First the participants must determine the amount of land and the type of land being developed (quantitative judgments), then they must determine the change in land type over time (determining trends and looking at relations on the land type attribute), and last they need to determine relative locations of redevelopment (spatial determinations). Figure 5 shows the breakdown of all the tasks into their information processing requirements.

Map Based Visualizations Discussions with the participants during and after the tests revealed that location of urban activity is extremely important for urban policy decisions. Figure 6 shows that visualizations with a map base type were preferred for the majority of tasks and processing activities. Because the map layout implicitly encodes important geographic location information and is the common display currently used by experts in urban planning, this was the expected result.

Processing Activity	Spatial	Comparison	Alternatives	Optima	Trends
Preferred Visualizations	2D Color Map Datum Progress 3D Col/Hgt Map 3D Col/Hgt Map 2D Intensity Map	2D Color Map Clustered Bar 3D Col/Hgt Map 3D Col/Hgt Map 2D Intensity Map	Clustered Bar 3D Col/Hgt Map 2D Intensity Map	3D Height Map	2D Color Map Datum Progress 3D Height 3D Col/Hgt Map 2D Marker Map
Favored Base Type	Map	Map	Map	*inconclusive*	Map
Favored Encoding Method	*N/A*	Distinct Color	*inconclusive*	*inconclusive*	Area Height/ Color

Processing Activity	Relations	Aggregation	Qualitative	Quantitative	Description
Preferred Visualizations	Datum Progress 3D Height 3D Col/Hgt Map 2D Intensity Map 2D Marker Map	3D Height Stacked Bar 2D Intensity Map 2D Marker Map	2D Color Map 3D Height 2D Intensity Map 3D Col/Hgt Map 2D Intensity Map	Clustered Bar Datum Progress Contribution Gr. Stacked Bar 2D Marker Map	2D Intensity Contribution Gr.
Favored Base Type	Map	Map	Map	Bar	*inconclusive*
Favored Encoding Method	Color to Area Height	Color	Intensity/ Area Height	Bar Size	*inconclusive*

Fig. 6. The preferred visualizations and the favored base type and encoding methods for each task that requires the corresponding processing activity.

We analyzed the activities where maps were the preferred base type to determine if there was a correlation between cognitive activity and preference for map encoding schemes. We determined that color distinctions and area height are the two most commonly preferred encoding methods for the processing activities. Due to the geographical nature of urban planning, the combination of color to distinguish values and map placement to determine relative locations was considered the most useful encoding for comparison, trend, and qualitative processing activities. The Favorite Encoding Method rows of Figure 6 show the results of this analysis.

There were three processing activities, Alternatives, Optima, and Description, for which there was no clearly preferred encoding method. Two of these also showed no significant base type preference. We attribute the lack of significant results for these three activities to the scarcity of tasks requiring them. Future tests will encompass the entire domain of activities that UrbanSim simulates, thus giving us a broader task base.

2D versus 3D Map Types Whether or not 3D is better than 2D is an open and very controversial question in information visualization. We designed our tests so that we could evaluate the differences between two- and three-dimensional map visualizations for this domain. For our analysis we considered two- and three-dimensional maps as separate visualization types rather than different encodings, because we believe there are fundamental differences in the expressive styles and capacities of two- and three-dimensional map visualizations. However, analysis shows that while there is a slight preference for two-dimensional maps for almost all of the processing activities, tasks that require the same processing activity show no bias to either two- or three-dimensional maps. This was a surprising result and is a further area of study that we plan to pursue.

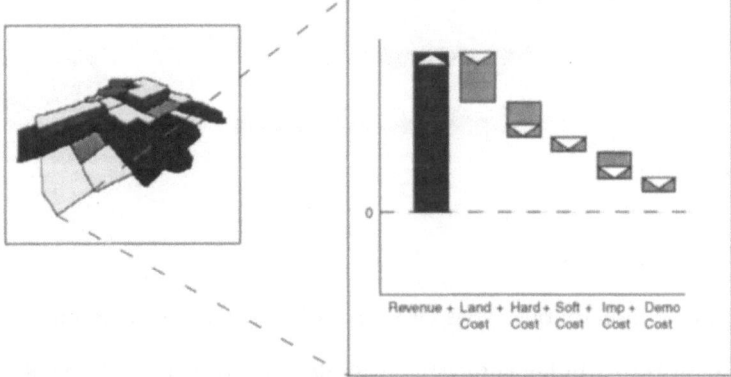

Fig. 7. Example contribution graphic from the user study. The large graphic shows why a particular land parcel was not developed due to the low total expected profit.

Companion Graphics Participant comments during the user tests revealed that in many cases one all-encompassing graphic is not sufficient. Participants liked the bar charts as companion and summary graphics for making quantity judgments and comparisons between planning attributes. For describing the internal logic of component models within a simulation system, the participants preferred a visualization that presented an overview of the actual process, such as the contribution graphic in Figure 7.

Other Preferred Visualization Types The one processing activity that revealed itself to be better represented by non-map visualizations involved tasks that required them to make quantitative judgments. For quantitative judgments, Figure 6 shows that even when presented with a majority of map type visualizations, participants preferred bar type graphics.

4.2 Task Based Analysis

We also performed an analysis based on the entire tasks rather than on the individual processing activities required by the tasks. The major result of this analysis was that map based visualizations were not considered useful for tasks that included a time dimension. In Tasks 3 and 6, described in Figure 2, users were asked to make judgments about the change in variable values over a time span. The preferred visualization types, shown in Figure 8, were datum progress visualizations, stacked bar graphs, and pie charts.

Initially we attributed a dislike of map based visualizations solely to the presence of quantitative reasoning required by both tasks. As discussed previously, when we broke tasks down into processing activities, we found map based visualizations were not the most preferred visualization type for quantitative processing activities. This alternate analysis based on entire user tasks suggests that it is really the time dimension in Tasks 3 and 6 that cause users to eliminate map based visualizations as a useful information display. To verify this hypothesis we plan to look at tasks that include a time dimension but do not

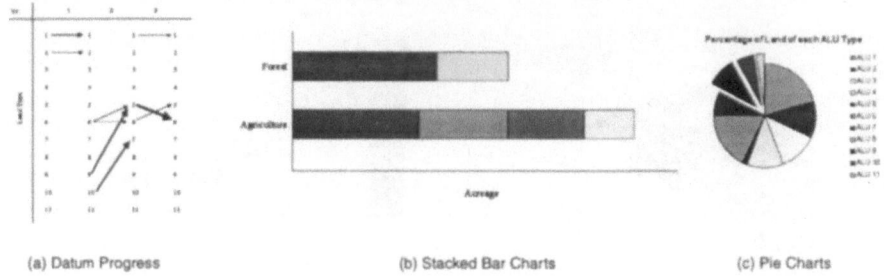

(a) Datum Progress   (b) Stacked Bar Charts   (c) Pie Charts

Fig. 8. Favored visualization types for tasks with quantitative processing activities.

require quantitative judgments. For example, we can extend Tasks 8 and 9 to include a time component and retest to see if users no longer prefer map based visualizations.

5 Conclusions

We found that for urban planning and analysis, map type visualizations provide the geographical information that plays a critical role in analysis of systems like UrbanSim. At the same time, we learned that for quantitative tasks bar charts and summaries better present the needed information. Of the encoding methods, analysts tended to like color and size encoding schemes.

User testing showed that there is great variance in which visualizations users consider effective. However, while there may not be one perfect way to design useful visualizations, we learned from this study that there are certain characteristics of visualization types that appear to be better for presenting the information necessary to help solve different types of tasks. This suggests that the visualization chooser can and should automatically present users with useful default visualizations based on their tasks. Our strategy is to provide a configurable but directed visualization system that integrates and facilitates the display of simulation information using many different types and styles of visualizations.

6 Future Work

There remain several problems for future research. Currently, UrbanView does not automatically choose visualization presentations; rather, the user specifies the visualization type and encoding methods and data to be viewed. Our research goal is the automatic generation of default visualizations. We hope to create a user interface where users indicate their task to the system and then are presented with default graphics.

Additional user studies include further investigation of the classification of user tasks by cognitive processing activities, increasing the variety of visualizations included in user studies, and expanding the domains of study beyond the visualization of the urban planning domain. We will also be performing user studies to determine the applicability of our exploratory results in static visualization to interactive model visualizations.

Finally, the focus of this paper has been the identification of useful visualizations for aiding urban planning professionals; but in the longer term we also want to aid citizens' groups and elected officials in understanding the component models and their outputs, thus opening the traditional black-box model to support informed civic deliberation and debate on issues of land use, growth, sprawl, and transportation choices [1].

7 Acknowledgments

This research has been supported in part by National Science Foundation Grants No. CMS-9818378 (Urban Research Initiative) and IIS-9975990, and in part by the PRISM project (University of Washington University Initiatives Fund). A.J. Brush supported in part by an NSF Graduate Research Fellowship.

We thank Paul Waddell for both participating in our user study and providing us with urban planning expertise, and Michael Noth for creating the UrbanSim system. We also thank Marina Alberti, John Caruthers, Kevin Krizek, and Hilda Blanco for participating in our user study. Finally, we would like to thank all of our reviewers for their helpful comments and suggestions.

References

1. A. Borning and P. Waddell. Participatory design of an integrated land use - transportation modeling system: First steps. In *Proceedings of the 1998 Participatory Design Conference*, pages 181–182, 1998.
2. Stephen Casner and Jill H. Larkin. Cognitive efficiency considerations for good graphic design. In *Proc. of the 11th Annual Conference on Cognitive Science Society*, pages 275–282, Ann Arbor, Mich., 1989.
3. Stephen M. Casner. A task-analytic approach to the automated design of graphic presentations. In *ACM Transaction on Graphics*, volume 10, pages 111–151, New York, NY, 1991. ACM.
4. Jill H. Larkin and Herbert A. Simon. Why a diagram is (sometimes) worth ten thousand words. *Cognitive Science*, 11(1):65–99, 1987.
5. Gerald L. Lohse, Kevin Biolsi, Neff Walker, and Henry H. Rueter. A classification of visual representations. *Com. of the ACM*, 37(12):36–49, 1994.
6. Jock Mackinlay. Automating the design of graphical presentation of relational information. *ACM Transactions on Graphics*, 5(2):110–141, 1986.
7. L. Denise Pinnel, Matthew Dockrey, and Alan Borning. Design and understanding of visualizations for urban modeling. Technical Report UW-CSE 99-12-01, University of Washington, December 1999..
8. P. Waddell. Simulating the effects of metropolitan growth management strategies. In *1998 Conference of the Association of Public Policy Analysis and Management*, 1998. Available from http://www.urbansim.org/papers.

ViSSh: A Data Visualisation Spreadsheet

Fabian Nuñez and Edwin Blake

Collaborative Visual Computing Laboratory
Department of Computer Science, University of Cape Town
Private Bag, Rondebosch 7701, South Africa
{fabian,edwin}@cs.uct.ac.za

Abstract. We describe a data visualisation system which uses spreadsheets as its user interface metaphor. Similar systems implemented in the past were hampered by the contradiction between an imperative formula language and the declarative spreadsheet framework. We have analysed spreadsheets from a data visualisation point of view, and built a system that is an improvement over past efforts. Our prototype combines the following three techniques: we store lists of values in each spreadsheet cell; we use a functional programming language as the formula language and we make use of lazy evaluation. The novel combination of these techniques makes our system consistently declarative in nature, and gives it several advantages such as small, uncluttered visual programs, the ability to deal with potentially infinite datasets and the use of advanced functional language features.

1   Introduction

Our prototype system for data visualization "ViSSh" (*Visualisation SpreadSheet*) uses a purely declarative spreadsheet paradigm for data visualization. ViSSh has been designed to explore the benefits to the user of a consistent application of a pure declarative programming paradigm. We note that spreadsheets are in essence declarative (since cells store either constants or side-effect-free formulas). One can also observe that the traditional programming systems for visualization (Modular Visualisation Environments or MVEs [2]) which use a visual data flow programming model also have a declarative flavour – data flow leads to pure declarative formalism.

A further insight on which our approach is based is that a spreadsheet is, of itself, a data visualization of sorts and as such its use in data visualization is a consistent extension of the basic spreadsheet paradigm.

Spreadsheets possess a number of useful properties for data visualization, *viz.*:

A. They are declarative in that cells store constants or functions without side-effects.
B. They are naturally tidy and uncluttered no matter how large they grow, while the same cannot be said for data flow graphs.
C. They derive a lot of power from their mutability: changes are easily made since all cells are available for editing.

However, they suffer from a number of drawbacks, mainly related to the way they scale with the size of the problem:

1. Large spreadsheets may be tidy but one can only view a small part of them.
2. Very large data sets are unwieldy since each cell may only contain a single value.
3. The large size of datasets commonly visualised would slow down recalculations.

We have developed extensions to the basic spreadsheet paradigm that overcome the above while retaining a pure declarative approach, and implemented them to demonstrate their viability. In essence, our extension of the spreadsheet paradigm consists of the following: storing lists of values in each cell, using a functional programming language and making use of lazy evaluation. Our prototype implements these extensions, as well as addressing some less fundamental usability problems we came across during its development. For point (1) above, we adopt an *overview window*, that acts as a map of the spreadsheet to aid navigation. Very large data sets (2) are naturally accommodated by using *lists* rather than scalars as the primitive data type in each cell and (3) is overcome by using lazy evaluation in the recalculation process. Sect. 5 further elaborates on these points. Additionally, we discovered that providing users with an *automated data flow view* of the corresponding spreadsheet (see Fig. 2) greatly helps users with the editing process. Since our spreadsheet is purely declarative there is a one-to-one correspondence between this dataflow view and the spreadsheet.

2 Previous Work

Most current data visualisation systems are based on the dataflow model. This includes systems such as *Iris Explorer*, *apE* and IBM's *Data Explorer* (see [2] for an overview). The dataflow paradigm is declarative in nature, and allows users to build non-trivial programs without extensive programming knowledge.

Spreadsheets have seldom been applied to data visualisation. Levoy [9] has described a system in which spreadsheet cells are populated not only by numbers and formulas (written in *Tcl*), but also by graphical objects (such as images or movies) and user interface objects (such as buttons or sliders). Levoy found that, compared to the commonly used dataflow model, spreadsheets are "more expressive, more scalable, and easier to program"[9]. However, Levoy's use of Tcl as a formula language leads to a mismatch in the programming paradigms used. Spreadsheets use a declarative programming paradigm, while Tcl is clearly imperative in nature. In addition to this, Levoy encourages the use of imperative programming and side effects (e.g., having cells directly modify other cells). Chi, Riedl, Barry and Konstan [3] extend the work of Levoy by not constraining the system to images, but instead allowing the user to visualise any type of information. Although this work does remove some of the limitations in Levoy's implementation, it still suffers from the same paradigm mismatch since it also uses Tcl.

The functional programming paradigm stipulates that functions may only read their arguments and generate only the result, without making any changes to their surroundings. Spreadsheet formulas on most commercial spreadsheets normally adhere to this paradigm, if one considers the formula to be the function, the cells referred to by the formula as the function's arguments and the value displayed in the cell occupied by the formula as the function's result. De Hoon [4] has shown that a spreadsheet can indeed be built from a purely functional perspective, and describes his implementation using the functional programming language *Clean*.

Since traditional spreadsheets make use of declarative formula languages, they do not suffer from the paradigm mismatch described above. However, it was not until the functional paradigm was used in conjunction with spreadsheets that the compatibility of spreadsheets and the functional paradigm was demonstrated. De Hoon [4] has constructed a text-based spreadsheet, which uses the functional language *Clean* [1] as its underlying formula language. This spreadsheet has some interesting properties, such as the ability to evaluate functional expressions symbolically. This depends on the choice of functional language, but does demonstrate the compatibility of both paradigms.

3 An Abstract View of Spreadsheets

Isakowitz *et al* [8] have analysed spreadsheets using concepts from the field of databases, and have come up with the concept of a *dual* view of a spreadsheet. On one side, there is the *physical* layout. This refers to the way cells are laid out, grouped, *et cetera*. On the other side there is a *logical* view of the spreadsheet. This handles abstract aspects of the spreadsheet, such as data dependencies between cells. The logical view could be likened to a database's schema, and in fact performs many of the same duties. Using an algorithm described in [8], one can break up a spreadsheet into several properties, called *schema*, *data*, *editorial* and *binding*. These four properties, taken as a whole, make up the entire spreadsheet, i.e. *spreadsheet = schema + data + editorial + binding*. These properties are defined as follows:

- The *schema* property stores a formal definition of the spreadsheet's formulas.
- The *data* property is the set of constants on which the *schema* property operates.
- The *editorial* property is what remains in the spreadsheet after the *schema* and *data* properties have been extracted: labels, comments, etc.
- The *binding* property is what defines the logical to physical mapping of the other three properties, using row and column addresses.

The logical structure of a spreadsheet is defined by the *schema* and *data* properties, while its physical structure is described by the *editorial* and *binding* properties.

It can be seen that Isakowitz's *schema* property can be likened to a dataflow system, since it keeps track of the dependencies between spreadsheet cells, and hence indicates the flow of data between these cells.

3.1 A Functional Look at the Logical Properties

Isakowitz and Schocken separate the concepts of *schema* and *data*, suggesting that functions and the data they operate on are different. Most functional languages, however, treat functions and data as equivalent entities. A re-examination of the logical structure of a spreadsheet, this time from a functional point of view, may reveal some interesting properties of spreadsheets.

In spreadsheets, cells can contain either constant values or formulas. A formula can be easily seen to be simply a different representation of a function. Any constant can also be trivially expressed in term of functions, simply by defining a function that always returns the same constant. Therefore it is possible to describe the contents of a spreadsheet cell entirely in terms of functions. From this, it follows that a spreadsheet can be seen as simply a grid of functions. For the sake of simplicity, each could in turn be called A1, A2, A3, ..., B1, B2, B3, ..., etc. If one considers a spreadsheet which contained a "1" in cell B1, a "2" in cell B2 and the formula "=SUM(B1:B2)" in cell B3, then these functions could be written (using the Scheme language syntax) as follows:

```
(define (B1) 1)
(define (B2) 2)
(define (B3) (+ (B1) (B2)))
```

The implications of this observation are not immediately obvious, yet they are quite fundamental. Since a spreadsheet can be completely defined in terms of named functions, it can be adequately described only in terms of functions.

Of course, some information is lost in the process, namely the spatial relationships between the cells. These relationships are used in two ways, namely to make relative references to cells (e.g., "add the contents of the cell just above this one to the contents of the cell just to the left of it"), and to manipulate groups of cells as ranges. However, this functionality can be implemented in other ways, for instance, consider the following implementations of relative references and cell ranges. Note that both of these are used only when *editing* the spreadsheet and do not affect *computations* in any way.

Relative References. Relative references can be implemented by applying simple transformations to the formulas as they are moved or copied to other cells. For example, a formula such as =sum(A1:A3) would become =sum(B1:B3) if copied one cell to the right of its current location. A mechanism would be required to "anchor" some cell references so they are not transformed (for example, the cell reference B4 in *Microsoft Excel* does not change when the cell referring to cell B4 is moved). This functionality would not be a part of the functional part of the spreadsheet, but of the cell editor.

Cell Ranges. Cell ranges can be implemented as compound types. For example, the cell range named B2:B4 could be expressed as the Scheme list (B2 B3 B4). Again, since this relies on cell adjacency, the responsibility lies with the cell editor.

4   A Redefinition of Spreadsheets

In Sect. 3.1 it was observed that a spreadsheet can be described as a rectangular grid of functions, instead of a mixture of functions and data. Therefore the logical view of a spreadsheet referred to by Isakowitz and Schocken [8] need not be broken down into *schema* and *data*, but can in fact be described as a coherent whole.

Since all inter-cell references are calls to named functions, spreadsheets have no direct need for a grid structure, and the grid organization of spreadsheet cells thus comes into question. It seems then that a spreadsheet can be described simply as a set of functions that are ordered on a rectangular grid. However, it should be remembered that the names that are given to functions are completely arbitrary (as long as they are consistent). This means that the grid layout of spreadsheets is in fact a user interface feature, and is not necessary for the spreadsheet's functioning. Note that this remark is not meant to demean the importance of the grid, but instead to highlight the separation that exists between the user interface and the underlying functional mechanism.

Isakowitz and Schocken's logical view of a spreadsheet has already been explained as a set of functions, while the physical view (i.e. the layout of the cells) has also been shown to be separate from the underlying functional structure. Therefore, since the physical layout is separate from the "real" structure, yet aids in its comprehension and manipulation, the physical layout itself can be classed as a form of data visualisation. Therefore, a spreadsheet can be defined as being a visualization of a finite set of functions, which taken as a whole solve a given problem. In other words, *a spreadsheet is an interactive system for manipulating and visualising declarative functions.*

This definition is quite useful, since it implies that since spreadsheets are already (simple) data visualisations, all that is needed to build a complex, interactive data visualisation based on spreadsheets is merely an extension of the basic spreadsheet paradigm.

5   The Extended Paradigm

We have extended the spreadsheet paradigm to address the needs of data visualisation systems, while maintaining the simplicity and declarative nature of traditional spreadsheets. Although some of the techniques described have been used individually in the past (e.g. Levoy's *Spreadsheets for Images* [9] and Eriksson's *Scheme in a Grid* [5]), by combining all of these features we have made an improved system.

Most programming systems deal with the concept of processing multiple data items by using ideas such as recursion or iteration, i.e. taking the elements of a set and processing them one after the other. While these methods have many advantages, they are ill-suited to the spreadsheet paradigm, where looping constructs cannot be elegantly expressed. Instead, numbers in traditional spreadsheets are manipulated individually, and where some form of grouping is desired, cell ranges can easily be formed by the user, and operations such as finding the average of a given set of numbers can be accomplished by idioms such as =average(B12:B72). Although this works quite well for financial statements, data visualisations normally deal with data sets containing tens of thousands of items. At this level, the range paradigm breaks down due to the unwieldiness of thousands of rows or columns containing thousands of items.

Our extended spreadsheet paradigm circumvents this problem by extending the spreadsheet paradigm to allow sets of items (in the form of lists) to be stored in each cell, as opposed to single items. Most non-circular data structures can be implemented using only lists; many programming languages provide lists as their only form of data abstraction. The concept of ranges is no longer needed – every single cell *is* a range.

In Sect. 4 we showed that a spreadsheet can be seen as an environment for the development of functional programs. Therefore the basic paradigm would not be altered if, instead of the formula language normally associated with spreadsheets, a full-featured functional language were used (since spreadsheet formulas do not have any side effects, they can be considered as a special-purpose functional language). By using a functional language the user benefits from its generality and extensibility, as opposed to being constrained by the primitives provided by the spreadsheet software.

Part of data visualisation is to eliminate information from a dataset, in order to obtain the underlying patterns in the data. Therefore, some data often does not make it all the way from the database to the screen. In those cases it would be useful to know what values are not going to be needed by later calculations, in order to save time by simply not calculating them. *Lazy evaluation* defers calculations until they are needed – for example, a lazy evaluation of $f(g(x))$ will pass "$g(x)$" to function f, instead of evaluating $g(x)$ and passing f the result; it will be f's responsibility to evaluate $g(x)$. If for some reason f does not need g's result, that value will not be calculated. Therefore, if datasets are processed using lazy evaluation, we will have the ability to deal with large datasets in an efficient manner (e.g., if a bitmap is generated and then scaled down to 25% of its size, only one out of every four pixels will actually be generated). However, this must be transparent to the user; recalculations must appear to happen as before.

To summarise, we have extended the spreadsheet paradigm by the addition of three techniques: storing lists of values in each cell, using a functional programming language and the use of lazy evaluation.

6 The Prototype

To test our ideas we have built a prototype, called ViSSh (see Fig. 1). A ViSSh spreadsheet consists of a set of typed cells, each of which is represented on the spreadsheet grid by a small user interface. Each type of cell performs a different function, taking as input a list of values and returning another list, the contents of which functionally depend on the input list and the type of the cell.

ViSSh attempts to provide users with a useful data visualisation system, using spreadsheets as its user interface metaphor. It uses the functional language Scheme (a dialect of the functional language Lisp) as its formula language.

6.1 Dealing with Large Datasets

The ability to deal with large datasets is crucial in a viable data visualisation system. ViSSh implements the extended spreadsheet paradigm described in Sect. 5, and as such is capable of handling potentially infinite datasets. Unlike Levoy's "Spreadsheets for Images" [9], ViSSh does not display the contents of each cell. This is both because of the potentially huge volumes of data (making the traditional "cells display their own contents" technique impractical in the general case), and because there is no single "right" way of representing an arbitrary data set stored in a cell. Instead, there are specialised cells that are used to view data in different ways. For example, in Fig. 1 the contents of cell D4 (a 3D surface) are displayed by cell E1 (a 3D renderer).

Functional Programming Model Whereas most spreadsheets use a fairly rudimentary language to express their formulas in, ViSSh uses the programming language Scheme. This gives users a lot more expressive power (as well as allowing the use of data abstractions, see Sect. 5), and allows extension of the system by writing new functions in Scheme. These are dynamically loaded into the spreadsheet at runtime and can be used by spreadsheet cells. To enhance usability we have extended the parser to allow expressions in infix form, as opposed to the native Scheme (e.g., m*x+c or (+ (* m x) c)).

One of the advantages of a functional programming system is its enhanced modularity [7], which from a user's point of view translates into a system that easily allows

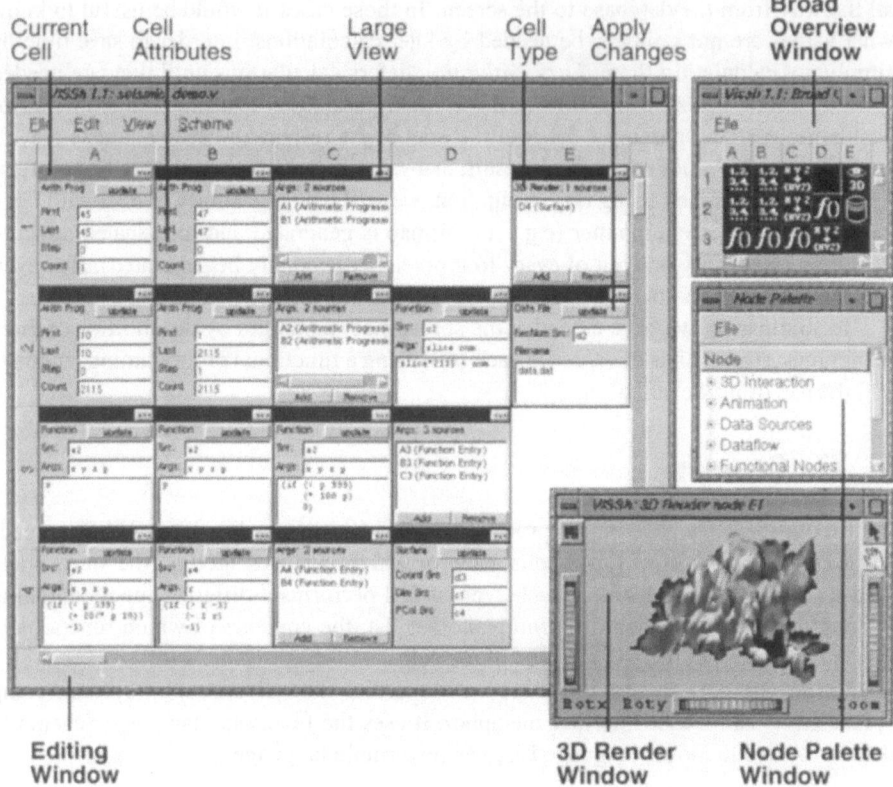

Current Cell Cell Attributes   Large View   Cell Type Apply Changes   Broad Overview Window

Editing Window   3D Render Window   Node Palette Window

Fig. 1. This illustrates ViSSh, our Data Visualisation Spreadsheet. The spreadsheet depicted here generates and displays an interactive 3D representation of seismic disturbance data read from a database. See Sect. 7 for a brief description of its operation.

code reuse. In many cases data visualisation benefits greatly by code reuse, and most data visualisation systems allow this practice by the use of some form of function libraries. ViSSh allows several spreadsheets to be linked together in a functional manner. This is done by the use of three specialised spreadsheet cells which, as a group, implement the idea of function calls in the context of spreadsheets. The user's view of functional linkage between two spreadsheets is quite simple: The *called* spreadsheet has an "Argument" cell, which collects the arguments that the function implemented by the spreadsheet takes (recall that each cell generates a list). It also has a single "Result" cell, which exposes the final result calculated by the spreadsheet. The *caller* spreadsheet has a "Subsheet" cell which behaves much like a function evaluation cell, but which "calls" the helper spreadsheet by evaluating its Result cell. This causes the normal evaluation process to propagate through the helper spreadsheet, until the Argument cell gets evaluated. This cell pulls in the arguments passed from the master spreadsheet, and the results then propagate forward until they reach the Result cell. The results then are given to the SubSheet cell, which returns them as its own results.

Navigational Aids A common problem with spreadsheets is that users can become "lost" when navigating through large spreadsheets which contain many similar subsections. This problem is especially acute in the case of ViSSh, since each cell is rather large and hence the number of them that can be displayed at any one time is fairly small (typically 7×5 cells in a 1024×768 display). This problem is tackled using two different tools, one for each aspect of the problem: grid navigation and dataflow.

Grid Navigation. To find out which part of the spreadsheet grid is being viewed though the editing window, as well as locating the current cell cluster in relation to the rest of the spreadsheet, the user can call up a window which contains a miniature version of the spreadsheet (See Fig. 1). Each cell contains a small (32×32 pixel) icon describing the function of the cell in the larger spreadsheet, wherein the area of the larger spreadsheet that is visible at any time is shaded in the small-scale spreadsheet. Since this can display a much larger portion of the spreadsheet (about 30×20 cells in a 1024×768 display), the spreadsheet navigation problem is greatly reduced by the use of this window. It can be scrolled independently of the main spreadsheet window, and clicking on one of the cells in the smaller window scrolls the main spreadsheet so that the corresponding cell becomes visible. The "Broad Overview" window acts as a road-map, with the icons representing the function of each cell being used as landmarks.

Dataflow. The other aspect of spreadsheets that can cause much frustration is keeping track of which cells depend on what other cells. ViSSh solves this problem by providing the user with a window which contains a dataflow diagram in which the nodes are iconic representations of the spreadsheet cells (see Fig. 2). The same icons that are used in the "Broad Overview" window to describe spreadsheet cells are used by this window. Like the "Broad Overview" window, the dataflow window can be used to navigate around the main spreadsheet window by clicking on a cell icon; the dataflow diagram is automatically kept synchronised to the main window. This dataflow diagram provides users with the advantages of the dataflow paradigm, while the spreadsheet editing environment shields them from the cluttering associated with medium to large dataflow editing environments.

7 A Practical Example

During its development, ViSSh has been tested with a variety of data sets. One of these consists of seismic disturbance data, recorded from locations in Southern Africa. Figure 1 illustrates the spreadsheet used to visualise this data. The flow of data through the spreadsheet is shown in Fig. 2. The cells in this spreadsheet can be classified into 3 functional groups, namely: extracting data from a file; generating a 3D surface with this data and displaying the 3D surface. The cells tasked with extracting the data are those in the range A2:E2. Cells A2:D2 generate record references (record-field pairs), which are used to extract data from the database by cell E2. The creation of the 3D surface is handled by two groups of cells. Cells A3:D3 form the 3D mesh using the data read in by cell E2, while cells A1, B1, A4, B4, C1 and C4 generate the vertex colours used by the surface, based on availability of data (the dataset has a number of "missing" data points, which must not be rendered) and the displacement value stored in the database. Rendering of the 3D surface is performed by cell E1.

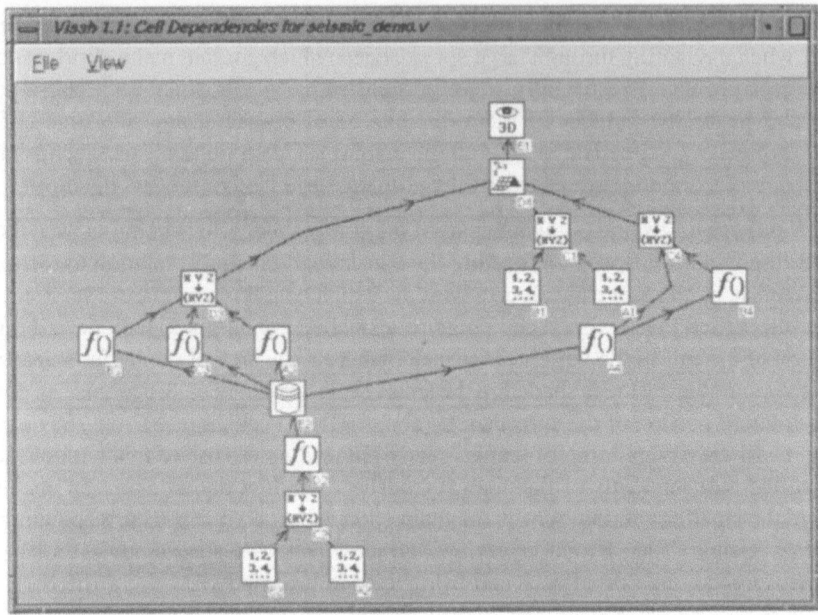

Fig. 2. This illustrates the intercell dependencies of the spreadsheet in Fig. 1. Each cell is represented by a small icon, using the same imagery as the "Broad Overview" window in Fig. 1. The arrows indicate the flow of data; clicking on an icon will make the corresponding cell be the current cell in the main spreadsheet window.

8 User Experiences

ViSSh is being currently tested by the authors and others. This testing has involved visualisations of the effects of different ATM network routing algorithms, as well as seismic disturbance data as described above. We have also compared ViSSh to Khoral Research's *Cantata*, and found that ViSSh was generally easier to use. We believe this is mainly because the spreadsheet layout made the visual program less cluttered than Cantata's dataflow model, thus easing the editing process. Additionally, the Scheme programming language made the formulas, especially those involving conditionals, compact and concise. Using the "Broad overview" and Dependencies windows (see Sect. 6.1 for details) also eased the visualisation task. The latter gives users the advantages associated with dataflow diagrams (see Fig. 2).

In future, in addition to doing further user testing, we will analyse ViSSh using Green's *Cognitive Dimensions Framework* [6]. This will give us a means with which to objectively compare ViSSh to other data visualisation systems.

9 Conclusion

In this paper we have looked at various attempts at using the spreadsheet paradigm for data visualisation. We have noted where this paradigm fails for data visualisation, and extended the paradigm by adding the following: firstly, each spreadsheet cell does not contain a single value, but a list of values. This allows us to store thousands of values in a spreadsheet, while keeping its size manageable. Secondly, we have used a full-featured functional programming language as the formula language. Using this instead of the formula languages used in most commercial spreadsheets provides the user with more expressive power, even allowing constructs such as high-order functions which are not expressible with most imperative languages. Finally, the lists of values are manipulated using lazy evaluation, which allows for the efficient manipulation of potentially infinite lists. In addition to this, we have added navigational and dataflow debugging facilities to circumvent problems inherent in the spreadsheet paradigm. Although others have implemented these separately, we believe that our main contribution is the combination of the three, together with the underlying theoretical framework for doing so. We have built a prototype to test the extended spreadsheet paradigm, and demonstrated its use for real-world data visualisation tasks. During our testing, we found that ViSSh spreadsheets were easier to edit than dataflow diagrams, mostly due to the absence of the clutter that normally occurs when dataflow diagrams are frequently modified. This result agrees with Levoy's findings regarding spreadsheet versus dataflow user interfaces [9]. Additionally, our dataflow debugging aids allow the transfer of skills from existing Modular Visualisation Environment (MVE) users.

10 Acknowledgments

We would like to thank Dr. Gary Marsden for his help and encouraging advice in the writing of this paper, Dennis Burford for the Kaap Vaal Seismic Experiment data and the South African National Research Foundation for their support.

References

1. T. Brus, M. van Eekelen, M. van Leer, M. Plasmeijer, and H. Barendregt. Clean - a language for functional graph rewriting. In *Proceedings of the Conference on Functional Programming Languages and Computer Architecture (FPCA '87)*. Springer, 1987.
2. G. Cameron. Special focus: Modular visualization environments (mves). *Computer Graphics*, 29(2), May 1995.
3. E. H. Chi, P. Barry, J. Riedl, and J. Konstan. Principles for information visualization spreadsheets. *IEEE Computer Graphics and Applications*, 18(4):30–38, July/August 1998.
4. W. de Hoon, L. Rutten, and M. van Eekelen. Implementing a functional spreadsheet in clean. *Journal of Functional Programming*, 5(3):383–414, July 1995.
5. U. Eriksson. Scheme in a grid. Online HTML document, 1999. "http://siag.nu/siag/".
6. T. R. G. Green and M. Petre. Usability analysis of visual programming environments: A 'cognitive dimensions' framework. *Journal of Visual Languages and Computing*, 7:131–174, 1996.
7. J. Hughes. Why functional programming matters. *Computer Journal*, 32(2):98–107, 1989.
8. T. Isakowitz, S. Schocken, and H. C. Lucas. Toward a logical/physical theory of spreadsheet modeling. *ACM Transactions on Information Systems*, 13(1):1–37, 1995.
9. M. Levoy. Spreadsheets for images. In *Computer Graphics Proceedings*, Annual Conference Series, pages 139–146. ACM SIGGRAPH, July 1994.

Fast Visualization of Special Relativistic Effects on Geometry and Illumination

Daniel Weiskopf

Institute for Astronomy and Astrophysics
University of Tübingen
Auf der Morgenstelle 10, D-72076 Tübingen, Germany
weiskopf@tat.physik.uni-tuebingen.de

Abstract. This paper describes a novel rendering technique for the special relativistic visualization of the geometry and illumination of fast moving objects. The physical basis consists of the relativistic aberration of light, the Doppler effect, and the searchlight effect. They account for changes of apparent geometry, color, and brightness of the objects. The rendering technique makes use of modern computer graphics hardware, in particular texture mapping and advanced per-pixel operations, and allows the visualization of these important special relativistic effects at interactive frame rates.

1 Introduction

Einstein's Special Theory of Relativity is widely regarded as a difficult and hardly comprehensible theory. One important reason for this is that the properties of space, time, and light in relativistic physics are totally different from those in classical, Newtonian physics. In many respects, they are contrary to human experience and everyday perception because mankind is limited to very low velocities compared to the speed of light. Therefore, computer simulations are the only means of visually exploring the realm of special relativity and can thus help the intuition of physicists and other people interested in the theory.

Hsiung and Dunn[5] were the first to use advanced visualization techniques for image shading of fast moving objects. They proposed an extension of normal three-dimensional ray tracing. Hsiung et al.[6] described a polygon rendering approach which is based on the apparent shapes of objects as seen by a relativistic observer.

In our previous work[13], we introduced texture-based relativistic rendering as a new means of visualizing the apparent geometry of fast moving objects. In this paper, the texture mapping approach is extended to the visualization of the special relativistic effects on both geometry and illumination. The physical basis consists of the relativistic aberration of light, the Doppler effect, and the searchlight effect. The novel rendering method makes extensive use of the functionality available on high-end graphics workstations[9], such as pixel textures and 3D textures. These features are exploited through OpenGL and its extensions.

Unlike the relativistic polygon rendering approach, texture-based relativistic rendering does not transform the coordinates or the color information of the vertices, but

220

transforms the images which are rendered in the normal non-relativistic way. Therefore, the standard rendering pipeline is not changed and only an additional step is added at the end of the rendering process. The relativistic transformation is performed on the image plane by texture mapping. This transformation is split in two phases. The first phase determines the geometrical effects by using standard 2D texture mapping. The second phase implements the Doppler and searchlight effects by using pixel textures and 3D texture mapping. In this way, interactive frame rates are achieved on modern computer graphics hardware.

2 Lorentz Transformation

This section describes only those terms and equations of special relativity which are relevant for rendering. A detailed presentation of special relativity can be found in [7, 8, 11]. The effects used from relativistic physics are the relativistic aberration of light, the Doppler effect, and the searchlight effect.

The relativistic aberration of light causes a rotation of the direction of light when one is changing from one inertial frame of reference to another. The aberration of light is sufficient to completely describe the apparent geometry seen by a fast moving camera.

The Doppler effect accounts for the transformation of wavelength from one inertial frame of reference to another and causes a change in color.

The searchlight effect is based on the transformation of wavelength dependent radiance from one inertial frame of reference to another. The transformation of radiance increases the brightness of objects ahead when the observer is approaching these objects at high velocity. The definition of wavelength dependent radiance can be found, e.g., in [3, Chapt. 13][14].

Let us consider two inertial frames of reference, S and S', with S' moving with velocity v along the z axis of S. The usual Lorentz transformation along the z axis connects frames S and S'.

In reference frame S, consider a photon with the wavelength λ and the direction determined by spherical coordinates (θ, ϕ). In frame S', the photon is described by the wavelength λ' and the direction (θ', ϕ'). These two representations are connected by the expressions for the relativistic aberration of light, cf. [8],

$$\cos \theta' = \frac{\cos \theta - \beta}{1 - \beta \cos \theta}, \qquad \phi' = \phi, \tag{1}$$

and for the Doppler effect,

$$\lambda' = \lambda D. \tag{2}$$

The Doppler factor D is defined as

$$D = \frac{1}{\gamma (1 - \beta \cos \theta)} = \gamma (1 + \beta \cos \theta'),$$

where $\gamma = 1/\sqrt{1 - \beta^2}$, $\beta = v/c$, and c is the speed of light.

Wavelength dependent radiance is transformed from one frame of reference to another according to

$$L'_\lambda(\lambda', \theta', \phi') = D^{-5} L_\lambda(\lambda, \theta, \phi). \tag{3}$$

A derivation of this relation can be found in [14].

3 Relativistic Rendering

The physical basis for texture-based relativistic rendering is the Lorentz transformation of properties of light. Suppose that the photon field is known at one point in spacetime, i.e. at a point in three-dimensional space and at an arbitrary but fixed time. This photon field is measured in one frame of reference which is denoted S_{obj}.

Now consider an observer, i.e. a camera, at this point in spacetime. In the following, the generation of a snapshot taken by this camera is investigated. The observer is not at rest relative to S_{obj}, but is moving at arbitrary speed relative to S_{obj}. However, the observer is at rest in another frame of reference, $S_{observer}$. The photon field can then be calculated with respect to $S_{observer}$ by the Lorentz transformation from S_{obj} to $S_{observer}$. Finally, the transformed photon field is used to generate the picture taken by the observer's camera.

Now we restrict ourselves to static scenes, i.e. all scene objects and light sources are at rest relative to each other and relative to S_{obj}. Here, all relevant information about the photon field—namely direction and wavelength dependent radiance of the incoming light—can be determined by standard computer graphics algorithms, since the finite speed of light can be neglected in all calculations for the static situation. With this information and with the use of the equations for the relativistic aberration of light and for the Doppler and searchlight effects, the picture seen by the relativistic observer can be generated. In this way, the relativistic effects on geometry, color, and brightness are taken into account.

3.1 Representation of the Photon Field

The relevant information about the photon field is the direction of the light and the spectral power distribution in the form of the wavelength dependent radiance. This is the information about the full plenoptic function[1] $P(\theta, \phi, x, y, z, t, \lambda)$ at a single point (x, y, z, t) in spacetime, effectively leaving a three-parameter function $\tilde{P}(\theta, \phi, \lambda)$. The polarization of light is neglected.

The function $\tilde{P}$ can be sampled and stored in the form of an image projected onto the unit sphere, with the camera being at the midpoint of the sphere. This image is called *radiance map*. The direction of an incoming light ray is determined by the spherical coordinates (θ, ϕ) of the corresponding point on the sphere.

The wavelength dependent radiance can be described by a vector with respect to a fixed set of n_L basis functions. Therefore, each point of the radiance map holds n_L components for the wavelength dependent radiance. Peercy[10] describes in detail how a finite set of orthonormal basis functions can be used to represent radiance. In this paper, no projection onto the basis functions is needed. Therefore, the orthonormality condition can be omitted and an arbitrary set of basis functions can be used. In this representation the wavelength dependent radiance is

$$L_\lambda(\lambda) = \sum_{j=1}^{n_L} \epsilon_j E_j(\lambda), \qquad (4)$$

with the set of basis functions, $\{E_j(\lambda)|j \in \mathbb{N}, 1 \leq j \leq n_L\}$, and the coefficients of the vector representation, ϵ_j.

In the literature, various sets of basis functions are described, such as box functions [4], Fourier functions[2], Gaussian functions[12], and delta functions[2]. As alternative basis functions, we propose the Planck spectral distributions at various temperatures with the wavelength dependent radiance

$$L_{\lambda,\text{Planck}}(\lambda, T) = \frac{2hc^2}{\lambda^5} \frac{1}{\exp\left(\frac{hc}{k\lambda T}\right) - 1},$$

where T is the temperature in Kelvin, h the Planck constant, and k the Boltzmann constant. The Planck distribution describes the blackbody radiation and thus has a direct physical meaning and a widespread occurance in nature.

For the display on the screen, three tristimulus values such as RGB have to be calculated from the wavelength dependent radiance. For example, the RGB values (c_R, c_G, c_B) can be obtained by

$$c_i = \int L_\lambda(\lambda) \bar{f}_i(\lambda) \, d\lambda, \qquad i = R, G, B, \qquad (5)$$

where $\bar{f}_i(\lambda)$ are the respective color matching functions for RGB, cf. [15].

How can the radiance map be generated for the non-relativistic situation, i.e. a camera at rest in the frame S_{obj} of the objects? Since perspective projection is restricted to a maximal field of view of 180°, the complete radiance map cannot be created in a single step by using standard computer graphics hardware. Therefore, the covering of the whole sphere is accomplished by projecting several images which are taken with differing orientations of the camera. A similar method is used for reflection and environment mapping.

Usually, images do not provide an arbitrary number of channels in order to store the n_L components for the photon field. However, several radiance maps each of which contain three standard RGB channels can be combined to represent the complete photon field.

3.2 Apparent Geometry

Let us assume that the photon field is known for an observer at rest in the frame S_{obj}. Now consider the same situation with the moving observer. The photon field has to be changed by the Lorentz transformation from S_{obj} to S_{observer}. This transformation is split in two parts, in the relativistic aberration of light and in the transformation of wavelength and radiance. Accordingly, the relativistic part of the rendering algorithm consists of two phases which determine apparent geometry and relativistic illumination.

The relativistic aberration of light yields a transformation of (θ, ϕ) to (θ', ϕ') according to (1). The resulting distortion of the image mapped onto the sphere is illustrated in Fig. 1.

Fig. 1. Effect of the relativistic aberration of light on the texture mapped onto the unit sphere. The left sphere shows the mapping without distortion, the right sphere illustrates the distortion for 90 percent of the speed of light. The direction of motion shows towards the north pole of the sphere.

Usually, the direction of motion is not identical to the z axis. Therefore, additional rotations of the coordinate system have to be considered. The complete mapping of the coordinates of a point (u, v) in the original image to the spherical coordinates (θ', ϕ') of the corresponding point seen by the moving observer is

$$T_{\text{mapping}} = T_{\text{rot, b}} \circ T_{\text{aberration}} \circ T_{\text{rot, a}} \circ T_{\text{proj}} \quad : \quad [0, 1] \times [0, 1] \longrightarrow S^2.$$

S^2 is the unit sphere. The map T_{proj} projects the rendered image onto the sphere and determines the corresponding coordinates on the sphere. The coordinates of the pixels are in the interval $[0, 1]$. The rotation $T_{\text{rot, a}}$ takes into account that the direction of motion differs from the z axis of the sphere. The actual relativistic transformation is accomplished by $T_{\text{aberration}}$ with the use of (1). Finally, $T_{\text{rot, b}}$ describes the rotation of the observer's coordinate system relative to the direction of motion.

The relativistic rendering process has to generate the texture coordinates for the mapping of the non-relativistic images onto the unit sphere which surrounds the moving observer. With the inverse map T_{mapping}^{-1}, the texture coordinates can be calculated from the spherical coordinates (θ', ϕ') in the coordinate system of the observer.

3.3 Relativistic Illumination

The Doppler and searchlight effects account for the relativistic effects on illumination. According to (2) and (3), both effects depend on the Doppler factor D. Therefore, in addition to the photon field, the information about the Doppler factors has to be known.

Here, we introduce the term *Doppler factor map*. Analogous to the radiance map, the Doppler factor map holds the Doppler factors for various directions. The Doppler factor map is a one-parameter map because, for a given velocity, the Doppler factor depends only on the angle θ'.

With (4) and (5), the tristimulus values (c'_R, c'_G, c'_B) for each pixel in the frame $S_{observer}$ can be calculated by

$$c'_i = \int \bar{f}_i(\lambda') L'_\lambda(\lambda') \, d\lambda' = \int \bar{f}_i(\lambda') \sum_{j=1}^{n_L} \epsilon_j E'_j(\lambda') \, d\lambda' = \sum_{j=1}^{n_L} X_{i,j}(\epsilon_j, D), \qquad (6)$$

with

$$X_{i,j}(\epsilon_j, D) = \int \bar{f}_i(\lambda') \epsilon_j E'_j(\lambda') \, d\lambda' \quad , \qquad i = R, G, B, \quad j = 1 \dots n_L.$$

The transformed $E'_j(\lambda')$ are computed from the original $E_j(\lambda)$ according to (2) and (3). The c'_i and $X_{i,j}$ can be combined to form the three-component vectors c' and X_j.

The function $X_j(\epsilon_j, D)$ can be represented by a three-component look-up table (LUT) depending on the two variables ϵ_j and D. This LUT can efficiently be implemented by using pixel textures. Pixel textures assign texture coordinates on a per-pixel basis instead of a per-vertex basis. Pixel textures are specified in the same way as normal 3D textures. With a pixel texture being activated, all pixels which are drawn from main memory into the frame buffer are interpreted as texture coordinates, i.e. each RGB color triple is mapped into the texture and then the interpolated texture values are actually drawn.

The LUT for each function $X_j(\epsilon_j, D)$ is stored in a 3D RGB texture. A 2D texture would be sufficient for a two-parameter function. OpenGL, however, does not support 2D pixel textures. Therefore, a third, dummy dimension which is set to one is included. The LUTs do not change for a fixed set of basis functions E_j. Therefore, the respective pixel textures can be built in a preprocessing step, thus not impairing rendering performance.

Finally, the relativistic transformation of wavelength and radiance is implemented as follows. Another sphere, now textured by the Doppler factor map, is rendered in addition to the sphere resulting from Sect. 3.2. The Doppler factors are stored in a previously unused, separate channel, such as the α channel. The final RGB values are evaluated according to (6) by iterating over the n_L channels which contain the vectors ϵ_j, by applying the corresponding pixel textures, and by adding up the results.

If n_L is greater than three, more than one radiance map is used to hold the n_L channels, and the whole process above has to be performed several times depending on the number of different radiance maps.

3.4 Rendering Process

Fig. 2 shows the structure of the relativistic rendering process. In addition to normal non-relativistic rendering, two phases for the relativistic transformations are appended to the rendering pipeline.

The geometric effects are taken into account in the first phase which consists of the white and gray boxes in the diagram. This phase resembles the implementation of reflection or environment mapping onto a sphere. The main difference is localized in the generation of the texture coordinates. The operations in the white boxes work on

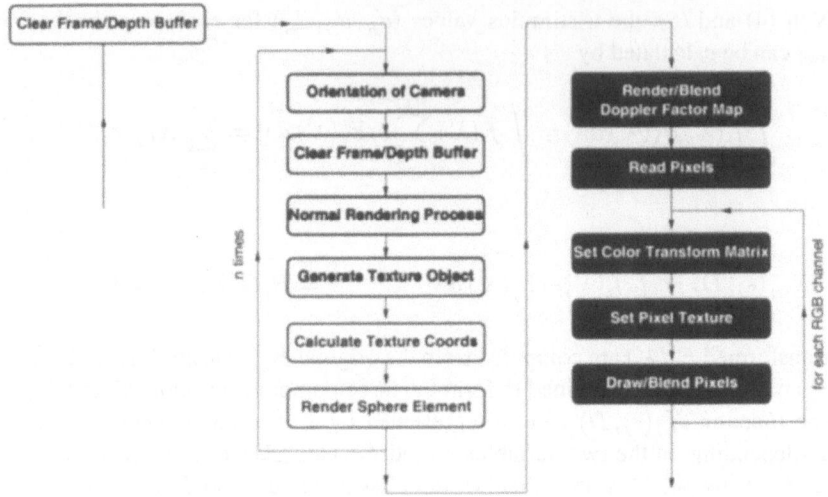

Fig. 2. Structure of the relativistic rendering process.

the back buffer of the normal frame buffer. The operations in the gray boxes work on an additional, off-screen frame buffer. In our implementation, pbuffers (SGIX_pbuffer extension) are used.

In the gray part, the textures for a spherical mapping are generated. Here, the standard non-relativistic rendering process is performed n times; n is the number of textures mapped onto the sphere and depends on the viewing angle and the orientation that are used for the rendering of the texture images. The OpenGL command glCopyTexImage2D transfers the results of non-relativistic rendering from the frame buffer to texture memory. Texture objects (glBindTexture) allow fast access to the stored textures.

In the white part, the texture coordinates are calculated with the use of T^{-1}_{mapping}. Then the textured sphere is actually drawn. The relativistic transformation is absorbed into the calculation of the texture coordinates. For the generation of the intermediate image which contains the visualization of apparent geometry, a picture is taken from inside the sphere. Note that the viewpoint has to be at the midpoint of the sphere, whereas the orientation is not restricted and allows for viewing into arbitrary directions.

The relativistic effects on illumination are taken into account in the second phase which consists of the black boxes. This phase works on the back buffer of the standard frame buffer. Another sphere, now textured by the Doppler factor map, is rendered into the α channel and blended with the result of the previous rendering steps. Then the complete RGBα buffer is read into main memory.

The last part is iterated three times, for the respective three color channels which hold the corresponding values ϵ_j. A color matrix (SGI_color_matrix extension) is set, which shifts the current color channel to the red channel and the α channel to the green channel. Now, the current values of ϵ_j are stored in the red channel and the corresponding Doppler factors in the green channel. Then the pixel texture which transforms

(ϵ_j, D) to displayable RGB values is applied. Finally, the operations in the rasterization stage are performed by drawing the image back into the frame buffer. These results are added up by a blending operation in order to obtain the final image.

So far, the rendering process supports up to three different basis functions E_j. A larger number of n_L can be implemented by iterating the whole rendering process several times with varying sets of basis functions and by adding up the results.

4 Implementation and Results

The relativistic rendering algorithm is implemented in C++ and runs on top of a standard OpenGL renderer. OpenGL version 1.1 with pbuffer (SGIX_pbuffer), pixel texture (glPixelTexGenSGIX), and color matrix (SGI_color_matrix) extensions is used. The implementation runs on an SGI Octane with Maximum Impact graphics and 250MHz R10000 processor.

An example of relativistic rendering can be found in Fig. 3. It illustrates the aberration of light, the Doppler effect, and the searchlight effect. The latter is very prominent and causes an extreme change of brightness.

In the current implementation, a constant number of six textures is used to cover the whole sphere. The six textures originate from the projection of the six sides of a cube onto the sphere. The texture coordinates are only recalculated if the velocity has changed. Three arbitrary basis functions E_j are supported. The following performance measurements are based on the test scene from Fig. 3 and a window size of 800*600 pixels. A frame rate of 7.6 fps is achieved for the relativistic visualization of apparent geometry and of 4 fps for the visualization of geometry and illumination. Approximately 30% of the total rendering time is used for the generation of the non-relativistic images, 10% for transferring these images from the frame buffer to texture memory, 40% for the pixel operations, and 20% for the remaining tasks.

5 Discussion

In this section, texture-based relativistic rendering is conferred to the well-known relativistic polygon rendering method.

Relativistic polygon rendering extends the normal rendering pipeline by a relativistic transformation of the vertex coordinates for the triangle meshes representing the scene objects. Since this transformation is non-linear, artifacts are introduced by the linear connection between the transformed vertices through straight edges. The error depends on the angular span under which each single triangle is viewed and might become very large for objects passing by closely. The artifacts can be reduced by a fine remeshing of the original objects in a preprocessing step or by an adaptive subdivision scheme during runtime. This is not needed for the texture approach, since the relativistic transformation is rather performed at the end of the rendering pipeline and affects every pixel in the image plane. Therefore, texture-based relativistic rendering does not need any modifications of the scene or the core rendering method. It does not increase the number of triangles of the scene objects to be rendered, it has no extra computational costs per vertex, and it does not introduce the geometric artifacts described above.

In the texture mapping approach, the sphere surrounding the observer is represented by a triangle mesh. The texture coordinates which are computed for the pixels inside each triangle by the usual perspective correction scheme differ from the true values. However, these errors do not impair the quality of the relativistic image as long as the angular span under which each single triangle is viewed is not too wide. The errors are independent of the geometry of the scene objects and can be controlled by choosing an appropriate size for the triangles representing the sphere, which is an important advantage over the relativistic polygon rendering approach. In the example depicted in Fig. 3, the whole sphere is tessellated by 5120 triangles, guaranteeing a good image quality for velocities as high as $\beta = 0.99$.

A very important issue is the physically correct handling of illumination. The reflection on a surface is usually described in terms of the angle between the light vector and the surface normal and the angle between the viewing vector and the surface normal. Relativistic polygon rendering transforms the coordinates of all vertices. Therefore, both angles are changed, which, in general, results in a wrong calculation of illumination. Texture-based relativistic rendering, however, completely transforms all relevant physical properties of the non-relativistic image to the frame of the fast moving camera, which brings about a physically correct visualization.

Rendering performance depends on different factors for relativistic polygon rendering and for texture-based relativistic rendering, cf. [13] for details. However, comparable frame rates are achieved by both rendering techniques on the used hardware.

One problem with texture-based relativistic rendering arises because of the properties of the aberration of light. The aberration equation does not conserve the element of solid angle. Therefore, the relativistic mapping does not conserve the area of an element on the sphere. The image is scaled down in the direction of motion, whereas the image gets magnified in the opposite direction. This magnification can reveal an inappropriate resolution of the texture. This issue could be solved by adapting the texture size to the relativistic distortion which depends on the observer's velocity and direction of motion.

Another problem could be the limited resolution of the RGBα channels, which might cause color aliasing effects whose extent depends on the chosen basis functions E_j and the interval of the used Doppler factors. These color aliasing effects are usually not very prominent for a depth of eight bits per channel, which is available on current hardware. They should completely disappear on future hardware supporting ten or twelve bits per channel.

6 Conclusion and Future Work

In this paper a texture-based approach to special relativistic rendering has been introduced. The physical basis consists of the relativistic aberration of light and the Doppler and searchlight effects. The algorithm uses texture mapping in order to transform the non-relativistic image into the frame of reference of a fast moving observer. Texture-based relativistic rendering allows for the fast visualization of special relativistic effects on both geometry and illumination.

There are several advantages compared to relativistic polygon rendering. There are no artifacts due to straight lines between vertices. There is no need for additional polygons and for the computation of the transformation of vertices. Most importantly, a physically correct model for the calculation of illumination is implemented.

In future work, the issue of rendering performance and image quality will be addressed. A limiting part in the rendering process is the pixel fill rate for the generation of the radiance map. The number of textures can be reduced if only that part of the sphere which is actually viewed by the relativistic observer is covered by textures. In addition, the resolution of the textures could be adapted to the magnification by the aberration of light. This will increase rendering speed and enhance image quality.

Acknowledgments

This work was supported by the Deutsche Forschungsgemeinschaft (DFG) and is part of the project D4 within the Sonderforschungsbereich 382. Thanks to B. Salzer and J. Schulze-Döbold for proof-reading, and to the anonymous reviewers for helpful comments.

References

1. E. H. Adelson and J. R. Bergen. The plenoptic function and the elements of early vision. In M. Landy and J. A. Movshon, editors, *Computational Models of Visual Processing*, pages 3–20, Cambridge, 1991. MIT Press.
2. A. S. Glassner. How to derive a spectrum from an RGB triplet. *IEEE Computer Graphics and Applications*, 9(4):95–99, July 1989.
3. A. S. Glassner. *Principles of Digital Image Synthesis*. Morgan Kaufmann, San Francisco, 1995.
4. R. Hall and D. P. Greenberg. A testbed for realistic image synthesis. *IEEE Computer Graphics and Applications*, 3(8):10–20, Nov. 1983.
5. P.-K. Hsiung and R. H. P. Dunn. Visualizing relativistic effects in spacetime. In *Proceedings of Supercomputing '89 Conference*, pages 597–606, 1989.
6. P.-K. Hsiung, R. H. Thibadeau, and M. Wu. T-buffer: Fast visualization of relativistic effects in spacetime. *Computer Graphics*, 24(2):83–88, Mar. 1990.
7. C. W. Misner, K. S. Thorne, and J. A. Wheeler. *Gravitation*. Freeman, New York, 1973.
8. C. Møller. *The Theory of Relativity*. Clarendon Press, Oxford, second edition, 1972.
9. J. S. Montrym, D. R. Baum, D. L. Dignam, and C. J. Migdal. InfiniteReality: A real-time graphics system. In *SIGGRAPH 97 Conference Proceedings*, pages 293–302, Aug. 1997.
10. M. S. Peercy. Linear color representations for full spectral sampling. In *SIGGRAPH 93 Conference Proceedings*, pages 191–198, Aug. 1993.
11. W. Rindler. *Introduction to Special Relativity*. Clarendon Press, Oxford, second edition, 1991.
12. Y. Sun, F. D. Fracchia, T. W. Calvert, and M. S. Drew. Deriving spectra from colors and rendering light interference. *IEEE Computer Graphics and Applications*, 19(4):61–67, July/Aug. 1999.
13. D. Weiskopf. A texture mapping approach for the visualization of special relativity. In *IEEE Visualization 1999 Late Breaking Hot Topics Proceedings*, pages 41–44, 1999.
14. D. Weiskopf, U. Kraus, and H. Ruder. Searchlight and doppler effects in the visualization of special relativity: A corrected derivation of the transformation of radiance. *ACM Transactions on Graphics*, 18(3), July 1999.
15. G. Wyszecki and W. S. Stiles. *Color Science*. John Wiley & Sons, New York, second edition, 1982.

Editors' Note: see Appendix, p. 293 for colored figure of this paper

AlVis - An Aluminium-Foam Visualization and Investigation Tool

Andreas H. König[1], Helmut Doleisch[1], Andreas Kottar[2], Brigitte Kriszt[3], and Eduard Gröller[1]

[1] Institute of Computer Graphics
{koenig|helmut|meister}@cg.tuwien.ac.at,
http://www.cg.tuwien.ac.at/home
[2] Institute of Experimental Physics
kottar@xphys.tuwien.ac.at
[3] Department of Material Science and Testing
bkriszt@mail.zserv.tuwien.ac.at

Vienna University of Technology, Karlsplatz 13, A-1040 Vienna, Austria

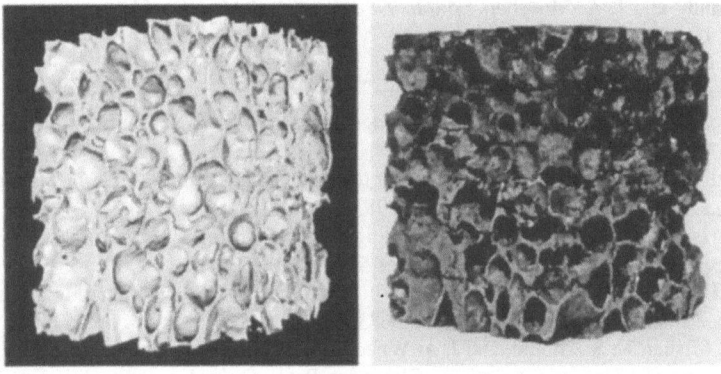

Fig. 1. Comparison of visualization (left hand side) and reality

Abstract. In recent years there has been an increased interest in metal foams in the field of material science. The stress absorbing potential is one of the most interesting properties for the application of aluminium foam (e.g. car manufacturing). Material scientists need to investigate the structure of metal foams in order to optimize their deformation behavior. An interactive tool for the investigation is presented in this work. Real-time surface rendering, automatic parameter determination, and display of local and global foam properties enable the user to understand the complex structure of the metal foam.

Keywords: aluminum foam investigation, surface extraction

1   Introduction

In recent years there has been a continuous increase in interest for metallic foams. Foams based on aluminium provide a number of advantages over solid materials: low specific weight, high specific rigidity, high energy-absorbing potential, reduced conductibility of sound and heat, non-combustibility, and the ability to be recycled. Metallic foams can be compacted at a relatively constant stress level which qualifies them for energy absorption applications (e.g. car manufacturing). Their mechanical behaviour is influenced by imperfections caused by the foaming process. Spatial variations in the foam cell-size distribution[1] and imperfect cell wall geometries are studied as examples for such imperfections. Small samples (about $2 \times 2 \times 2$cm) are deformed in a repetitive way. Pressure is applied to the sample in one direction, until cells start deforming and the sample loses some small amount of length in this direction. Then the deformation characteristics of the sample is investigated. This procedure is repeated until all the cells are entirely collapsed and the sample is completely compacted. In order to aid these investigations, a tool is needed, which allows the non-destructive investigation of samples during the single steps of the deformation procedure. The next section discusses the demands material scientist expect such a tool to meet.

2   Requirements for the investigation of aluminium foam

The demands for the investigation of aluminium foam risen by material science experts are manifold. Due to the repetitive nature of the investigation of deformation results, the foam sample must not be destroyed for the investigation. Thus, cutting techniques cannot be employed. The approach presented in this work utilizes a non-immersive way of gathering information about the local properties of the foam sample: The sample is scanned with the usage of a computer tomograph (CT) modality. Whereas medical CT modalities are limited in terms of scanning resolution (due to the radiation harm for the patient), industrial CT scanners feature resolutions of about 0.04mm.

The visualization of the scanned volumetric data sets shows the requirements of classical volume visualization problems as well as special demands:

- **Interactivity**: As the investigation procedure includes zooming in on different cells of interest, an interactive visualization approach is crucial. The manipulation of the viewpoint (movement of the camera) as well as manipulations of the data set (cutting planes, etc.) should be possible with interactive frame rates.
- **Accuracy**: As the relation of cell size to wall thickness is of interest to the investigation, high accuracy has to be ensured. The visualization should

[1] The field of computer graphics often refers to the term *cell* as a sub-cubiod of a volume data set with eight data values at its corners. In the field of metal foam research the term cell refers to the air filled bubbles within the solid material. In the presented work the term cell is used in the later meaning.

represent the the structure and topology of the real sample as accurately as possible (with respect to the limitations of the scanning procedure).

- **Measurements**: Tools for investigating the size of cell features have to be provided.
- **Shape criteria**: Different metrics for the description of cell characteristics are needed.
- **Statistical evaluations**: Besides the investigation of local cell properties, also information about global foam sample properties and the entirety of cells is desirable. Visual aids for the user to distinguish between regions of the foam sample with different characteristics should be provided.

Existing commercial as well as academic visualization systems cannot be employed for the investigation task. Due to the high spatial resolution of the CT scanners, data sets of enormous size are generated. Existing systems are not capable of rendering these data sets in real-time. Furthermore the display of special properties of foam cells as required by the cooperating material scientists (like shape and size criteria) is not supported by existing systems.

The approaches employed in AlVis to deal with these requirements are described in the following sections.

3   Interactive Rendering

In order to meet the requirements of interactive rendering, a surface shaded display approach is used. Modern hardware graphics accelerators are capable of rendering large amounts of polygons in real-time frame rates. The OpenGL [5] graphics library is employed to generate the visualization.

The segmentation of the volume data set into a surface representation is done with the aid of an isosurface extraction approach. The well known *Marching Cubes* [2] method is utilized. To overcome the drawbacks of ambiguity artifacts [3,6] an approach by Montani [4] is used.

Three problems have to be overcome with the usage of isosurface extraction: First, a suitable threshold for the surface generation has to be defined. In order to allow the display of foam cell properties, individual cells are identified. Then a way of dealing with the huge number of generated triangles has to be found.

3.1   Threshold definition

The aluminium/air boundary in the sample is represented by a highly wrinkled convoluted surface due to the high frequency nature of the data. Isosurfaces as used in medical visualization applications are usually smoother. For the presented application the specification of an appropriate threshold is crucial in order to meet the accuracy requirements of the investigation procedure. The threshold has to be chosen in a way such that the volume of the foam sample is preserved. Errors in threshold definition have a cubical influence to the volume of foam cells. A rough approximation of a cell is a sphere. A small change in threshold changes the radius of the sphere. This change has a cubical influence on the volume of the sphere. The appropriate threshold for the extraction of the aluminium/air boundary is dependent on the type of modality and external influences. For these reasons, the utilized threshold has to be adaptively derived for every sample to be investigated.

The following procedure yields the optimal threshold:

- The real sample is weighted using high precision scales. Let W_S be the weight of the sample.
- The specific weight of aluminium is known: W_{Al} is 2.7kg/dm^3 (2.7g/cm^3)
- The volume of aluminium in the sample resolves to $V_{Al} = \frac{W_S}{W_{Al}}$
- The volume of the sample cuboid can be calculated using the measured dimensions of the sample: $V_S = x_S * y_S * z_S$
- The ratio R of aluminium to air in the sample resolves to:

$$R = \frac{V_{Al}}{V_S}$$

- Given the histogram of the volume data set $H = \{h_0, \dots, h_{max}\}, h_x = |\{v|value(v) = x\}|$, R_t is calculated as

$$R_t = \frac{\sum_{i=t}^{max} h_i * i}{\sum_{i=0}^{max} h_i * i}$$

R_t describes the ratio of aluminium to air in case t is selected as threshold. Selecting the threshold t so that $R_t \leq R \leq R_{t-1}$ guarantees that the apparent volume in the computed sample closely corresponds to the real world sample. This is very important as the lacunarity (size and distribution of foam cells) is crucial for judging structural characteristics.

Visual comparisons and spatial relation measurements on the real sample as well as in the geometry representation of the visualized sample have affirmed the validity of the derived threshold (see Figure 1).

3.2 Foam cell identification

Before the isosurface is extracted, single cells are identified. The volume data set is interpreted as a binary image, separated into voxels having lower or higher values than the threshold derived in the last section. A *two-pass labeling* approach yields the number of distinct foam cells as well as the different sets of voxels contributing to the single cells. Based on this information, the isosurface is extracted on a cell by cell basis. The generated triangles are stored indexed with accordance to the cell they represent. Special care is taken about additional closing triangles at the cut up faces of the foam sample.

3.3 Rendering acceleration

The extraction of the isosurface from the volume data set yields an enormous number of triangles. Even for up to date hardware rendering accelerators, special strategies have to be employed in order to provide real-time frame rates.

Two cases have to be distinguished for rendering: The viewpoint might be inside the boundaries of the foam sample, investigating inner cell structures. It might also be outside the sample, when the user wants to gain an overview of cell size and shape distribution.

The viewpoint will be most of the time during the examination inside the cave-like structure of the foam. As the maximal viewing distance is very restricted, only a small part of the geometry data is visible. A special subdivision scheme is applied for rendering. Similar to the proposals of other authors [1, 7] a regular partitioning of the display geometry is employed. The volume data set is subdivided into sub-cuboids of user defined size. These subregions are called *macrocubes*. The surface portions generated during the marching cubes procedure are stored separately for each macrocube in an optimized OpenGL triangle-strip data-structure. The rendering procedure employs an progressive approach:

- **Interactive phase**: As long as the user moves the viewpoint, the rendering budget is limited by the performance of the used graphics hardware. A limited set of macrocubes within the viewing frustum is determined. This set is rendered front to back, until no more time for this frame is available (see figure 2, left hand side). Thus the maximal amount of information for real time frame rates is rendered.
- **Completion phase**: When the user stops moving the camera, more rendering time can be invested. The OpenGL z-buffer is checked for regions, where no triangles have been rendered possibly due to not yet rendered macrocubes. Using the camera location and direction, the according set of macrocubes is derived, which is rendered front to back, either until the z-buffer is completely covered with triangles or until no more macrocubes inside the viewing frustum are left to render. The later case might emerge, if the user has positioned the camera facing from within an open cell to the outside of the sample.

Un-filled z-buffer regions can only emerge in situations, when the camera faces structures of connected foam cells, which tunnel into the depth of the visualization. This happens rarely during the investigation. Usually just the interactive phase has to be performed.

When the camera is outside the limits of the data set (to gain an overview of the entire foam sample), the macrocubes rendered in the interactive step are those that lie in the three faces of the data set facing the camera.

In addition to the gain of rendering speed, the usage of macrocubes has an additional benefit. When the triangle sets of different macrocubes are colored individually, the size of foam cells can be quickly visually derived by the user. Due to the usage of perspective rendering the perception of depth also gains a

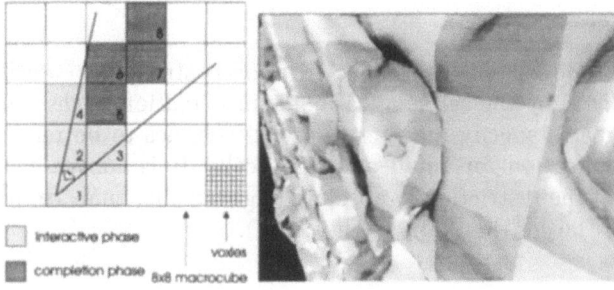

Fig. 2. Left: The rendering subdivision scheme using macrocubes. In the interactive phase, macrocubes 1-4 are rendered. In the completion phase, macrocubes 5-8 are rendered in order to patch an un-filled region. Right: Individually colored macrocubes allow a quick perception of feature size and depth information. The size of one macrocube in this image is $0.32 \times 0.32 \times 0.32$mm.

lot by the different size of near and far macrocubes (see figure 2, right hand side).

4 Measurements

The size of foam cells is one of the most important properties for the investigating material scientists. The size of cells in relation to the thickness of the surrounding walls is characteristic for the deformation behaviour. In order to allow spatial measurement a metaphor based on the analogy of a rubber band (elastic) has been integrated. Figure 5 (color-plate section) shows the usage of this tool for the determination of the diameter of a very small foam cell. The length of the rubber band being manipulated by the user is displayed in real time.

5 Focused Views

Occlusion problems, which cannot be dealt with using cutting planes, have to be solved with special techniques. In order to extract the important visual information, an approach well established in the field of information visualization is employed: *Focused Views*. Not the entire foam sample is rendered, but only cells which meet certain criteria. Additionally, colors can be used to illustrate criteria characteristics of certain cells. Figure 6 (color-plate section) demonstrates the usage of a simple focused view, where only cells in the inner part of the sample are rendered. This allows the material scientist to gain an overview on the size and shape distribution of cells, which determines the inner stiffness of the aluminium foam. Useful criteria to restrict the rendered information include the size and the shape of single cells. These criteria are discussed in detail in the following sections.

5.1 Cell size

The evaluation of cell size as a focus criterion is straight forward. The number of voxels enclosed by the cell is a feasible approximation to the volume of a cell. The user is free to specify an interval of cell volumes to be rendered. Figure 3 shows the application for small and large cells. It is convenient to allow the specification of the restriction in absolute (e.g. "smaller than 7mm^3") or relative numbers (e.g. "larger than 80% of the cells").

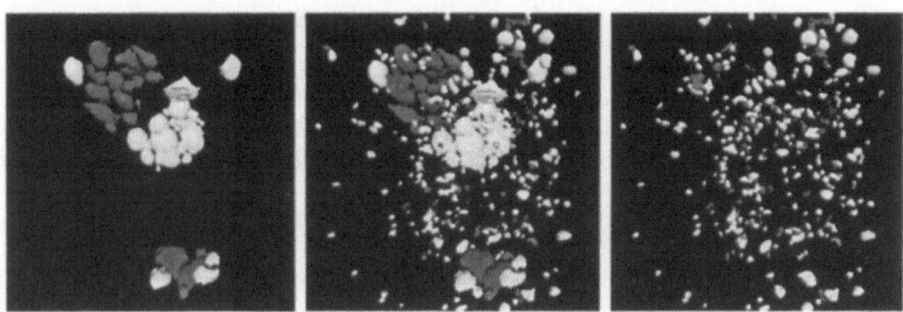

Fig. 3. Focused view based on size of cells (left: largest 10% of all foam cells, right: smallest 20%, middle: both)

5.2 Cell shape

One of the requirements of the investigation of the foam sample was the determination of the shape of the foam cells. The uniformity of cells guarantees an optimal stress absorbance potential. A metric for the judgement of cell topologies is defined. The desirable shape of a foam cell is a sphere. A shape criterion f based on the comparison of the cell shape to a sphere can be defined as follows: Let B be the set of voxels, which are enclosed by the surface of a certain cell C. $B = \{v_i | v_i \text{ inside cell } C, i = 0 \ldots n - 1\}$ Let $\mathbf{M}$ be the center of mass of B:

$$\mathbf{M} = \frac{1}{n} \sum_{i=0}^{n-1} \mathbf{location}(v_i)$$

A distance function evaluates the absolute distance of certain voxels to the center of mass:

$$\text{distance}(v_i) = |\mathbf{location}(v_i) - \mathbf{M}|$$

The average distance D_{avg} evaluates to:

$$D_{\text{avg}} = \frac{1}{n} \sum_{i=0}^{n-1} \text{distance}(v_i)$$

Two different shape characteristics are of interest to material scientists investigating aluminium foam:

- foam cells closely resembling spheres, only having a small number of high frequency outlying peaks
- foam cells deviating from the shape of a sphere by a lot of minor dents and bumps

Figure 4 gives a graphical representation of these two cell types. Material scientists want to investigate, if one of these common cell types is responsible for high cell collapsibility during the deformation process. Now two different criteria for

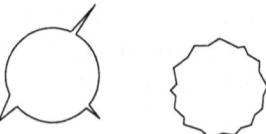

Fig. 4. Cells closely resembling spheres although having single peak-like deformations (left hand side) feature high f_{max} and low f_{avg} values. Cells with a lot of minor dents and bumps (right hand side) vice-versa.

the cell can be calculated: the average and the maximal distance of voxels to the center of mass:

$$f_{max} = \max_{i=0..n-1} |\text{distance}(v_i) - D_{avg}|$$

$$f_{avg} = \frac{1}{n} \sum_{i=0}^{n-1} |\text{distance}(v_i) - D_{avg}|$$

It is convenient to have these absolute f_{max} and f_{avg} criteria scaled to relative values on the basis of all cells in the foam sample. Furthermore the two values can be combined into a weighted sum for the ease of user interaction. The specification of the weights w_{max} and w_{avg} decides which kind of shape anomaly shall be emphasized or ignored when a focused view is applied.

$$f = w_{max}\frac{f_{max}}{m_{max}} + w_{avg}\frac{f_{avg}}{m_{avg}}$$

where m_{max} (respectively m_{avg}) is the maximal value of f_{max} (respectively f_{avg}) for all cells of the sample. Using this approach is is for instance possible for the user to have all cells with an shape criterion f in the best percentile displayed by focusing the view to cells with $f < 0.10$. If cells deviating with single peaks from the shape of a sphere are of interest, w_{max} should be assigned a high value, whereas if w_{avg} is emphasized, cells with a lot of minor deviations are favored.

6   Results

The presented approach was implemented using Microsoft Visual C++ (Developer Studio ver. 6.0), MS Windows NT 4.0. AlVis was tested on an Intel PIII based system (450 MHz, 512 MB RAM, 3DLabs Oxygen GVX1 graphics accelerator). The data set used to generate the results presented in the work featured the characteristics shown in table 1. High resolution images of the results presented in this work are to be found on the WWW:
http://www.cg.tuwien.ac.at/research/vis/Miscellaneous/alufoam. The presented system has been developed in cooperation with a group of material scientists. User studies involving four aluminium foam experts and three different data sets (each with a number of deformed versions) were made. Permanent feedback during the development has assured that the possibilities of the system meet the requirements of the experts. Interactivity has shown to be the most crucial point for daily work. Only the usage of the presented focused views techniques enabled the quick perception of local and global homogeneity properties of the foam structure. As a first research result AlVis has already enabled the material scientists to distinguish different classes of foam cells, which might help to optimize the deformation behaviour of the foam.

size of sample	22.3 × 22.0 × 30.0mm (14.718 cm^3)	relative weight of sample	0.499g/cm^3
weight of sample	7,43 g	fill ratio	18.5%
scanner resolution	0.04 × 0.04 × 0.04mm	value of aluminium	220
data value resolution	8 bit	optimal threshold	35
data set resolution	626 × 626 × 760	# generated triangles	9112628
size of data set	298 MByte	# triangles per voxel	0.04
scanning time	5 hours 37 minutes	# ambiguity cases	11077 (0.04%)
		macrocube size	8 × 8 × 8 voxel
		# identified cells	1150

Table 1. The statistics at the left hand side of the table were known before the investigation. Values at the right hand side were determined by the system.

7   Conclusions and Future Work

In close cooperation with material scientists a system for the investigation of aluminium foam has been developed. Between iterated steps of deformation CT scans of a foam sample are acquired. Due to the highly complex and wrinkled structure of the aluminium/air boundary, the visualization of these data sets is difficult. As high accuracy is necessary, the threshold value for the employed isosurfacing technique has to be volume preserving. It is derived using the weight and size of the real sample. Due to the structure of the foam, enormous amounts of triangles have to be rendered. A sub-division scheme utilizing macrocubes guarantees real-time frame-rates. In order to aid the investigation task, statistical properties of individual foam cells can be displayed. Focused views enable the user to restrict rendering to cells with certain size and shape criteria. User studies with material science experts proved the usefulness of the integrated tools. As a first research result, certain types of foam cells have been identified, which might influence the deformation behaviour of the aluminium foam.

As future work the system will be extended to handle multiple data sets resulting from the different steps of the deformation analysis. Animating the changes of foam cell size and geometry will yield in deeper insights to the complex relation of foam structure and deformation behaviour.

8 Acknowledgements

The authors would like to thank Armin Kanitsar and Andre Neubauer for the implementation. The work presented in this publication has been supported by the $V^{is}M^{ed}$ project. $V^{is}M^{ed}$ is supported by *Tiani Medgraph*, Vienna, http://www.tiani.com, and the *Forschungsförderungsfonds für die gewerbliche Wirtschaft*, Austria, http://www.telekom.at/fff/.

References

1. R. Grzeszczuk, Ch. Henn, and R. Yagel. *SIGGRAPH 98 "Advanced Geometric Techniques for Ray Casting Volumes" course #4 notes*. July 1998.
2. W. E. Lorensen and H. E. Cline. Marching cubes: A high resolution 3D surface construction algorithm. *Computer Graphics*, 21(4):163–169, July 1987.
3. S. V. Matveyev. Approximation of isosurface in the marching cube: Ambiguity problem. In R. Daniel Bergeron and Arie E. Kaufman, editors, *Proceedings of the Conference on Visualization*, pages 288–292, Los Alamitos, CA, USA, October 1994. IEEE Computer Society Press.
4. C. Montani, R. Scateni, and R. Scopigno. A modified look-up table for implicit disambiguation of Marching Cubes. *The Visual Computer*, 10(6):353–355, 1994.
5. J. Neider, T. Davis, and M. Woo. *OpenGL Programming Guide*. Addison-Wesley, Reading MA, 1993.
6. G. M. Nielson and B. Hamann. The Asymptotic Decider: Removing the Ambiguity in Marching Cubes. In *Visualization '91*, pages 83–91, 1991.
7. Orion Wilson, Allen VanGelder, and Jane Wilhelms. DIRECT VOLUME RENDERING VIA 3D TEXTURES. Technical Report UCSC-CRL-94-19, University of California, Santa Cruz, Jack Baskin School of Engineering, June 1994.

Editors' Note: see Appendix, p. 294 for colored figures of this paper

WWW-based Visualization of the Real Time Run of a Space Weather Forecasting Model

Sergei Maurtis[1], Jeff McAllister[1], Brenton Watkins[2]

[1]Arctic Region Supercomputing Center, University of Alaska Fairbanks, AK 99775, USA
maurits@arsc.edu   jeff_mcallister@mail.com
[2]Geophysical Institute, University of Alaska Fairbanks, Fairbanks, AK, 99775, USA
ffbjw@aurora.alaska.edu

Abstract. As use of earth-orbiting space technology increases, so does the need to understand and forecast the space weather. The University of Alaska Fairbanks Eulerian Parallel Polar Ionosphere Model (UAF EPPIM) is a first principles three-dimensional time-dependant simulation, applied for real-time ionospheric forecasts. Its network-based continuous run uses on-line remote inputs, including current satellite data. Forecasted conditions are continuously visualized and disseminated via two mirrored WWW-sites www.arsc.edu/ SpaceWeather and dac3.gi.alaska.edu/~sergei in various formats such as GIF-files, including their JavaScript animation; NCAR Graphics metafiles; and Vis5D databases. This paper describes practical approaches to issues such as synchronizing the model run to the real-time inputs and achieving the highest possible resolution on a variety of computational platforms. Experience running the model with network inputs continuously for nearly three years is presented and summarized. Examples of remote use of model forecasts are discussed together with new opportunities this real-time approach provides for the U.S. National Space Weather Program.

1 Introduction (Modeling Space Weather Phenomena)

The current 11-year solar activity cycle is expected to peak in 2000-2001. Combined with the dramatically increased dependence on space-based technologies, this upcoming solar maximum is attracting new attention to space weather phenomena. The U.S. National Space Weather Program has formulated a list of priorities in creating a capability to mitigate potential consequences of large-scale solar disturbances in Earth's close vicinity (www.ofcm.gov/nswp-ip/text/cover.htm). The need for operational real-time prognostic models with the capability to immediately disseminate forecasts is especially urgent. Simulations of the terrestrial ionosphere are an important part of this effort —particularly of the North Polar Region as satellite communication traffic is especially dense there. This paper summarizes experience creating the WWW-based real-time forecasting run of the University of Alaska Fairbanks Eulerian Parallel Polar Ionosphere Model (UAF EPPIM).

One of many peculiarities of the polar terrestrial ionosphere is a prominent advective motion in the horizontal direction at altitudes 150 *km* and up. The pattern of this motion is centered around the geomagnetic pole, though its mid-latitude edge regularly reaches as far as 50 degrees of geomagnetic latitude. The motion speed reaches 1 *km/sec* and up, which makes the process capable of significantly modifying all plasma parameters, including the electron density and its vertical profile. The plasma advection is governed by electromagnetic **ExB**-drift in the crossed Earth's magnetic field and the electric field of magnetospheric origin. While the Earth's main magnetic field can be assumed constant in time, the electric field is highly variable. It is controlled by a stream of energetic particles from the Sun, known as solar wind. Influence via the Interplanetary Magnetic Field (IMF), a field "frozen" into the moving stream, is the most pronounced. Significant variations of the IMF, including its drastic reversals on a scale of a few minutes, are a common occurrence [2].

Consequently, the advective motion pattern is in constant fluctuation. Combined with the considerable lifetime (several hours) of upper ionosphere plasma, this variability introduces a "memory" to the ionospheric system. Similarly to meteorological weather, the current space weather state depends not only on the immediate distribution of the forcing inputs, but also on their prior development in time. Such period-specific history is largely unique, and following its progression in the real-time simulation appears a feasible way to introduce prior influences into the forecast.

Another important characteristic of the terrestrial ionosphere (as well as the upper atmosphere in general) is that it responds quickly to changes in the forcing parameters. The significant variability of these parameters results in serious limitations in how far "weather-scale" ionospheric conditions can be predicted in advance. Currently, the practical limits of forward simulations cannot exceed one-two hours, the time scale at which the forcing parameters are predictable or measurable.

The EPPIM obtains forecasts of solar wind parameters via input from satellites placed at upstream positions between the Sun and the Earth. The probing satellite, ACE (Advanced Composition Explorer), is located near the gravity-free Lagrangian point (L1) at a distance of ~1.5 million kilometers (www.srl.caltech.edu/ACE/ASC). As the solar wind's velocity ranges from 250 to 800 *km/sec*, while the data downlinking is practically instantaneous, the satellite provides real-time round-the-clock forecast of the solar wind parameters from 100 to 30 minutes in advance.

The L1 gravity-free location is uniquely advantageous for a long-term placement of the probing satellite. L1 can be orbited for years without depleting propellant supply. A global chain of several tracking stations is required for round-the-clock data downlinking from this type of trajectory. The real-time data online depository is located at the NOAA's Space Environment Center (ftp://solar.sec.noaa.gov/pub in Boulder, Colorado). This new practically non-interrupted availability of geophysical inputs makes the continuous real-time forecasting run of the EPPIM feasible.

2   The UAF EPPIM and its Real-Time Forecasting Applications

The University of Alaska Fairbanks Eulerian Parallel Polar Ionosphere Model is a first principles ionospheric model [3], which solves the equations of continuity, motion, and energy balance for seven ion species, electrons, and a few minor neutral atmosphere components important for the ionization balance. The model geographic domain covers a significant portion of the Northern Hemisphere (9,200x 11,000-*km* area), and extends vertically from 80 to 500 *km*. The model uses the Azimuthal Equidistant Projection to apply a Cartesian frame to the Earth's spherical geometry (Fig.1).

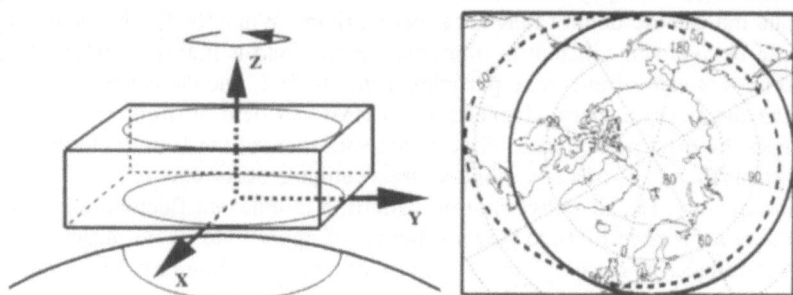

Fig. 1. Schematic representation of the UAF EPPIM domain. The Earth's spherical geometry (*left panel*) is mapped into a Cartesian frame using the Azimuthal Equidistant Projection (AEP), while the AEP metrics is used to compensate for resulted distortions. Right panel shows extent of the geographic coverage, the solid line circle represents 50°N geographic latitude, while the dashed circle corresponds to 50°N geomagnetic latitude

The UAF EPPIM flow chart at dac3.gi.alaska.edu/~sergei/Introduction.htm depicts the model structure. Its modularity allows variable approaches depending on input availability. For instance, the model interchangeably uses several electric field modules to generate the advective drift pattern. Among them is a statistical model [6], driven by the Interplanetary Magnetic Field parameters; or a combination of a statistical pattern and the current measurements from the polar chain of Doppler radars SuperDARN [5]; or, finally, a period-specific reconstruction of the electric field by Assimilative Mapping of Ionospheric Electrodynamics technique [4].

The UAF EPPIM code consists of ~25,000 Fortran lines structured into about a hundred subroutines. It supports the MPI-parallelization based on the domain decomposition as well as a single processor environment. Multiple porting and extensive scalar optimization developed the code's robust performance. This performance is well scalable both with an increase of processor power in a single CPU workstations and, in case of a MPP supercomputer environment, with the number of the assigned processors. One of the code environment variables is the model horizontal resolution, which sets the cell size from 220 *km* up to 10 *km* as a uniform mesh throughout the entire domain. Color Plate 1 demonstrates the resolution-related increase of the solution fidelity for the electron density. It shows consecutive magnifications of portions of the horizontal cross-section of the 3-D model domain, cut near the vicinity of the main maximum of the ionospheric layer *F2* at 290 *km* altitude. The left column represents the model sub-domain of ~2000x1500 *km* size and the right column shows ~1000x750 *km* area near the center, magnified. The otherwise geophysically identical runs differ only in horizontal resolution, namely ~220x220 *km*, ~ 55x55 *km*, and ~14x14 *km*, or factor of 16 for each transition.

It is evident from the picture that the relatively smooth solution obtained using a horizontal resolution of 220x220 *km* is replaced by an increasingly structured pattern as resolution increases. For applications related to satellite communications, it is important to simulate not only the electron density, but also electron density gradients. Gradient magnitudes are critically important for forecasting the growth rates of the ionospheric plasma instabilities, as these occurrences are known to degrade communication signal quality. Hence, high horizontal resolution can add significantly to ionospheric forecast comprehensiveness.

The EPPIM's time-splitting algorithm allows for varying the model's time resolution without jeopardizing its numerical stability. Using high time resolution (tens of seconds) provides the means to trace interesting short-lived developments, while low time resolution (tens of minutes) allows the same computational resources to cover extended periods. Frequently, a five minute time step is chosen for the EPPIM runs as a good compromise between reasonable time coverage and the computational effectiveness of the model. Time and space resolution can be adjusted to fit the performance of a given computational platform.

Table 1. Characteristics of several computational platforms and the UAF EPPIM horizontal resolutions, which they support for real-time simulations with 5 minutes time stepping

Platform, year	DEC Alpha 3000/300x, 1994	SGI O2, 1997	SGI Octane, 1999
SPECfp95	~2.6*	9.7	20.3
Horizontal resolution	80 x 80 km	50 x 50 km	30 x 30 km
Problem size	131 x 113 x 43	217 x 181 x 43	361 x 301 x 43
Memory needed	110 MB	290 MB	800 MB

* - Estimate from DEC Alpha 3000/300x value of benchmark SPECfp92 = 100

Generally, for a given computational platform the criterion of real-time compatibility of a time-dependent application can be formulated as follows: the elapsed time required for completion of the application time advance (the response time) should not exceed the time advance value (synchronization interval). Provided the EPPIM time advance is chosen at 5 minutes, it defines the highest possible resolution or the problem size for a given platform performance. (In all cases, the memory is assumed adjustable to accommodate any problem size.) Table 1 summarizes the achievable in real time horizontal resolutions of the EPPIM for several model ports.

For all combinations of platforms and resolutions in Table 1, the average wall-clock time to advance the model for one time step of 300 sec (5 minutes) was ~230-250 sec. Thus, the EPPIM code response time comprises ~80% of the real-time synchronization interval. By any means, this is a tight arrangement for a real-time run. To maintain synchronization it dictates a full dedication of system resources. The next question is whether a 20% of idling overhead is too wasteful. Even on a fully dedicated platform, a computation/idling ratio as 80/20% is justified. It accommodates possible fluctuations in computational time related to seasonal variability, quiet vs. disturbed geophysical conditions, etc. Next, allowing for a small idling time at the end of every time step is convenient for synchronization purposes. "Borrowing" this time for computations related to the next time step provides a flexibility to control dynamic shift between the model time and the current clock time during the run, as described below.

As shown in Table 1 and Color Plate 1, the UAF EPPIM can reach useful horizontal resolutions using workstation-class computational platform. High-end workstations support horizontal resolutions of 30x30 *km* and better. It is adequate for ionospheric density forecasting and providing reasonable density gradient estimates, an important goal for communication-related applications. Even though a real-time setting requires full platform resources for extended periods, as long as a dedicated workstation is significantly cheaper than a supercomputer, it is an economically feasible arrangement. Although dedication of expensive supercomputers is not unheard of in real-time meteorological forecasting, space weather programs are still in their infancy. It is important to bridge the gap between the real-time needs and current funding realities by developing scalable approaches, capable of effortless porting to available computational resources.

Finally, the real-time mode of the UAF EPPIM naturally creates a capability to continuously track the history of geophysical developments. To preserve model real-time performance for weeks- and longer runs, it is necessary to arrange a series of network exchanges to update the model inputs in a time-effective manner. The next chapter describes the networking solutions, which maximized the forecasting abilities of the EPPIM.

3 WWW-based Ionosphere Forecasting

As discussed above, the UAF EPPIM has a modular structure. Many modules exist in several interchangeable versions. The EPPIM supports simulations with real data but also can be switched for modules comprised of statistical description of various geophysical parameters. The statistical modules are driven by a limited number of geophysical indices. Among these indices are the solar activity index ($F_{10.7}$); the geomagnetic activity index (Kp); and the magnitude and orientation of the Interplanetary Magnetic Field (IMF). This combination is sufficient to invoke the statistical modules of the EPPIM to describe the current geophysical situation with a reasonable level of accuracy. This approach reflects the current stage of ionospheric modeling. Generally, an inclusion of real data for simulation is limited to post-analysis of specific space weather events. For the real-time prognostic run, when data availability is extremely limited, the EPPIM resorts to statistical descriptions driven by a combination of current or anticipated values of geophysical indices, available online for remote access.

The EPPIM network-based real-time run is hosted at the University of Alaska Fairbanks. The model time is dynamically shifted forward with respect to the current time and update of geophysical indices is arranged as a series of automatic network exchanges. The exchange frequency varies for various geophysical parameters. The slow varying solar activity index $F_{10.7}$ is assumed constant for the entire model day. During the run, its value is updated once a day from on-line depository at the Space Environment Center (SEC) ftp://solar.sec.noaa.gov/pub/latest/SRFF.txt in Boulder, Colorado. The synoptic nowcast of the geomagnetic activity index Kp is generated

244

by the USAF Space Weather Squadron in Colorado Springs, Colorado and disseminated through the SEC depository ftp://solar.sec.noaa.gov/pub/latest/Mahr.txt. The file with the established Kp value for the previous hour is updated every hour at its 27th minute. Thus, the Kp index is updated in the model with a noticeable delay between one and two hours. This arrangement limits the precise timing for the Kp surges, but generally allows for trend tracing. In the future, we plan to use the Kp forecast for this input, which recently became available at the SEC.

The most variable parameter for updating during the real-time run is the Interplanetary Magnetic Field (IMF) magnitude and orientation, measured by the ACE satellite at 1.5 million kilometers from the Earth. It is available online also at the SEC depository ftp://solar.sec.noaa.gov/pub/lists/ace_mag_1m.txt. The file is updated every minute and includes the latest information on Earth-bound solar wind. Data is available in less than a few minutes after the instant the measurements are taken, which is a remarkable achievement of space technology. The EPPIM reading script applies the running averaging algorithm to cover the latest 10 minutes, which filters out the higher-frequency random component and leaves the IMF variation trends intact, from which statistically realistic ionospheric drift patterns can be reconstructed.

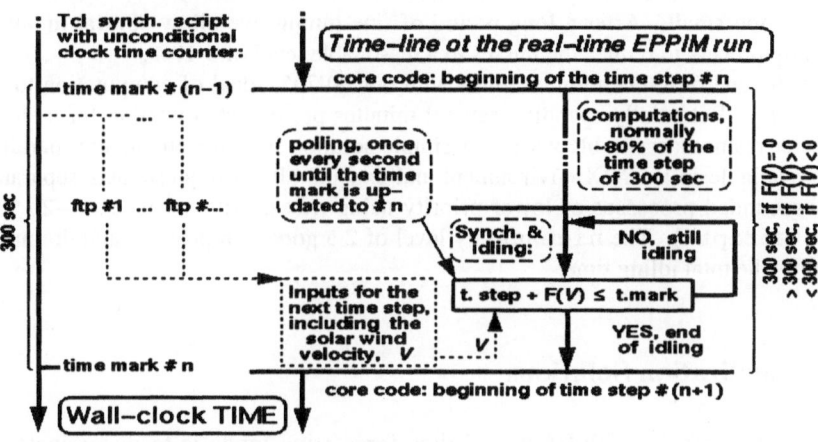

Fig. 2. Parallelization of the EPPIM UNIX processes during a real-time run. The Fortran computational part of the code upon completing each time advance checks the absolute timing marks from the Tcl script and makes a decision whether to continue idling or to proceed to the next time step. The decision depends upon many factors, including the current shift of model time vs. wall-clock time and the solar wind anticipated delay, expressed as $F(V)$

Due to the upstream position of the ACE satellite in the solar wind, it takes a certain amount of time for a sampled solar wind parcel to propagate to the Earth's vicinity from the measurement point. Since the information about incoming variations is available in some advance, the model time of the run is shifted forward to accommodate this advance. This shift facilitates the model's forecasting capability and it is essential to arrange its dynamic update during real-time runs of the EPPIM, which may extend for weeks. The advance time varies depending on the current solar wind velocity, while the distance to ACE can be assumed constant. Another file from the SEC depository is read to find the current velocity of the solar wind. Compared to the IMF, this is a slowly varying parameter. For the ACE location at L1 point, the advance time is in the range 30 to 100 minutes with an average value at about one hour. The dynamic adjustment of the model time shift with respect to the current clock time is schematically shown in Fig.2 as elimination—or increase—of idling time at the end of each time advance.

As follows from the discussion above, the real-time ionospheric model run requires a frequent repetition of many ftp-requests to maintain the inputs updated. This can severely degrade the code's performance if the transfers are invoked sequentially. A network transfer of a few kilobytes is a very fast process, but with several files the accumulated wait for connections, etc. can significantly increase the time required to gather necessary data. Consideration of the worst case scenario is the only feasible way to create a stable application with operational capabilities. During a continuous real-time run of the EPPIM it is necessary to perform about 2000 time advances weekly, which means roughly ~5000 ftp-exchanges. A practical approach to recover the code from the inevitable ftp failure is to impose a hard time limit for each ftp-attempt. If ftp-ing fails, the run continues using the input values from the previous time step. However, the percentage of misses can be unacceptably high if the time limit is too small. After a long period of fine-tuning, we allow several minutes for each ftp process to reduce the misses to a reasonable level of 5-10%.

Such generous allocation conflicts with our 80/20% ideal of computation to idling time ratio. Potentially spending several minutes per timestep waiting for a slow file transfer is an unacceptable waste. A simple solution is to use the natural parallelization available in a UNIX environment and execute all ftp requests as a separate and backgrounded processes of lower priority using `nice` utility (`>nice -2 "ftp-ing script"`). The `nice` priority level of 2 a good compromise and dramatically reduces the total idling time.

4 Visualization Solutions

The goal of running EPPIM in real-time forecasting mode is to disseminate space weather information as widely and rapidly as possible through its WWW-site. The site is designed to be useable both for fast reference to ionospheric conditions (pre-prepared 2-D "ionospheric maps"), and for research collaboration, such as during ionospheric measurement campaigns. These users, who can be called "high-end users", are interested in model outputs in formats suitable for real-time or near real-time downloading to remote powerful resources for in-depth interactive analysis.

Color Panel 2 shows the basic format adopted at the EPPIM Web-site for 2-D maps to demonstrate both the ionospheric developments and variation of input parameters. The visualization is performed at the end of each model time step by the NCAR Graphics package. The resulting NCAR Graphics metafile is a convenient cross-platform vector format, supporting both direct viewing and/or batch-mode rasterization into widespread formats such as GIF. The adopted "geographic latitude vs. local time" frame allows for permanent positioning of the local noon (upper side of the plot), midnight (lower side), and, correspondingly, morning (right) and evening (left), independent of Earth's rotation. The input values are also placed on the map, either as textual information (time and date stamp, solar activity, etc.) or in graphical representation (Kp, the IMF magnitude and orientation). Several operations are performed to simplify the picture (generalizing contouring) and its interpretation (adoption of discreet color code, geographic coordinate mesh). This approach enables a user to attribute certain geographic locations with the current values of ionospheric parameters and to evaluate the regional ionospheric conditions in general.

The EPPIM WWW-site is hosted on a server, which is not involved in computations. This separation facilitates handling of the WWW-requests much faster than from the dedicated platform under heavy computational load. Additionally, under this mini-cluster arrangement the WWW-handling does not interfere with computations, making the run timing more predictable. The clustering allows the computational processor to assign certain visualization tasks to the WWW-server. When the NCAR Graphics metafile is ready, the clustering arrangement uses an *ssh*–connection to the WWW-server to trigger the pull of metafile from the computational platform and rasterize it subsequently. After the copying is completed, the computational processor continues with the next modeling task, while the WWW-server uses a batch mode of the NCAR Graphics utility `ctrans` and software rendering to rasterize the metafile. The limited number of colors facilitates an effective transformation to the economical 8-bit GIF format. Two images are prepared at each time step, one at resolution 511x431 with a general view of the ionospheric map and the geophysical inputs, and another with a magnified map (431x431). The metafile itself is made directly accessible to WWW-users. It can be ftp-ed for local viewing by the native NCAR Graphics viewer `idt` with non-lossy zooming capabilities, including an option of animating a magnified area of interest if a sequence of metafiles is downloaded and appended with NCAR Graphics utilities.

For the horizontal resolution of the current run at 80x80-*km*, the metafile size is in the range of 60-80 kB and the two GIF-views require 20-25 kB each, depending on image complexity. Thus, the model generates about 30MB of graphical information daily. This imposes the need to arrange an expiration of the outdated information at the WWW-site to maintain its size within reasonable limits. Currently, the site stores ionospheric maps covering the last 24 hours. This collection of single frames creates opportunities for the animation of ionospheric developments. However, for this setting deciding what method of animation to use is not trivial. Formats like QuickTime or MPEG exploit interframe coherence and can noticeably compress the resulting animation. The price paid is a heavy CPU load at the WWW-server. Since the ionospheric map is permanently updated, the life span of a pre-prepared animated sequence will be just 5 minutes before it is outdated. Thus, it is not practical to constantly modify these movies, as this would overload the WWW-server's processor.

In addition to the obvious solution of only generating animation on demand at the WWW-server, the task of developing animation can also be delegated to the viewer's workstations. The site provides a JavaScript-based animator to download the desired number of GIF frames and combine them into an animated sequence on the user's screen. The script supports playback both in forward and in reverse direction with variable speed, frame by frame advance, and pause. It requires a transfer of ~0.5, ~3.0, and ~8 MB of data and 16, 32, and 64 MB of free RAM to animate 3, 9, and 24 hours of the ionospheric developments represented by 36, 108, and 288 frames respectively. Some intermediate combinations with skipping of every second or third frame are also supported, which allow covering the same period of modeled time with a smaller data transfer. A strong side of this approach is an immediate response to the WWW-request and the animation "previewing" while downloading. By contrast, a generation of compressed movie on demand requires several minutes only to produce it without counting time for downloading. Admittedly, the adopted approach assumes that the client workstations are powerful enough to handle the animation load, which is generally true for the site's target "high-end users". Several representative pre-compressed animations in conventional formats are available at the site for re-playing.

Still, animated or steady ionospheric maps are just 2-D cross-sections of the 3-D multi-variable simulation domain. Volume rendering is still a challenging task for remote visualization. We approach its solution by using Vis5D format for the iono-spheric datasets. The volume rendering package Vis5D [1], freely available for downloading from its home site (www.ssec.wisc.edu/~billh/vis5d.html), is a powerful graphical tool for exploration of complex 3-D time-dependent datasets. It is compatible with a variety of UNIX platforms and has been ported to Linux, WindowsNT, and OS/2 environments. Cross-platform capabilities of Vis5D allow it to function as a WWW-browser plug-in. To facilitate this capability during the real time run, at every time step the EPPIM outputs a 3-D distribution of the essential ionospheric parameters from the entire modeling domain or from its representative part into a Vis5D database. The Vis5D format applies powerful data compression up to one byte per node per variable, reducing the overall size of the data file to a manageable level. For the current real-time run dimensions of the EPPIM grid (113x135x43), the size of volumetric data for one scalar variable is reduced by this compression to 0.65 MB. Output of the every second node in the representative altitude sub-range of 200-400 km reduces this number to ~0.1MB. The EPPIM real-time users can download these files —either with full or with reduced resolution— and open them locally with the Vis5D viewer. Additionally, a simple script placed in a *crontab* shell allows for automatically appending the Vis5D database by ftp-ing the current data every 5 minutes. Thus, the site's "high-end users" can acquire both animation and volume-rendering capability for their ionospheric data analysis (Color Plate 3).

4 Conclusions and Future Work

This paper presented a case study of computational, networking, and visualization arrangements to maintain a synchronized real-time run of the ionospheric model on a practically permanent basis. This study directly relates to space weather, however its findings are useful for development of network-based applications in various fields. From a computer visualization standpoint, an aggressive use of such elements as run-time visualization and animation, networking, clustering, and computational steering made the scope of this study more universal than geophysical modeling.

Our immediate future plans include an upgrade of the run's computational platform with an order-of-magnitude more powerful workstation SGI Octane, which was provided by the ARSC for this goal. It will allow us to dramatically increase the run horizontal resolution to a 30x30-*km* cell and to consider more geophysical parameters, such as electron density gradients and growth rates of plasma instabilities. New computational and graphical power will also allow us to proceed to an interactive 3-D representation of the ionospheric fields, both locally and remotely using Java- and JavaScript-based approaches.

The availability of a second computational platform for dedicated real-time runs will help us to utilize alternative approaches to the polar ionosphere nowcasting, such as using the direct measurements of the plasma motion from the SuperDARN Doppler radar network [5]. Additionally to a run with IMF-driven statistical advection patterns, it opens numerous opportunities for comparisons of modeling results between different runs, different empirical and database-type ionospheric models, and with the data from the global ionosonde network now available on-line. It is important to note that space weather real-time capabilities are relatively recent developments with a long-term potential. The space weather research community expects a continuation of monitoring at the solar wind upstream locations. The next generation satellite design (2003) includes a solar sail to shift the equilibrium point further sunward to increase the time advance of the measurements and, consequently, the forecasts.

Acknowledgements. The authors would like to acknowledge the contribution from Don Rice (Geophysical Institute, UAF) related to the networking solutions on earlier project stages and from Shawn Houston (ARSC, UAF), who provided Perl scripts to circumvent certain bugs in the latest versions of Netscape precluding from the graphics regular refreshing for WWW-sites users.

This research was supported in part by Cray Research and the University of Alaska Fairbanks Research and Development Grant Program, by the NSF grant ATM-9523818, and by a grant of HPC and visualization resources from the Arctic Region Supercomputing Center.

References

1. Hibbard, W.L., B.E. Paul, D.A. Santec, C.R. Dyer, A.L. Battaiola, and M-F. Voidrot-Martinez, Interactive Visualization of Earth and Space Science Computations, *Computer*, **27**, No.7, pp. 65-72, July 1994.
2. Kelley, M.C. with contributions from Roderick A. Hellis, *The Earth's Ionosphere (Plasma Physics and Electrodynamics)*, Academic Press, Inc., p.10-21, 1989
3. Maurits, S. and B. Watkins, UAF Eulerian Model of the Polar Ionosphere, *STEP Handbook of Ionospheric Models*, p. 95-123, 1996.
4. Richmond, A.D. and Y. Kamide, Mapping electrodynamic features of the high-latitude ionosphere from localized observations: technique, *J.Geophys. Res.*, **93**, 5741, 1988.
5. Ruohoniemi, J. M., and R. A. Greenwald, Statistical patterns of high-latitude convection obtained from Goose Bay HF radar observations, *J. Geophys. Res.*, **101**, 21743-21763, 1996.
6. Weimer, D. R., A flexible, IMF dependent model of high-latitude electric potentials having "space weather" applications, *Geophys. Res. Lett.*, **23**, 2549, 1996.

Editors' Note: see Appendix, p. 295 for colored figures of this paper

Towards visual matching as a way of transferring pre-operative surgery planning

Stijn De Buck, Johan Van Cleynenbreugel,
Guy Marchal, and Paul Suetens

Faculties of Medicine and Engineering
Medical Image Computing (ESAT and Radiology)
University Hospital Gasthuisberg
Herestraat 49
B-3000 Leuven
stijn.debuck@uz.kuleuven.ac.be

Abstract. This paper presents a case study towards intra-operative visualization of pre-operative medical images and surgery planning information. We describe a two stage procedure to provide a surgeon with valuable information during surgery, by augmenting live video with a 3D visualization of surgery planning.
A first stage consists of automatic determination of the intrinsic parameters of the video camera by means of one image of a calibration object. It comprises an automatic ellipse extraction and a solution to the 2D-3D correspondence problem. With the intrinsic camera parameters known, we perform a second stage to compute the extrinsic camera parameters with respect to the patient under surgery, allowing us to position and visualize the medical images in the right place.
Techniques are investigated to visualize and manipulate the pre-operative medical images in order to cope with changes in the surgery scene.

1 Introduction

In an increasing number of surgeries, planning based on three dimensional reconstruction and visualization techniques is used. However intra-operatively such planning is shown on a monitor or as printouts besides the operation table. To benefit from this information the surgeon will have to transfer his field of view systematically back and forth between the images and the patient, thereby trying to integrate both. This way he tries to place his instruments as accurate as possible. Although other solutions do exist (navigation, mechanical drill guides), an integrated visualization of pre-operative and intra-operative images can alleviate this problem. This integration of real and virtual images is a typical augmented reality approach (e.g. [1],[2]). Augmented reality systems aim to place virtual objects in real space in such a way the user can not distinguish them from the real ones. This requires a modelling of the camera, a registration of the real world and a natural integration of the virtual objects (e.g. lighting).

Three different steps have to be attended to in an augmented reality system:

- First segmentation is needed to identify real objects in the scene. Possibly several real objects have to be distinguished (e.g. in surgery of joints, one may want to keep track of the two bone structures involved). Segmentation can be simplified through the use of markers allowing to search the image for known geometries, color, ...
- Next, the segmentation results (e.g. markers positions, corresponding points in 2 images, ...) are employed to obtain calibration and registration of the camera.
- Finally a virtual camera is fed with the 'real' camera and object parameters thus obtained in order to render the virtual objects on the actual camera image.

Other work on transfer of planning by visualization has been reported in [3] and [4]. In their work, it is assumed that patient and intra-operative video camera(s) are not able to move relatively to each other. The approach discussed here differs in a number of technical realisations on automatic camera calibration and holds the potential to release the assumption. Indeed, *visual matching* is an approach to overlay synthetic and real images in an interactive way, thus coping with (small) movements of the patient with respect to the camera.

In this paper we will assume that the intrinsic camera parameters do not change during surgery. This allows to separate the calibration and registration part. During calibration, performed in a automatic way previously to the actual "surgery", we determine the parameters that are proper to the camera (section 2). These parameters are used for rendering the virtual scene, and for the registration of the pose of the camera relative to the patient (section 3). In section 4 we will highlight the visualization and user interaction aspects of our implementation.

2   Automatic camera calibration

Basically, camera calibration is the determination of the relation between 3D points and image coordinates observed by a camera. We can model this relation through several different models of which the complexity depends on the problem at hand and the limitations that are imposed when solving it. We will first discuss the calibration algorithm we used and next the procedure to gather 2D image pixels and corresponding 3D points, used by the algorithm.

2.1   Camera calibration algorithm

From the many algorithms for camera calibration already existing in literature ([5], [6], [7], [8], [9] and [10] with [7], [8], [9] and [10] even as freeware implementations available) we are applying the algorithm developed by R. Tsai ([7], [8]) and implemented by Willson [11].

However our camera model differs a bit, so we first discuss this issue together with our modifications to the algorithm.

Camera models. Mathematically, the pinhole camera, modelling ideal perspective projection, is expressed by an equation of the following form:

$$w_i = K(R^t W_i + T) \tag{1}$$

where w_i is an image point $\left[\mu u_i \ \mu v_i \ \mu \right]^t$, W_i a 3D point, K an upper triangle matrix, R a rotation matrix and T a translation vector. The matrix

$$K = \begin{bmatrix} \frac{f}{s_x d_x} & s & C_x \\ 0 & \frac{f}{d_y} & C_y \\ 0 & 0 & 1 \end{bmatrix} \tag{2}$$

is called the calibration matrix and summarizes the intrinsic parameters of a camera. For a given camera and lens system these parameters are constant: when using zoom and focus they are likely to change. Not only f and s_x, which model the focal length and aspect ratio, but also C_x and C_y will change (cfr Willson [12]). The s appearing in K is a skew factor for the pixel skew and is mostly close to zero and further will be assumed zero.

The matrix R and vector T are determined by the cameras' extrinsic parameters and relate only to the position and orientation of the camera corresponding to the registration part. Zooming and focussing also would affect position and orientation of the camera (cfr Willson [12]) so we will not use these features.

The model of Tsai extends the pinhole model with radial distortion which is a non-linear distortion in the image plane proportional to the square of the distance of a pixel to the center of the image (C_x, C_y) and modelled by two radial distortion coefficients k_1 and k_2 .

Algorithm. Our renderer uses ideal perspective projection to visualize 3D objects. If we would fit our data with Tsai's method the rendering would show not only the normal fitting error but also an error due to the neglection of the radial distortion coefficients.

Tsai's method uses a two-stage calibration with an analytical calculation followed by non-linear optimisation. Since the analytical step is independent of the value of k_1 or k_2, we use the same analytical solution and omit the radial distortion coefficients in the non-linear optimisation function. This results in a best fit to the pinhole-model parameters.

2.2 Elliptical region detection and 2D-3D correspondence

For automatic calibration we use the calibration jig shown in figure 1. It consists of two perpendicular white planes in which circular objects are laid out in a regular pattern. These objects are "closed" black discs, except for the square configurations of four "open" black discs that appear on each plane near the lower corner of the jig. The color of these the open discs is used to distinguish the planes they belong to. In order for our method to be applicable these eight open discs should be visible to the camera.

Complying to the pinhole model, each disc is projected as an elliptical region. Next, we describe how to extract the centroids of the elliptical regions and how to find the correspondence between the centroids and the 3D locations of the discs on the calibration jig.

Fig. 1. The calibration jig consisting of two planar patterns

Elliptical region detection. One approach to the problem of recognizing ellipses is applying a Hough transform [13]. This would lead to a five dimensional search space without the guarantee that all the detected features are real ellipses in the image nor that all ellipses are found. Therefore we chose for a two step implementation. First we look for all the objects that could correspond to an ellipse; second, we filter all detected objects such that only the projections of circles remain. This second step is integrated in the 2D-3D correspondence solution.

To start the first stage the image histogram is normalized in order to be robust to variations in lighting. After thresholding and erosion/dilation, a region-based segmentation is done. For every region the bounding box, minor axis, major axis and Euler number are calculated. Only the regions of which the minor axis and major axis are between certain boundaries, are retained. We can employ such a constraint because the distance between camera and object is between known boundaries in a surgery environment.

The coordinates $(EC_x(i), EC_y(i))$ of centriod i are the first order moment divided by the area A of the centriods bounding box [2]:

$$EC_x(i) = \frac{\sum_{boundingbox} \rho(u,v)u}{A} \tag{3}$$

$$EC_y(i) = \frac{\sum_{boundingbox} \rho(u,v)v}{A} \tag{4}$$

with

$$\rho(u, v) = 1 - \frac{\text{the intensity of pixel}(u, v)}{\text{the maximum pixel intensity}}$$

We identify the eight centers originating from the open discs by their Euler number equal to 0. For each of these the pixels around the center are converted to the HSV-colourspace. Based on the hue-component and sufficient saturation we can separate the white from the yellow ones. We then know 4 points from each plane enabling us to solve the 2D-3D correspondence problem.

2D-3D correspondence. In order to solve the 2D-3D correspondence, we consider the image plane and the two planes of the calibration jig as 2D projective planes $\mathcal{I}$, $\mathcal{P}_1$ and $\mathcal{P}_2$ respectively. Then the pinhole projection of each $\mathcal{P}_j$ onto $\mathcal{I}$ is modelled by a homography H_j of projective planes. We show that determining H_j for $j = 1, 2$ will solve the 2D-3D correspondence. Suppose the detected centers of the elliptical regions are described in $\mathcal{I}$ as $m_k = \begin{bmatrix} u_k & v_k & 1 \end{bmatrix}^t$ for $k = 1 \ldots N$, where N is the total number of detected centers. Suppose the known centers of the circular objects on calibration jig plane j are described in $\mathcal{P}_j$ as $m_k^j = \begin{bmatrix} x_k^j & y_k^j & 1 \end{bmatrix}^t$ for $k = 1 \ldots N_j$, where N_j is the total number of objects on plane j. It is good to keep in mind that to each m_k^j does correspond a known 3D point also.

If m in $\mathcal{I}$ corresponds to m^j in $\mathcal{P}_j$ then H_j satisfies

$$\lambda m = H_j m^j \tag{5}$$

So if we know "enough" corresponding pairs (m, m^j), we can solve these equations for H_j. Suppose there are r such pairs. Without loss of generality we can assume them to be (m_i, m_i^j) for $i = 1, \ldots, r$. This leads to a linear system of the following form (in which arbitrarily $\lambda_r = 1$) with $H_j = [h_{kl}]$ and $k, l = 1, \ldots, 3$:

$$
\begin{bmatrix}
E_1 & -m_1 & 0 & \cdots & 0 \\
E_2 & 0 & -m_2 & \cdots & 0 \\
\vdots & & & & \vdots \\
E_{r-1} & 0 & 0 & 0 & -m_{r-1} \\
E_r & 0 & 0 & 0 & 0
\end{bmatrix}
\begin{bmatrix}
h_{11} \\
\vdots \\
h_{33} \\
\lambda_1 \\
\vdots \\
\lambda_{r-1}
\end{bmatrix}
=
\begin{bmatrix}
0 \\
\vdots \\
\vdots \\
0 \\
m_r
\end{bmatrix}
\quad \text{with } E_i =
\begin{bmatrix}
m_i^{jt} & 0^t & 0^t \\
0^t & m_i^{jt} & 0^t \\
0^t & 0^t & m_i^{jt}
\end{bmatrix}
$$

So what is the minimum r and how can the corresponding pairs be obtained? After all this whole procedure is meant to find correspondences, but it seems to be dependent on such correspondences. First, $r \geq 4$, since four non-collinear projective points span a basis of a projective plane and the mapping of a basis onto a basis determines the homography. Second due to our design of the calibration jig and the elliptical region detection, for a given plane j four corresponding pairs are readily determined. So we start from these four and obtain an estimate for H_j. This H_j is used to project an other known m_k^j onto $\mathcal{I}$. By taking the elliptical region that is closest to this estimated image point, we associate m_k^j with the center of this region and hence with an m_k. The more pairs that are obtained in this way the better the approximation of H_j from iteratively solving the linear system.

Theoretically this solution should be optimized further by a non-linear scheme ([14], [15]). The linear solution is however sufficient for our purpose of establishing 2D-3D correspondences as input for a subsequent camera calibration. Also, false responses to the elliptical region detection procedure are filtered out in this stage because no world plane coordinates are corresponding to them.

2.3 Accuracy

To determine the overall accuracy of our method we generated a sequence of 30 synthetical images of our calibration jig. After adding an amount of Gaussian noise, these images are input to our routine yielding the results in Table 1. The error depicted here is the difference between a projection of a set of 300 random points with the estimated and original camera parameters. The fact that we even have an error when no noise is added can be explained by discretisation errors of the software renderer (Open Inventor) that generates the synthetical images.

Table 1. Accuracy of the calibration method.

	Variance of the noise		
	No noise	0.01	0.1
Mean projection error (in pixels)	0.743	0.745	0.813
Standard Deviation (in pixels)	0.155	0.188	0.182

2.4 Conclusion

The elliptical region detection and the correspondence deliver us with pairs of 2D-3D coordinates. These are used as input for the modified calibration algorithm giving the camera's intrinsic parameters and its extrinsic parameters with respect to the calibration jig. We only retain the intrinsic parameters which we suppose to be constant throughout the sequel of the approach and use them to model a software camera.

3 Registration

In order to progress towards visual matching, we give some details on a laboratory setup (cfr fig. 2) enabling us to temporarily solve the initial registration problem required.

A plastic phantom skull, on which a number of radio-opaque markers are attached, was CT-scanned. From this CT-volume a 3D surface representation (mimicking the pre-operative planning) was derived (see figure 2). The same skull is then imaged by a camera intrinsically calibrated as described above. For the world 3D coordinate system of the skull, we use the one defined by the CT-image volume.

We need to know the relative position of the skull in the reference frame of the camera or, vice versa, the position of the camera with respect to the skull. This pose determination problem can be seen as an extrinsic camera calibration. The implementation by Willson provides in such a procedure. It requires 7 point matches between world and image coordinates. These are gathered manually by pointing out the radio-opaque markers which are very well visible on the video image also. By corresponding the 2D image coordinates thus obtained to the 3D coordinates of the respective markers from the CT, the camera pose is determined.

Although the procedure is rather coarse, the quality of the results obtained can be judged from fig. 4 (Left).

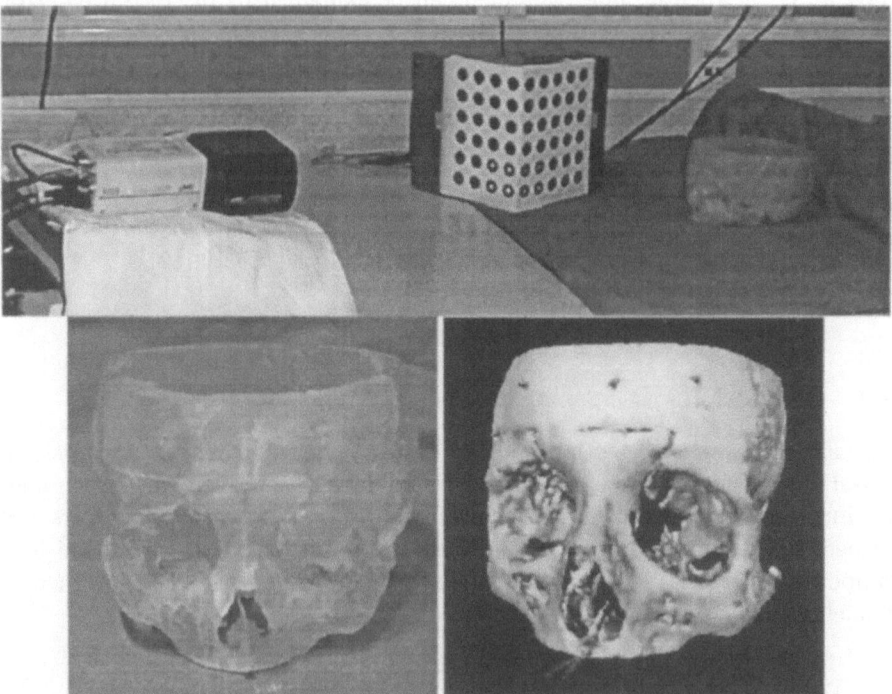

Fig. 2. Top: the laboratory setup we used to test the integrated visualization; Bottom left: the skull phantom; Bottom right: the 3D surface representation of the CT-volume of the skull phantom

4 Intraoperative visualization and visual matching

The visualization itself is developed in Open Inventor (on SGI Octane). The intrinsic and extrinsic parameters of the video camera (JVC, KY-F55B) are fed into a software camera. The viewport of the latter is overlayed on live-video input.

The system can be useful to surgeries where certain pre-operatively determined distances, angles, etc. have to be measured intra-operatively. The measures the surgeon needs during surgery are then planned in the CT-image volume and transferred to the intra-operative video.

We illustrate this procedure by showing a simulation on inserting zygomatic oral implants (osseo-integrating screws). This system is an extension of the work presented in [16]. Pre-operatively the surgeon simulates on the CT-image volume different positions to place the implant and selects the most optimal one (cfr fig. 3). The surgeon has to position his drill at the precise angle and location determined during planning, in order to insert the zygomatic implant. During surgery an integrated view of live video and the rendering of the virtual images by the software camera (cfr fig. 3) is shown on a workstation display. The surgeon can interactively decide which virtual images he wants to see: the CT-image volume, the planned drill axis, ... (cfr fig. 3 and fig. 4). Visual matching capabilities are provided: by means of a 3D-mouse the surgeon can realign the virtual images with his patient after a small movement. This is shown in figure 4 (Right). Here the skull is slightly rotated and the virtual objects are realigned, making use of the 3D-mouse.

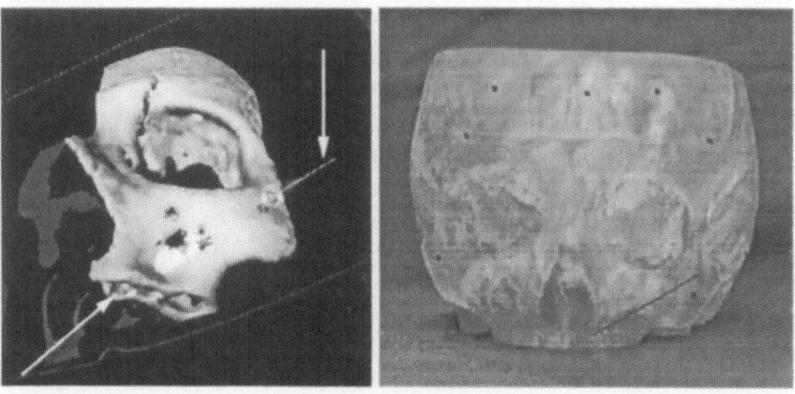

Fig. 3. Left: the planning of the placement of a zygomatic oral implant; Right: Transfer of the planned drill axis to the surgery environment

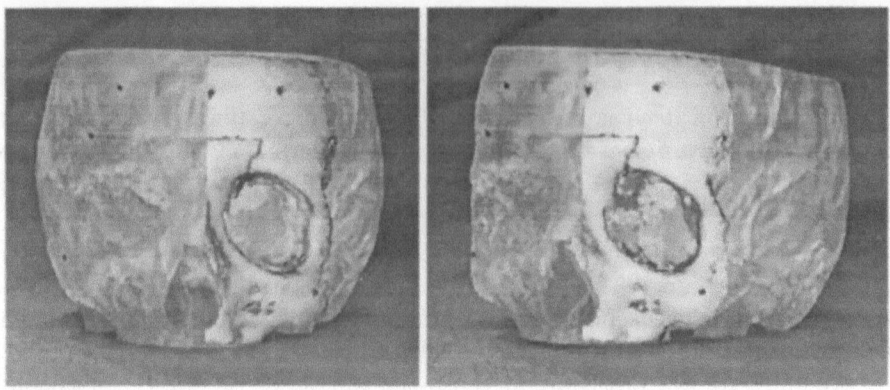

Fig. 4. Left: an integrated view of the skull, a partial CT-surface and markers after the calibration and registration procedure. Right: after the skull is slightly rotated, the virtual objects were realigned by visual matching

5 Discussion

In this paper we propose a procedure towards visual matching for intra-operative transfer of planning data.

Visual matching can only be applied when the virtual world is perceived in the same way as the real world. Modelling of the video camera is a means to fulfill this requirement. The automatic camera calibration procedure we propose offers a robust, fast solution to this problem. So it can be applied without much overhead and specific knowledge in a surgery environment. Furthermore we obtain subpixel accuracy on synthetical images, which is sufficient for our application, and a low sensitivity to noise.

The seperation of extrinsic and intrinsic parameters makes visual matching capabilities like a 3D-mouse possible. In turn, this technique together with the surgeons' clinical experience permits him to adapt the integrated view to his needs in an intuitive manner.

At the moment we are further investigating how depth impression can be induced in order to give the surgeon a better perception of the scene recorded by the camera(s). Techniques like stereo vision and stereoscopic display of the augmented intra-operative video images look promising towards that goal.

Acknowledgements

This work is part of SuperVisie, ITA-II/980302, an IWT ITA-II project sponsored by the Flemish Government.

References

1. J. Vallino. Introduction to augmented reality. http://www.cs.rit.edu/~jrv/research/ar/introduction.html.
2. J.P. Mellor. Enhanced reality visualization in a surgical environment. In *MIT AI-TR*, 1995.
3. Colchester et al. Development and preliminary evaluation of VISLAN, a surgical planning and guidance system using intra-operative video imaging. *Medical Image Analysis*, 1(1):73–90, March 1996.
4. Grimson et al. An automatic registration method for frameless stereotaxy, image guided surgery and enhanced reality visualization. *IEEE Transactions on Medical Imaging*, 15(2):129–140, April 1996.
5. D. Vandermeulen. *Methods for registration, interpolation and Interpretation and of Threedimensional medical image data for use in 3D-display, 3D-modelling and therapy planning*. PhD thesis, KUL, 1991.
6. J.P. Tarel and A. Gagalowicz. Calibration de caméra à base d'ellipses. In *9ème congrès Reconnaissance des Formes et Intelligence Artificielle*, Paris, France, 1994. AFCET.
7. R.Y. Tsai. An efficient and accurate camera calibration technique for 3D machine vision. In *CVPR86*, pages 364–374, 1986.
8. R.Y. Tsai. A versatile camera calibration technique for high-accuracy 3D machine vision metrology using off-the-shelf tv cameras and lenses. *RA*, 3(4):323–344, 1987.
9. J. Heikkilä and O. Silvén. A four-step camera calibration procedure with implicit image correction. In *CVPR97*, pages 1106–1112, 1997.
10. J. Heikkilä and O. Silvén. Camera calibration and image correction using circular control points. In *SCIA97*, 1997.
11. R.G. Willson. Freeware implementation of R. Tsai's camera calibration algorithm. http://www.cs.cmu.edu/afs/cs.cmu.edu/user/rgw/www/TsaiCode.html.
12. R.G. Willson. Modeling and calibration of automated zoom lenses. In *CMU-RI-TR*, 1994.
13. Van Gool L. and Proesmans M. *An introduction to image processing and analysis*. 1998. Course text ESAT KULeuven.
14. Z. Zhang. A flexible technique for camera calibration. Technical Report MSR-TR-98-71, Microsoft, 1998.
15. R. Horaud and G. Csurka. Self-calibration and euclidean reconstruction using motions of a stereo rig. In *ICCV98*, pages 96–103, 1998.
16. Verstreken K., Van Cleynenbreugel J., K. Martens, G. Marchal, D. van Steenberghe, and P. Suetens. An image-guided planning system for endosseous oral implants. *IEEE Transactions on Medical Imaging*, 17(5):842–852, October 1998.

A Case Study of Isosurface Extraction Algorithm Performance

Philip M. Sutton[1], Charles D. Hansen[1], Han-Wei Shen[2], and Dan Schikore[3]

[1] Department of Computer Science
University of Utah
Salt Lake City, UT 84112
[psutton | hansen] @cs.utah.edu

[2] Computer and Information Science Department
The Ohio State University
Columbus, OH 43210

[3] CASC
Lawrence Livermore Laboratories
Livermore, CA 94550

Abstract. Isosurface extraction is an important and useful visualization method. Over the past ten years, the field has seen numerous isosurface techniques published, leaving the user in a quandary about which one should be used. Some papers have published complexity analysis of the techniques, yet empirical evidence comparing different methods is lacking. This case study presents a comparative study of several representative isosurface extraction algorithms. It reports and analyzes empirical measurements of execution times and memory behavior for each algorithm. The results show that asymptotically optimal techniques may not be the best choice when implemented on modern computer architectures.

1 Introduction

Researchers in many science and engineering fields rely on insight gained from instruments and simulations that produce discrete samplings of three-dimensional scalar fields. Visualization methods allow for more efficient data analysis and can guide researchers to new insights. Isosurface extraction is an important technique for visualizing three-dimensional scalar fields. By exposing contours of constant value, isosurfaces provide a mechanism for understanding the structure of the scalar field. These contours isolate surfaces of interest, focusing attention on important features in the data such as material boundaries and shock waves while suppressing extraneous information. Several disciplines, including medicine [10, 17], computational fluid dynamics (CFD) [4, 5], and molecular dynamics [8, 12], have used this method effectively.

The original Marching Cubes algorithm [11, 20] for isosurface extraction examined all cells in the data set. A tremendous amount of research has focused on reducing the number of cells visited while constructing an isosurface [1, 2, 9, 15, 19]. These methods utilize auxiliary data structures to examine only those cells that contain a portion of the isosurface. While the search structures introduced by many of these methods increase the storage requirements, the acceleration gained by the isosurfacing technique offsets this overhead.

Algorithms that use data structures to accelerate isosurface extraction generally provide lower latency than simple marching methods in visualization applications. In the context of isosurface extraction, latency is defined as the elapsed time between receiving a query and returning a complete isosurface. Reducing latency greatly improves interactivity, providing researchers with a better understanding of their data. The visualization literature lacks studies or surveys comparing the latency and overall performance of the many different three-dimensional isosurface extraction algorithms. Authors of isosurfacing papers usually compare their algorithm's performance only with that of Marching Cubes. Analysis of theoretical average- and worst-case efficiency also plays a large role in the literature. Unfortunately, different implementations and different platforms make objective, empirical comparisons between algorithms difficult. Memory behavior on modern computer architectures, for example, plays a crucial role in an application's performance, but an analysis of this important factor rarely appears.

This paper presents a comparative study of several representative isosurface extraction algorithms. Each algorithm uses a different data structure to accelerate the search for cells containing an isosurface, then computes the surface using a Marching Cubes-style interpolation. This paper reports and analyzes empirical measurements of execution times and memory behavior for each of these algorithms. Section 2 describes the algorithms tested, along with implementation details of each. Section 3 describes the experiments and presents the results. Section 4 summarizes the paper and draws conclusions.

2 Isosurface Extraction Techniques

Visualization applications in many fields [5, 8, 10] use the Marching Cubes [11, 20] algorithm to extract isosurfaces from volumetric data. Marching Cubes and other algorithms use a voxel representation of the volume, considering each data point as the vertex of some geometric primitive, such as a cube or tetrahedron. These primitives, or cells, subdivide the volume and provide a useful abstraction for computing isosurfaces. The Marching Cubes algorithm tests each cell in the volume for intersection with the isosurface. By visiting cells in an order based on their position, this method can exploit the spatial coherence of the isosurface by reusing interpolation calculations along edges shared by two or more cells. However, the Marching Cubes method spends a high percentage of time visiting cells that do not contain portions of the isosurface.

Researchers have introduced a number of techniques to increase the efficiency of isosurface extraction over the linear search proposed in the Marching Cubes algorithm. These methods fall into two general categories, characterized by the criteria used to partition the cells. Geometric techniques retain the original representation of the volume and partition along divisions in the geometric mesh. Span space decomposition techniques create and manipulate abstract representations of the cells. Both geometric and span space methods can extract isosurfaces from unstructured grids as well as regular grids (see [13, 16] for geometric decomposition techniques for unstructured grids). Sections 2.1 and 2.2 describe representative methods from these two categories.

2.1 Geometric Decomposition Techniques

Wilhelms and van Gelder [19] describe the branch-on-need octree (BONO), a space-efficient variation on the traditional octree. Their data structure partitions the cells in the data based on their geometric positions. Extreme values (minimums and maximums) propagate up the tree during construction, enabling the extraction phase to prune branches of the tree. The extraction algorithm traverses only those nodes whose values span the isovalue, i.e. those with $minvalue < isovalue < maxvalue$. The branch-on-need strategy partitions the volume such that the "lower" subdivision in each direction covers the largest possible power of two cells. This results in fewer nodes when the dimensions of the volume do not equal powers of two, making the tree traversal more efficient. Leaf nodes in the branch-on-need octree generally reference eight cells (nodes may reference fewer cells along the edges of the volume). This greatly reduces the memory required, as one pair of extreme values covers eight cells. In the original paper, a hash table of edges was used to exploit spatial coherence. After the initial interpolation of a point along an edge, cells that share that edge access its hash entry to avoid recomputing the interpolation.

Another technique involves propagating the isosurface from a set of seed cells. This method combines aspects of both geometric decomposition techniques and span space algorithms[1]. A seed set must contain at least one cell per connected component of each isosurface. The algorithm groups seed cells into a hierarchical search structure, then traverses that structure to find all seeds that intersect the current isosurface. Construction of the isosurface begins at these seeds and propagates through neighboring cells using adjacency and intersection information. The difficult portion of the surface propagation algorithm lies in locating and selecting the seed cells. Itoh et al. [6, 7] find the local extremum points in the data and connect them with a graph in the spatial domain. The seed set consists of the cells containing extremum points, plus all cells intersected by the arcs of the graph and some cells along the boundaries if the volume has "holes". A thinning algorithm, commonly used in image processing, can then generate a skeleton of the seed set that connects all extremum points, yet contains fewer cells. van Kreveld et al. [18] also use a graph of local extremum points, but add saddle points to create a contour tree. Bajaj et al. [1] use set theory to find seed cells and a segment tree to organize and traverse them. Both structured and unstructured meshes can utilize these techniques, which theoretically provide near-optimal worst-case time complexity. However, noisy data may disturb the complicated seed set construction process. Measurement data such as MRI and CT scans can cause these algorithms to produce large numbers of seed cells, causing slower preprocessing time.

---

[1] Seed sets contain aspects of span space techniques, but surface propagation requires information about the structure of the volume, establishing it as an inherently geometric technique.

2.2 Span Space Decomposition Techniques

Span space techniques partition cells based on their extreme values. Livnat et al. [9] introduce the span space, where each cell maps to a point in 2D space. The cell's minimum value defines the x-coordinate of the point, and the maximum value determines the y-coordinate. All points in span space lie above the $y = x$ line. For a given isovalue v, the points representing cells which intersect the isosurface have $y \geq v$ and $x \leq v$. The NOISE algorithm described in [9] overlays a kd-tree on the points. This structure organizes the points such that during traversal, the algorithm needs to test only one of the two extreme values at each node in the tree. The authors use a pointerless representation of the kd-tree to avoid the additional overhead of pointer traversal. Constructing this array involves sorting the cells based on their extreme values. Sorting in a preprocess minimizes the effect on isosurface extraction performance.

Shen et al. [15] use a lattice subdivision of span space in their ISSUE algorithm. The user defines a lattice resolution L and the algorithm divides the span space points into one of the $L \times L$ lattice elements. Given an isovalue v, the ISSUE method assigns each lattice element to one of five categories. The algorithm trivially excludes cells in region 1 and trivially includes those in region 2. Cells in region 3 require a test against their maximum value, those in region 4 require a test against their minimum. Only region 5 requires a full min-max search of its cells. Shen proposed a modified version of the ISSUE algorithm [14] which creates search structures only at the lattice elements along the diagonal region 5 and coalesces lattice elements at other regions into sorted linear arrays. These modifications are implemented by building a kd-tree for each lattice element along the $y = x$ line, since only these elements may fall into the region 5 classification. This method simplifies and accelerates the search phase of isosurface extraction, but this acceleration comes at the cost of a higher memory requirement than the NOISE algorithm. The lattice structure itself requires additional memory and by creating a search structure in lattice elements that may require a full min-max search, further memory overhead is introduced. However, simple division of lattice elements among parallel processors makes this algorithm easily parallelizable. The authors report near-linear speedups using this parallel algorithm [15].

The Interval Tree technique introduced by Cignoni et al. [2, 3] guarantees that the worst-case efficiency is asymptotically optimal. This algorithm groups cells, represented by the intervals defined by their extreme values, at the nodes of a balanced binary tree. Each node contains two lists, one sorted in ascending order of cell minima, the other sorted in descending order of cell maxima. For any isovalue query, the algorithm traverses at most one branch from a node after scanning through one of the lists. The number of nodes created depends on the number of distinct interval extremes, usually much smaller than the number of cells in the volume. The memory requirements for representing intervals and for the two lists at each node dominate. The authors propose improvements specific to the underlying geometry (structured or unstructured mesh). A hash table of edges exploits spatial coherence in unstructured meshes, while regular grids can utilize a form of local surface propagation.

3   Experimental Results

Both geometric and span space techniques accelerate isosurface extraction by limiting the number of cells examined. This acceleration is usually described in terms of average- and worst-case algorithm complexity. The analysis of asymptotic complexity given by various authors [1, 2, 9, 15] shows that in the limit, the Interval Tree [2, 3] and seed set [1] algorithms guarantee worse-case optimal efficiency while the NOISE [9] algorithm provides near-optimal complexity. However, no quantitative performance comparison between the different algorithms exists, since most authors compare their technique only with Marching Cubes. This section describes the comparative performance of various three-dimensional isosurface extraction techniques, each implemented and tested using the same hardware and software framework. Marching Cubes [11, 20], the branch-on-need octree (BONO) [19], and a surface propagation algorithm using seed sets [1] represent the geometric decomposition techniques, while NOISE [9], ISSUE [15], and the Interval Tree [2, 3] represent span space algorithms. The NOISE implementation uses a pointerless representation of the kd-tree. A similar data structure could be used by the other span space algorithms, but since span space algorithms must always index cells, a certain amount of memory overhead is unavoidable. Each implementation includes the optimizations given in the paper, with the exception of techniques explicitly designed to exploit spatial coherence, such as those given in [19] and [3]. Every algorithm would benefit from these improvements, so fairness dictates their omission. Each algorithm performs both the isosurface query and triangle construction and thus are representative of execution times for the isosurface generation process.

The test data consists of both a noisy, measurement data set and a simulation data set that contains a continuous scalar field. The Head128 data set contains results from a CT scan of a human head and consists of $128 \times 128 \times 128$ points. The Rage256 data set represents a CFD simulation of the classic Rayleigh-Taylor hydrodynamic instability, in which two fluids of differing densities mix. This data set contains $256 \times 256 \times 256$ points. Figure 4 (see color plate) shows a sample isosurface from each data set. The strikingly different characteristics of these data sets can be seen in Figure 5 (see color plate). These images show the density of points in span space for each data set. The Head128 data set shows a nonuniform density, characteristic of noisy measurement, with large empty areas gradually coalescing into two hot spots. The Rage256 data set contains much more uniform data spread over the entire range of values, with very localized hot spots at the extremes.

Experimental results allow comparison of both execution time and memory system behavior from execution on a single dedicated processor of an SGI Origin 2000 with 8GB of memory. The Origin 2000 is a commonly used visualization system with a memory hierarchy typical among high end systems. Table 1 shows execution times for each algorithm using the Head128 data set. Table 2 displays results from the Rage256 data set. Results in Table 1 represent averages from ten repeated executions of ten representative isovalue queries. Table 2 represents averages from ten executions of five representative isovalue queries.

All algorithms exhibit significant speedup over Marching Cubes. Most algorithms perform similarly for the Head128 data set, as shown in Table 1. However, for the

Table 1. Experimental results from the Head128 data set.

Type	Algorithm	Average execution time (s)
Standard	Marching Cubes	2.13
Geometric	BONO	0.58
	Seed Set	0.66
	ISSUE	0.57
Span Space	Interval Tree	0.56
	NOISE	0.51

Table 2. Experimental results from the Rage256 data set.

Type	Algorithm	Average execution time (s)
Standard	Marching Cubes	17.31
Geometric	BONO	4.92
	Seed Set	6.67
	ISSUE	5.60
Span Space	Interval Tree	8.84
	NOISE	5.04

Rage256 data set, large disparities in performance exist. Table 2 shows that BONO, IS-SUE, and NOISE present the largest speedups, with the Branch-on-Need Octree providing the best performance. The surface propagation code used for the seed set technique contains a large number of branches. Since the processor cannot readily predict the outcome of these branches, this algorithm performs poorly. The Interval Tree technique, although provably optimal in the limit, actually executes slower than every algorithm but Marching Cubes for the Rage256 data set. To discover the causes of this result, each implementation used the performance counters on the R10000 to track the number of clock cycles, TLB misses, and L1 and L2 cache misses during execution. Figure 6 shows the experimental results for the Head128 data set. Figure 7 shows the results obtained from the Rage256 data set. The clock cycles charted in Figures 6(a) and 7(a) correspond closely to the execution times given in Tables 1 and 2. Figure 6(b) shows that the span space techniques have a high TLB miss rate for the Head128 data set, but Figures 6(c) and 6(d) demonstrate no such distinctions in L1 and L2 cache behavior. Figures 7(b), 7(c), and 7(d) uncover the reason for the Interval Tree's low performance in the Rage256 data set — the algorithm's poor memory behavior. The large number of TLB and cache misses imply that this algorithm visits data in a different order than that adhered to by the data in memory. The severity of this difference in marching order requires processing to stall repeatedly as the operating system swaps information in and out of these hardware structures. In contrast, the BONO and Marching Cubes algorithms visit data in an order similar to that of the data in memory, since they traverse the geometric volume. These two algorithms have low instances of TLB and L2 cache misses, which incur high penalties.

To demonstrate the tradeoff between performance and storage space, Figure 8 shows the amount of memory overhead required by each algorithm for both test data sets. These figures do not include the memory required to store the data set, nor the mem-

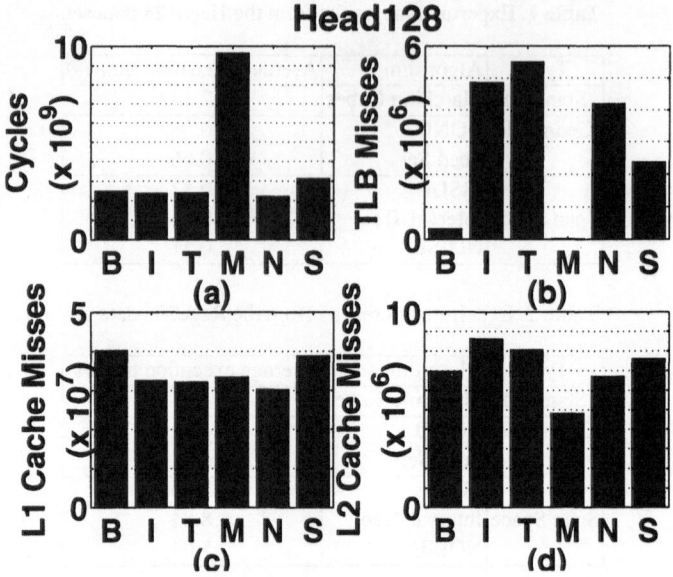

Fig. 6. Experimental results from the Head128 data set. B = BONO, I = ISSUE, T = Interval Tree, M = Marching Cubes, N = NOISE, S = Seed Set/Surface Propagation.

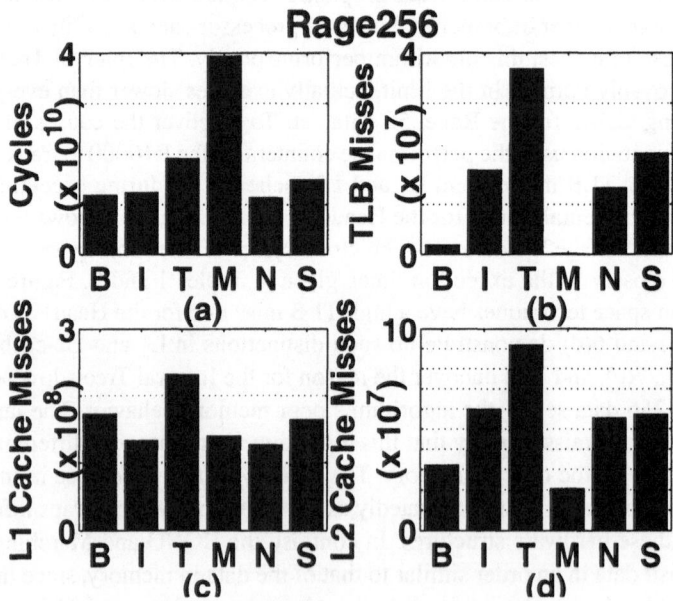

Fig. 7. Experimental results from the Rage256 data set. B = BONO, I = ISSUE, T = Interval Tree, M = Marching Cubes, N = NOISE, S = Seed Set/Surface Propagation.

266

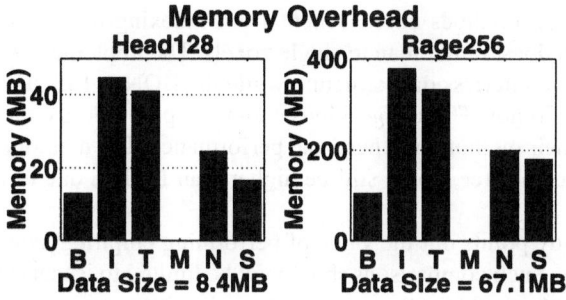

Fig. 8. Memory overhead requirements for the Head128 and Rage256 data sets. B = BONO, I = ISSUE, T = Interval Tree, M = Marching Cubes, N = NOISE, S = Seed Set/Surface Propagation.

ory consumed by storing the triangles that compose the isosurface, since all algorithms require this memory. BONO uses the least amount of additional memory, which is to be expected since it retains the original representation of the data volume. The seed set method has nearly as much memory overhead as the NOISE algorithm because a extrema data structure is used. In these implementations, the ISSUE and Interval Tree algorithms consume large amounts of memory, approximately five times the amount required to store the original data. In fairness, this is due to implementation details. Both ISSUE and the Interval tree could use a pointerless data structure which would reduce the amount of memory overhead. Also, the span space methods as implemented use indices to individual cells while BONO indexes eight cell simultaneously. An interesting comparison would be to raise the index to groups of cells rather than individual cells.

These experiments show that data structures increase the memory required to extract isosurfaces, but allow the computation to execute more quickly than in a simple marching method. However, algorithms that consume larger amounts of memory do not necessarily yield better performance.

4   Conclusion

The comparison of multiple isosurface extraction algorithms has not been previously available. This case study performs such a comparison with several data sets and attempts to show the empirical performance differences on both measurement data and simulation data. Based on the data structures required, the large constant for provably optimal algorithms is amplified by modern computer architectures where cache and page misses induce large performance penalties. With such architectures being the prevalent compute platform, the theoretical gains do not appear in practice.

While this case study relates a relatively fair comparison of different isosurface techniques, several enhancements can be made. Most importantly, this comparison should be repeated for more data sets and on different compute platforms. This would better demonstrate the variations caused by diverse data sets and by other memory hierarchy designs.

The span space methods could benefit from indexing to groups of cells, as the BONO algorithm does, rather than to single voxels. The implementation of the NOISE algorithm uses a pointerless data structure while the BONO, ISSUE and Interval Tree implementations do not. These algorithms could use pointerless data structures which would improve their memory overhead and performance. The memory overhead for ISSUE and the Interval Tree would still be higher than NOISE due to the required data structures.

This case study points out the value of performing empirical comparisons of published algorithms. Such comparisons should not be limited to isosurface techniques but should also be performed for other visualization algorithms to examine the practicality of such techniques.

5 Acknowledgements

The authors would like to thank the reviewers for their insightful comments which improved the final version of this paper. This work was supported in part by the C-SAFE DOE ASCI Alliance Center, the DOE Advanced Visualization Technology Center (AVTC), and a grant from LLNL. A portion of this work was performed under the auspices of the U.S. Department of Energy by Lawrence Livermore National Laboratory under contract no. W-7405-Eng-48. The Rage data set is courtesy of Robert Weaver (Los Alamos National Laboratory).

References

1. Chandrajit L. Bajaj, Valerio Pascucci, and Daniel R. Schikore. Fast isocontouring for improved interactivity. In *1996 Symposium on Volume Visualization*, pages 39–46. IEEE Computer Society, IEEE Computer Society Press, Los Alamitos, CA, October 1996.
2. P. Cignoni, C. Montani, E. Puppo, and R. Scopigno. Optimal isosurface extraction from irregular volume data. In *1996 Symposium on Volume Visualization*, pages 31–38. IEEE Computer Society, IEEE Computer Society Press, Los Alamitos, CA, October 1996.
3. Paolo Cignoni, Paola Marino, Claudio Montani, Enrico Puppo, and Roberto Scopigno. Speeding up isosurface extraction using interval trees. *IEEE Transactions on Visualization and Computer Graphics*, 3(2):158–170, April 1997.
4. Jean M. Favre. Towards efficient visualization support for single-block and multi-block datasets. In Roni Yagel and Hans Hagen, editors, *Proceedings of Visualization 1997*, pages 423–428. IEEE Computer Society, IEEE Computer Society Press, Los Alamitos, CA, October 1997.
5. Philip D. Heermann. Production visualization for the asci one teraflops machine. In David Ebert, Hans Hagen, and Holly Rushmeier, editors, *Proceedings of Visualization 1998*, pages 459–462. IEEE Computer Society, IEEE Computer Society Press, Los Alamitos, CA, October 1998.
6. Takayuki Itoh and Koji Koyamada. Automatic isosurface propagation using an extrema graph and sorted boundary cell lists. *IEEE Transactions on Visualization and Computer Graphics*, 1(4):319–327, December 1995.
7. Takayuki Itoh, Yasushi Yamaguchi, and Koji Koyamada. Volume thinning for automatic isosurface propagation. In Roni Yagel and Gregory M. Nielson, editors, *Proceedings of Visualization 1996*, pages 303–310. IEEE Computer Society, IEEE Computer Society Press, Los Alamitos, CA, October 1996.

8. Marco Lanzagorta, Milo V. Kral, J.Edward Swan II, George Spanos, Rob Rosenberg, and Eddy Kuo. Three-dimensional visualization of microstructures. In David Ebert, Hans Hagen, and Holly Rushmeier, editors, *Proceedings of Visualization 1998*, pages 487–490. IEEE Computer Society, IEEE Computer Society Press, Los Alamitos, CA, October 1998.

9. Yarden Livnat, Han-Wei Shen, and Christopher R. Johnson. A near optimal isosurface extraction algorithm using the span space. *IEEE Transactions on Visualization and Computer Graphics*, 2(1):73–84, 1996.

10. William E. Lorensen. Marching through the visible man. In Gregory M. Nielson and Deborah Silver, editors, *Proceedings of Visualization 1995*, pages 368–373. IEEE Computer Society, IEEE Computer Society Press, Los Alamitos, CA, October 1995.

11. William E. Lorensen and Harvey E. Cline. Marching cubes: A high resolution 3d surface construction algorithm. In Maureen C. Stone, editor, *Computer Graphics*, volume 21, pages 163–169. ACM SIGGRAPH, ACM SIGGRAPH, July 1987.

12. Colin R. F. Monks, Patricia J. Crossno, George Davidson, Constantine Pavlakos, Abraham Kupfer, Cláudio Silva, and Brian Wylie. Three dimensional visualization of proteins in cellular interactions. In Roni Yagel and Gregory M. Nielson, editors, *Proceedings of Visualization 1996*, pages 363–366. IEEE Computer Society, IEEE Computer Society Press, Los Alamitos, CA, October 1996.

13. S. Parker, M. Parker, Y. Livnat, P. Sloan, C. Hansen, and P. Shirley. Interactive ray tracing for volume visualization. *IEEE Transactions on Visualization and Computer Graphics*, 5(3):238–250, July 1999.

14. Han-Wei Shen. Isosurface extraction in time-varying fields using a temporal hierarchical index tree. In David Ebert, Hans Hagen, and Holly Rushmeier, editors, *Proceedings of Visualization 1998*, pages 159–164. IEEE Computer Society, ACM Press, New York, NY, October 1998.

15. Han-Wei Shen, Charles D. Hansen, Yarden Livnat, and Christopher R. Johnson. Isosurfacing in span space with utmost efficiency (issue). In Roni Yagel and Gregory M. Nielson, editors, *Proceedings of Visualization 1996*, pages 287–294. IEEE Computer Society, IEEE Computer Society Press, Los Alamitos, CA, October 1996.

16. Philip M. Sutton and Charles D. Hansen. Accelerated isosurface extraction in time-varying fields. *Submitted to IEEE Transactions on Visualization and Computer Graphics*, 2000.

17. Ulf Tiede, Thomas Schiemann, and Karl Heinz Höhne. High quality rendering of attributed volume data. In David Ebert, Hans Hagen, and Holly Rushmeier, editors, *Proceedings of Visualization 1998*, pages 255–262. IEEE Computer Society, IEEE Computer Society Press, Los Alamitos, CA, October 1998.

18. Marc van Kreveld, René van Oostrum, Chandrajit Bajaj, Valerio Pascucci, and Dan Schikore. Contour trees and small seed sets for isosurface traversal. In *Proceedings of the Thirteenth Annual Symposium on Computational Geometry*, pages 212–220. ACM Special Interest Groups for Graphics and Algorithms and Computation Theory, ACM Press, New York, NY, June 1997.

19. Jane Wilhelms and Allen van Gelder. Octrees for faster isosurface generation. *ACM Transactions on Graphics*, 11(3):201–227, July 1992.

20. Geoff Wyvill, Craig McPheeters, and Brain Wyvill. Data structure for soft objects. *The Visual Computer*, 2(4):227–234, 1986.

Editors' Note: see Appendix, p. 296 for colored figures of this paper

Case Study: Resource Steering in a Visualization System

Ed H. Chi*, John Riedl

Computer Science and Engineering Department, University of Minnesota
4-192 EE/CS Building, Minneapolis, MN 55455
[echi,riedl]@cs.umn.edu

Abstract. Visual computational steering environments extend traditional visualization environments by enabling the user to interactively steer the computations applied to the data. In this paper, we develop a new type of computational steering. "Resource steering" extends current visual steering techniques by providing machine resource estimation and control to the user. With resource steering, the user controls the execution of the computation on a parallel or distributed computer based on experimentally or theoretically derived estimates of the parallel performance of the computation. We demonstrate this extended steering model by applying it to an information visualization system that analyzes genetic sequence similarity reports. We show how our extended steering model enhances the user's ability to control visualization computations.

1 Introduction

The process of discovery includes trial-and-error followed by deductive insight. Computational steering in visualization enables users to interactively control the computation as it is being visualized, enabling us to apply human intuition and experience to the analysis of data sets. Although existing visual steering environments enhance computational interactivity, users do not receive feedback on computational resource usage, and have no control over how resources are used.

It is important to be able to view and interact with the data as it is generated [10, 13]. Early work include Interactive Zoner, which is an interactive grid generator running on a workstation that is connected to a 2D flow simulation running on a supercomputer [2]. Marshall et. al.ś system enable users to change the parameters of an algorithm as the computation proceeds [9, 15]. The user has two types of control. She uses the interactive control to direct the simulation by changing variables or the simulation condition. She can also manipulate the visualization, such as rotating, translating, zooming, and scaling. (Figure 1)

Recent research has refined computational steering. Jablonowski et. al. described a system where a program could be partially rewritten to support steerable visualization in IRIS Explorer [7]. They provided a framework for a computation written in FORTRAN to include directives for steering the algorithm. Mulder et. al. discussed a customizable widget system where attributes of the widgets are linked to steerable algorithmic variables, and direct manipulations of the widgets result in state changes in the variables [11]. Brodlie et. al. implemented steering along with a history tree that allowed the user to return to previous computation results [3]. Many other systems for visual steering have been developed, such as CUMULVS (Oak Ridge), VIPER (TU

---

* Work done while at U of M, current contact: chi@acm.org, Xerox PARC, Palo Alto, CA 94304

270

Muenich), SCIRun (Utah), Progress and Magellan (Georgia Tech). Several of these pay attention to distributed computing. The SCIRun system provide some interactive visual and quantitative feedback of performance metrics [12]. Pablo and Falcon are some recent work on parallel performance metrics for steering systems provide library routines for instrumenting source code to extract performance data [14].

Visualization applications should extend computational steering to provide resource usage feedback and control to the user for two reasons. First, visualization systems utilize expensive and scarce computing resources, which means users must carefully control resource usage. Visualization applications should provide ways to manage these resources during the entire user session. Second, users interacting with visualization environments can initiate long computations with simple manipulations of the interface. When users initiate computation, they often cannot predict the amount of system resources needed or the length of time required to complete the computation.

We introduce an extended model that incorporates several types of computational steering. Previous steering techniques can be classified into two types: (1) Parameter steering is the specification of initial parameters to the algorithm, and (2) algorithm steering is changing some aspect of the algorithm during computation. We introduce a new steering technique called *resource steering* that allows the user to control the amount of computational resources utilized. Based on resource limits chosen by the user, the model predicts how long the computation will take, and how efficiently it will make use of computing resources.

We tested our idea by extending a computational biology visualization system [5, 4] to including these types of computation steering. The visualization application helps biologists analyze genetic sequence similarity. Previously, biologists used the system in batch mode by creating data sets off-line. In the new system, biologists initiate and steer computations. The computations use an appropriate number of processors, based on the user's selection of desired response time or efficiency.

2   Resource Steering under Our Model

Our steering model is shown in Figure 2, which depicts the interactions between the user and the visualization analysis loop. First, the user initiates an analysis session by

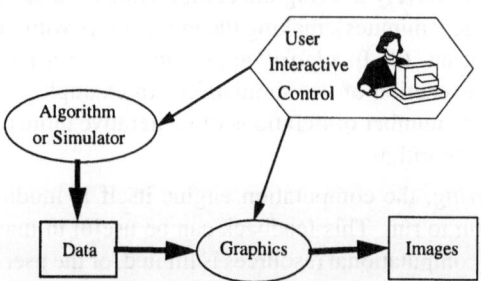

Fig. 1. Existing steering model. The simulation generates data that is given to a visualizer to generate the images on the screen. The user can interact with parameters of the algorithm and manipulate the objects on the screen.

specifying the initial parameters and conditions. Then the parameters are fed into the simulation algorithm to start the simulation process. The parallel machine, in response to the request, starts multiple processors on the simulation, and then generates the data. The data is then processed and given to the visualizer to present to the user.

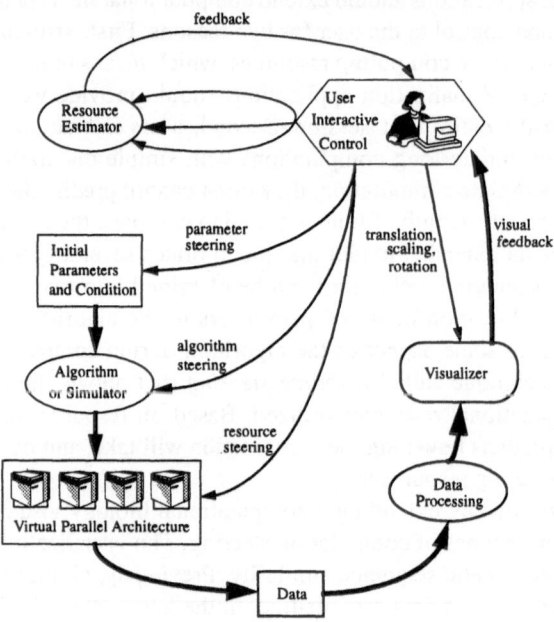

Fig. 2. Extended steering model. Thin arrows represent flow of control, while wide arrows represent the flow of data.

Previous computational steering can be described in two ways: (1) In *parameter steering*, the initial input parameters and conditions are specified for every run of the algorithm. Interactive manipulation of initial input parameters to the algorithm is often mentioned as a simple form of computational steering. In some cases, the computation can be parallelized effectively to bring the computation time down to interactive speed (tens of seconds or a few minutes), making the interactivity with initial input parameters or conditions significant. (2) In *algorithm steering*, some aspect of the algorithm is changed during the execution of the simulation. An example of algorithm steering is the specification of the number of iterations of an iterative numeric method, or perhaps an entirely different algorithm.

In *resource steering*, the computation engine itself is modified either during the execution or from run to run. This feedback can be useful in many situations, such as when the amount of computational resources is limited, or the user and the administrator of the system is interested in obtaining good efficiency. If the response time is too slow, the user might want to increase the number of processors working on the computation. The user may also be interested in the efficiency of the system, especially if she is paying for supercomputing time.

We can provide this feedback by studying the performance of the parallel system under different conditions, and then predicting the performance. For example, given some initial parameters, the current state of the algorithm, and the machine characteristics desired (the number of processors), the estimator returns the expected response time. Alternatively, given some desired response time, the estimator returns the number of processors needed to meet that requirement.

In our model, the three types of steering parameters are given to a resource estimator as constraints. The resource estimator then returns an estimation of the amount of resources needed. The resource estimator uses a *resource estimation function* that maps from the steering parameters to the amount of resources needed. Here we will first describe how we can obtain a performance model for the resource estimation function. Then we discuss how to use the model in a resource estimator.

2.1 Establishing a performance model

The *resource estimation function* is a function that takes the parameter space as input, and returns the runtime (e.g. runtime = r.e.f.$(X_1, X_2, X_3, \ldots)$, where X_i is the input parameters). To predict performance, we can obtain the resource estimation function by using either a theoretical model or an experimental model.

Theoretical Model To develop a theoretical model, we can derive the parallel runtime T_p from the algorithm that solves the problem. The parallel runtime is a function of the parameters and the number of processors. Theoretical derivations have been verified and shown to model the parallel system reasonably in many cases [8]. For example, consider the problem of adding n numbers on a p-processor machine arranged on a hypercube network [8]. Let us assume that it takes T_a units of time to add two numbers and T_c units of time to communicate one number between two directly connected processors. Each processor is assigned n/p numbers initially, and this will take $(n/p - 1)T_a$ to sum at each processor. After the local results are known, it takes $\log p$ steps to add these partial sums together, where each step takes one addition and one communication. Thus, knowing n and p, we can compute the parallel runtime. The parallel runtime is:

$$T_p = (n/p - 1)T_a + (T_a + T_c) \log p \tag{1}$$

The theoretical derivation for many types of algorithms is available in the parallel computing literature, such as sorting, graph algorithms, dynamic programming, fast Fourier transforms, numerical algorithms, and discrete optimization algorithms [8]. If the scalability of a parallel algorithm is known, we can use it to predict the parallel runtime, given the initial parameters and the number of processors.

Experimental model We can also use an experimental model of the algorithm for establishing the performance model. Performing experiments to obtain the resource estimation function is often necessary for a variety of reasons. For example, (a) the theoretical derivation might be difficult; (b) there are unknown variables in the computation; and (c) there is a need to establish the validity of a theoretical model or a need to be more accurate. Therefore, often an experimental model is desirable.

The major drawback of experimental models is that they are relatively difficult to obtain accurately. However, the difficulty can be reduced with carefully planning of experiments. We can establish an experimental performance model by first systemati-cally listing all the steering parameters, and then varying the steering parameters and studying the effect on the parallel runtime. Imagine the parameter space coupled with an axis measuring runtime. We can sample this parameter space on a grid, and measure the runtime at each point on the grid. Then we obtain the runtime characteristic as a multi-dimensional surface in this parameter space. Given this rough sketch of the esti-mation function, we can interpolate between the points to obtain an runtime estimation. This will work well if there are not too many sample points (e.g. there are only a hand-ful of parameters, and the ranges of each parameter is relatively narrow.) In our case study, we used this method to study the effect of each of the steering parameters on the parallel runtime. The limitation of this approach is that this parameter space may not be continuous, since other factors such as number of page faults can result in substantial jumps in runtime. Experimental modelling works well if the steering parameters are mostly independent from each other, and do not affect the runtime in some unexpected way when combined.

2.2 Resource Estimator: Using the performance model to estimate resources

We can use the performance model in a variety of ways, since given all except one vari-able of the resource estimation function, we can determine the value of that last variable. For example, (1) We can use the performance model straightforwardly to obtain an es-timated runtime. Since the user often cannot predict the length of a computation, this provides the necessary feedback to the user. (2) We can use the model to predict the number of processors p needed when given the desired response time. In order to do this, we need to invert the estimation function with respect to p. We can obtain the in-verted function directly if we know the precise runtime equation for the function. Or we can compute p implicitly by using a simple for-loop going from $p = 1$ to the maximum number of processors available, and stopping when p can satisfy the required runtime. (3) We can use the estimation function to predict the speedup and the efficiency. In an experimental model, given T_p, we can directly compute speedup and efficiency as $S = \frac{T_s}{T_p}$ and $E = \frac{S}{p} = \frac{T_s}{pT_p}$, respectively. Given the runtime equations, we can directly compute the algebraic representation of speedup and efficiency. For example, in the example of parallel addition above:

$$S = \frac{nT_a}{(n/p - 1)T_a + (T_a + T_c)\log p} \tag{2}$$

$$E = \frac{nT_a}{(n - p)T_a + (T_a + T_c)p\log p} \tag{3}$$

So given the number of processors p, we can obtain the parallel runtime and the effi-ciency of the system. We can again invert the efficiency estimation function to calculate how many processors we can utilize when we allow the efficiency to drop.

The above examples show that the user can specify any one of the resource steering parameters (e.g. runtime, number of processors, and efficiency), and get an estimation of the other parameters. In high resource load situations, users can use resource steering to obtain efficient performance. If the user is paying for the computing time, this method gives them the power to manage the classic tradeoff between efficiency and speed. This

274

allows the users of the system to have strict control over the utilization of computing resources.

3   An Application of Resource Steering

Here we describe a problem solving environment for genetic sequence similarity computation that incorporates the above ideas and illustrate the techniques for controlling resource usage.

Computational Model Molecular biologists utilize known sequence data in large genomic databases and similarity algorithms to help determine the function of new sequences. BLAST [1] is a well-known similarity algorithm. We first derived a rough theoretical derivation of the BLAST parallel performance model. The BLAST sequence comparison algorithm first serially computes some hashing information, then the database is partitioned into many pieces, which are then given to each of the processors on-the-fly to compare. Thus, the parallel runtime is the time for the serial component plus the parallel database searching component. We expect the parallel component to be nearly linearly scalable, since the database is large compared to the size of the query sequence. However, because the database content is not uniform, and we are not confident whether dynamic load-balancing contributes a large overhead, an experimental study is needed to help us refine and validate the theoretical model. Such validation is required in many applications.

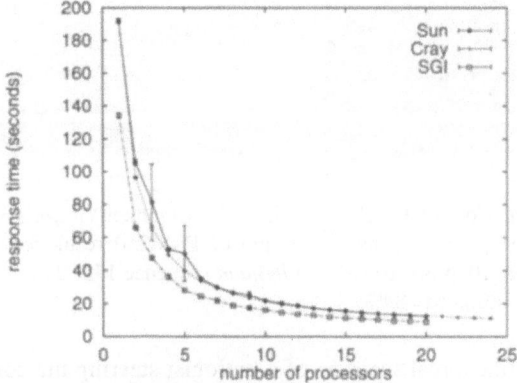

Fig. 3. Response time for SGI Challenge, Sun SparcCenter 2000, and Cray CS6400 with a 500 base sequence using parallel BLASTX SMP algorithm.

For BLAST, we obtained and evaluated the experimental model using several different length sequences on three Shared-Memory Parallel machines (SGI Challenge XL, Sun SparcCenter 2000, Cray CS6400). The parallel response time curve is plotted in Figure 3. We refined the performance model to include other parameter of the algorithm. Sequence comparison is linearly scalable up to tens of processors, so we can parallelized heavily to bring the computation close to interactive speeds (tens of seconds or several minutes). For example, a BLAST process that took nearly 1.3 hours on a single processor completed in only 3.5 minutes with 24 processors on Cray CS6400 [6]. Short sequences that used to take 5–10 minutes can be done in tens of seconds.

Application Usage Walk Through Previously, we presented a system called Alignment Viewer for visualizing similarity information between a single DNA sequence and a large database of other DNA sequences [5]. Each report consists of an input sequence and many *alignments*. An alignment indicates a region of similarity between two sequences. Each alignment has a matching vector and twelve variables. AlignmentViewer uses three spatial axes and one temporal axis. Any of twelve variables can be mapped to any of the four axes. The temporal axis allows the user to construct animations with respect to the temporal variable [5]. The matching vector is represented by a comb-like glyph. For more details on the visualization representation, please see our previous papers.

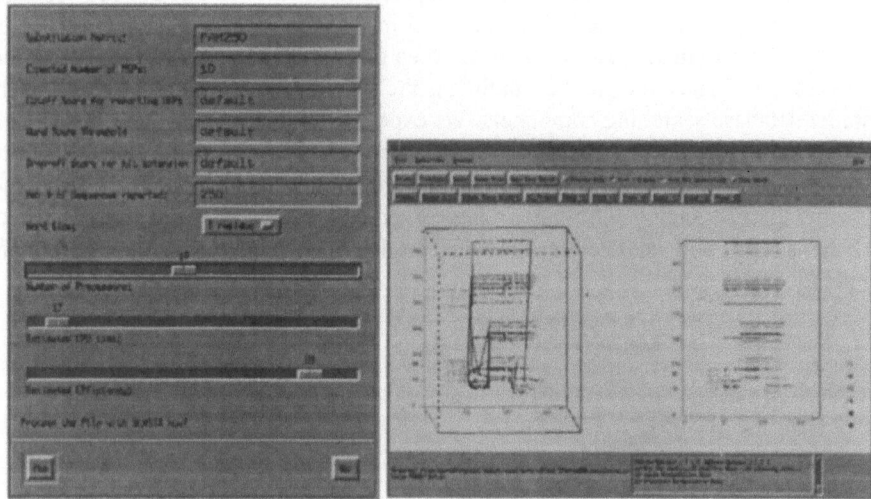

Fig. 4.1: Steering Control Panel: For response time ¡ 20 sec., this sequence requires 10 processors. The efficiency is estimated =89%.

Fig. 4.2: AlignmentViewer's visual representation of PAM250 result for an *Arabidopsis thaliana* sequence 172C2T7.

Let us demonstrate a real scenario of a biologist steering the computation of a sequence. The system asks the user how to process the input sequence using the steering control panel (Figure 4.1, initial parameters in the top half, resource steering controls in the bottom half). The default of one processor is predicted to require 152 seconds, but the biologist wants the response time to be shorter. She drags the response time slider to 20 seconds. The estimating function then computes the required number of processors and the resulting efficiency to meet that request. As shown in Figure 4.1, this sequence requires 10 processors to get the response time under 20 seconds on a SGI Challenge XL machine. The efficiency will be an estimated 89 percent, which is above the 60 percent minimum specified by the system administrator. The user then presses the "Yes" button to start the algorithm running with 10 processors. The result of the computation is visually presented to the user (Figure 4.2).

This sequence report has many good alignments, represented by the abundance of red teeth on the combs. This suggests that the database sequences we found are all closely related. After some analysis, the biologist decided to re-run the algorithm with parameters of closer evolutionary distance. The biologist obtains a new visualization by running the sequence algorithm again with parameter matrix=PAM60 and asking for at least 90 percent efficiency.

A new feature of our system implements the subtraction of two visualized data sets, showing alignments found by one data set but not the other. By using "detail-on-demand", the biologist select an alignment and then view the details of that alignment in a separate window. The result is that we have found "motifs" of a protein called "peroxidase." Motifs are short regions that have been preserved with little change over evolution, presumably because their existence is important to the function of the protein.

The above example showed that the resource steering works well in practice. Our molecular biology application incorporates computational steering in a visualization system that analyzes genetic sequence similarity reports. Using this model, the user is able to change the initial parameters of the algorithm and estimate what effect the parameters will have on the performance of the system. Biologists have been regularly using this system since late 1996. The feedback we have received from them is that using this visualization system has increased their ability to analyze large amounts of similarity data by at least two or three folds.

4   Conclusion

Existing systems do not provide feedback on the efficiency of the computational engine. This feedback is important for two reasons. First, if the user initiates a simulation, she should have information that tells her how long the simulation will take. Second, since computing resources can be scarce, users need to be able to specify the upper limit on the usage of resources. In this paper, we described a particular approach to computation steering that enable users to closely monitor resource usage. Resource steering can be used in many situations, offering the user the ability to fine-tune the efficiency and response time. Sometimes the system administrator may impose minimum efficiency on the computing resources. The system may even provide an estimate of the amount of dollars a particular computation may cost. As depicted in the case study, the new combined approach enhances our ability to integrate computation with visualization.

Acknowledgments This work has been supported in part by the National Science Foundation under grants BIR9402380 and CDA9414015. We wish to thank members of the Arabidopsis sequencing group at Michigan State University and the genomic database group at the University of Minnesota for their advice and suggestions.

References

1. S. Altschul, W. Gish, W. Miller, E. Myers, and D. Lipman. Basic Local Alignment Search Tool. *Journal of Molecular Biology*, 215:403–410, 1990.
2. G. Bancroft, T. Plessel, F. Merritt, and V. Watson. Tools for 3d scientific visualization in computational aerodynamics at NASA Ames Research Center. In *Proc. SPIE 1083: Three-Dimensional Visualization and Display Technologies*, pages 161–172, 1989.

3. K. Brodlie, A. Poon, H. Wright, L. Brankin, G. Banecki, and A. Gay. Grasparc — a problem solving environment integrating computation and visualization. In *IEEE Visualization '93*, pages 102–109. IEEE CS Press, 1993.

4. E. H. Chi, P. Barry, J. Riedl, and J. Konstan. A spreadsheet approach to information visualization. In *Proceedings of the Symposium on Information Visualization '97*, pages 17–24,116. IEEE CS, 1997. Phoenix, Arizona.

5. E. H. Chi, J. Riedl, E. Shoop, J. V. Carlis, E. Retzel, and P. Barry. Flexible information visualization of multivariate data from biological sequence similarity searches. In *Proc. IEEE Visualization '96*, pages 133–140, 477. IEEE CS, 1996. San Francisco, California.

6. E. H. Chi, E. Shoop, J. Carlis, E. Retzel, and J. Riedl. Efficiency of shared-memory multiprocessors for a genetic sequence similarity search algorithm. Technical Report TR97-005, University of Minnesota Computer Science Department, 1997.

7. D. Jablonowski, J. Bruner, B. Bliss, and R. Haber. VASE: The visualization and application steering environment. In *IEEE Visualization '93*, pages 560–569. IEEE CS Press, 1993.

8. V. Kumar, A. Grama, A. Gupta, and G. Karypis. *Introduction to Parallel Computing*. Benjamin Cummings, 1994.

9. R. Marshall, J. Kempf, S. Dyer, and C.-C. Yen. Visualization methods and simulation steering for a 3D turbulence model of Lake Erie. In *Proceedings of Symposium on Interactive 3D Graphics*, pages 89–97, 264. SIGGRAPH, 1990.

10. B. McCormick et al. Visualization in scientific computing. In *Computer Graphics*, volume 21. ACM Press, November 1987.

11. J. Mulder and J. van Wijk. 3D computational steering with parametrized geometric objects. In *IEEE Visualization '95*, pages 304–311. IEEE CS Press, 1995.

12. S. Parker, D. Weinstein, and C. Johnson. The SCIRun computational steering software system. In E. Arge, A. Bruaset, and H. Langtangen, editors, *Modern Software Tools in Scientific Computing*, pages 1–44. Birkhauser Press, 1997.

13. D. Reed, C. Elford, T. Madhyastha, E. Smirni, and S. Lamm. The next frontier: Interactive and closed loop performance steering. In *Proceedings of the 25th Annual Conference of International Conference on Parallel Processing*, 1996.

14. J. Vetter and K. Schwan. Progress: A toolkit for interactive program steering. In *Proceedings of the 24th Annual Conference of International Conference on Parallel Processing*, pages 139 – 142, 1995.

15. C.-C. Yen, K. Bedford, J. Kempf, and R. Marshall. A three-dimensional/stereoscopic display and model control system for great lakes forecasts. In *IEEE Visualization '90*, pages 194–201. IEEE CS Press, 1990.

Authors Index

Color Plates

Melançon and Herman (pp. 3–12)

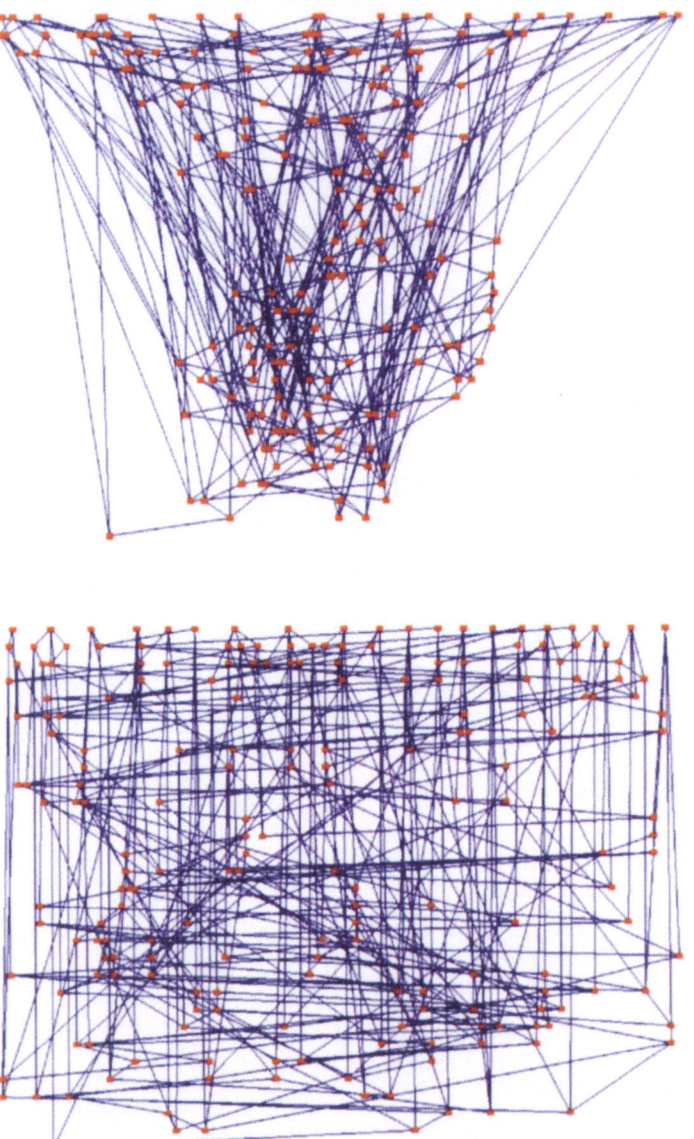

Fig. 4. Layout examples. Barycenter method (above) and spanning tree layout adapted from Reingold and Tilford (below)

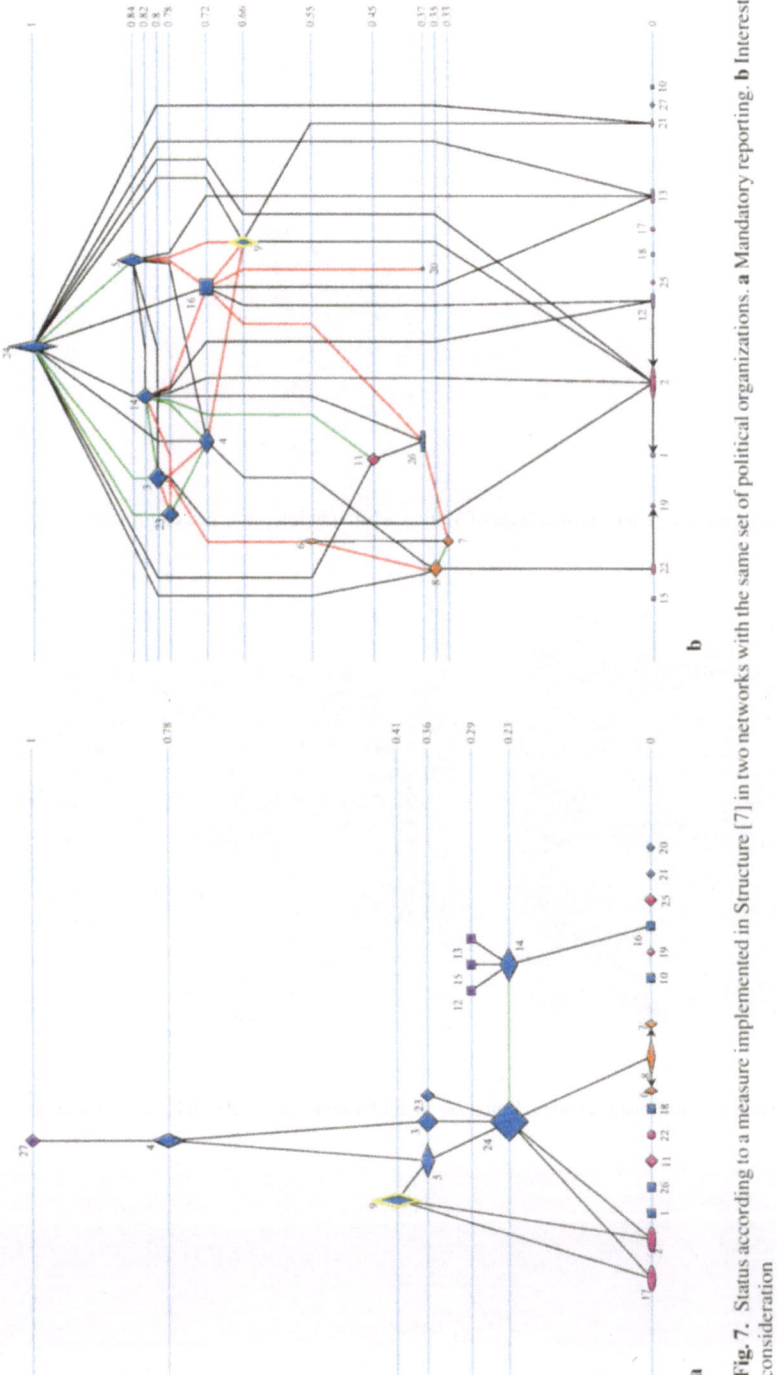

Fig. 7. Status according to a measure implemented in Structure [7] in two networks with the same set of political organizations. **a** Mandatory reporting. **b** Interest consideration

Brandes et al. (pp. 23–32)

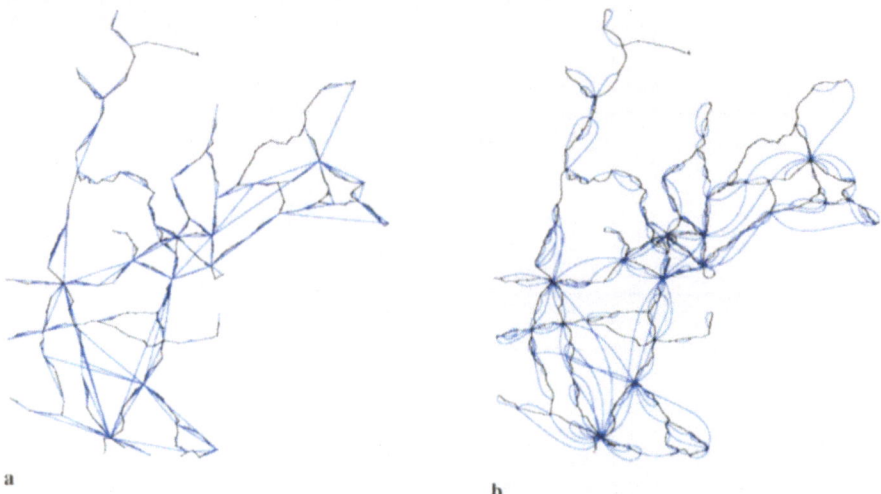

a
b

Fig. 7. Medium size time table graph: surroundings of Venice. **a** Straight lines, **b** balanced rotation

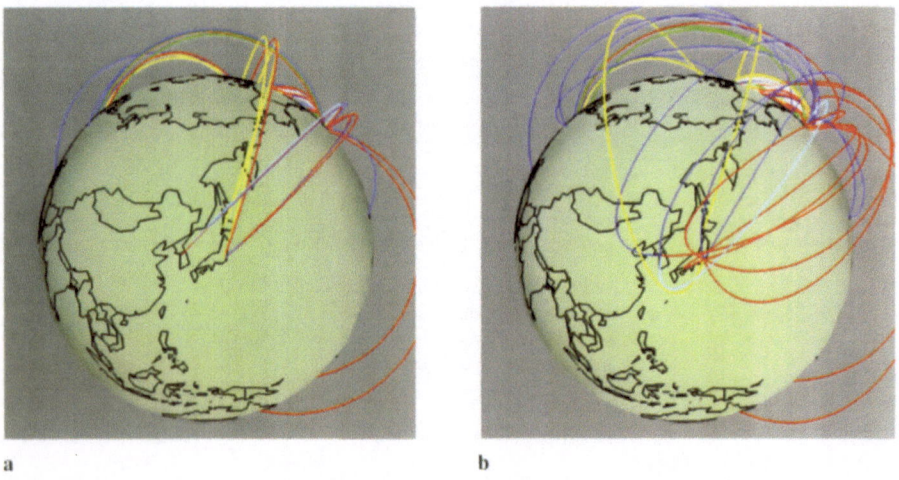

a
b

Fig. 8. Part of the Internet's multicast backbone: Korea/Japan. **a** Elevated great circles, **b** balanced rotation

Pu and Pečenović (pp. 43–52)

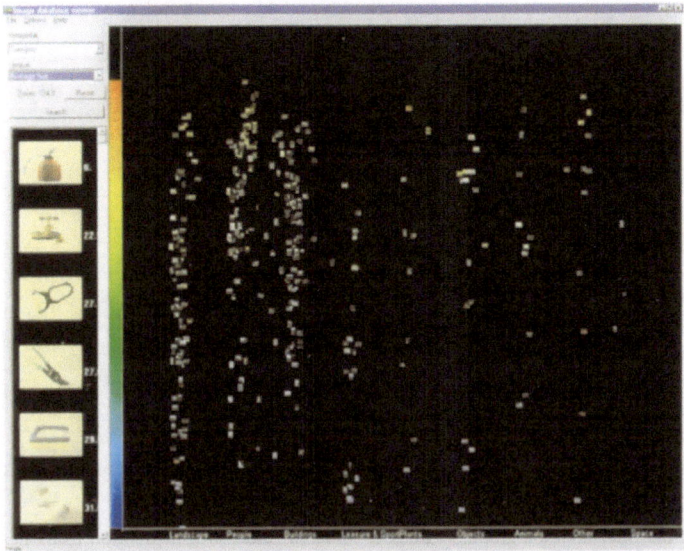

Fig. 9. A galaxy of images spread by Category and Name on the X axis, and by Hue on the Y axis

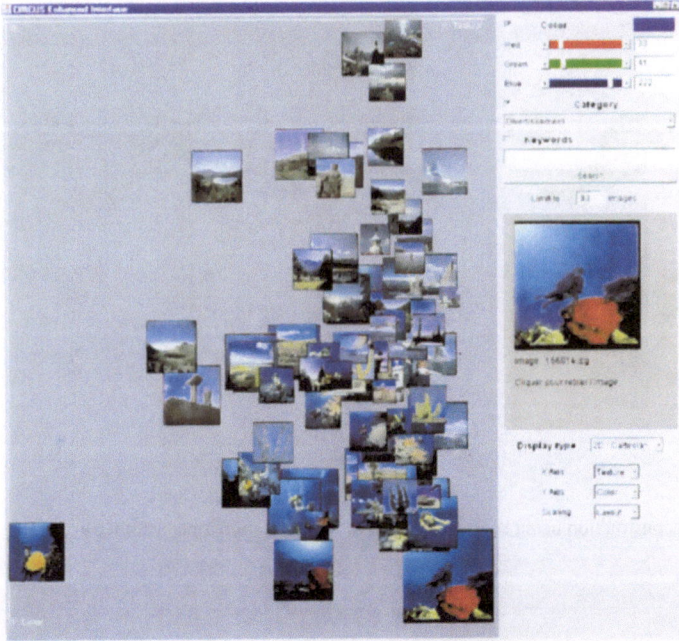

Fig. 11. Blue "sports and entertainment" images in Cartesian coordinates. The query image is in the lower-right corner, we show *similarity* on Color on the X and Texture on the Y axis. The size represents similarity on Layout

Reinders et al. (pp. 73–82)

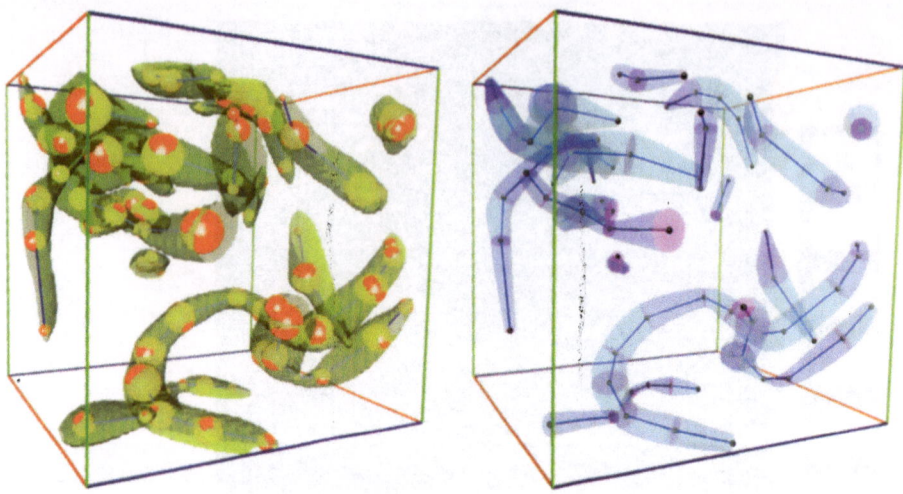

Fig. 7. A visualization of the final skeleton graph. The nodes and edges are shown by spheres and small tubes, and the iso-surface of the object is shown transparently

Fig. 9. Skeleton surface reconstruction using hermite tube icons

Fig. 8. Skeleton surface reconstruction using solid spheres and cones, or a more open geometry

Holliday and Nielson (pp. 83–92)

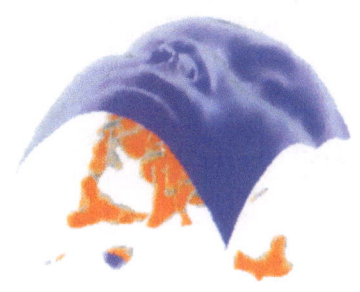

Fig. 5. An adaptively approximated approximated MRI data set. The tetrahedrization resulting from a fit and the isosurfaces computed from the model are shown for various approximations

LaMar et al. (pp. 105–114)

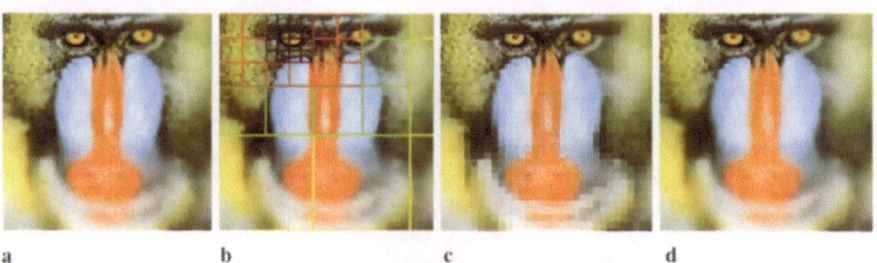

a   b   c   d

Fig. 6. Multiresolution Mandrill: **a** without blending; **b** node boundaries high-lighted; **c** blended nearest-neighbor; **d** blended, bilinearfiltering

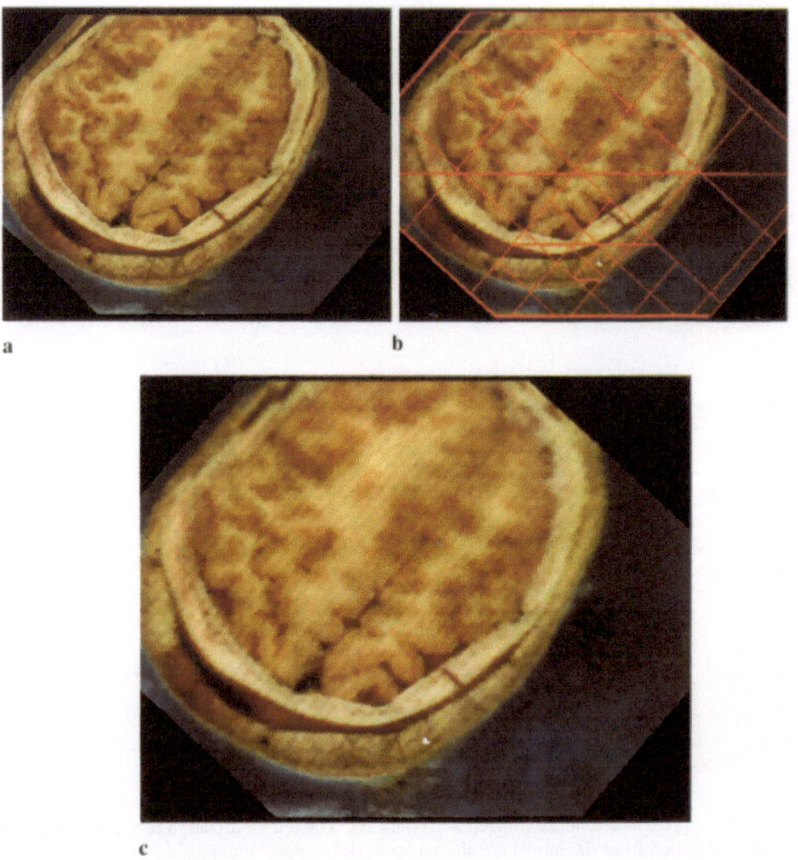

Fig. 7. Multiresolution cutting plane of the Visible Female data set: **a** fixed resolution; **b** blending with node boundaries high-lighted; **c** MR, blending

Hanson et al. (pp. 115–124)

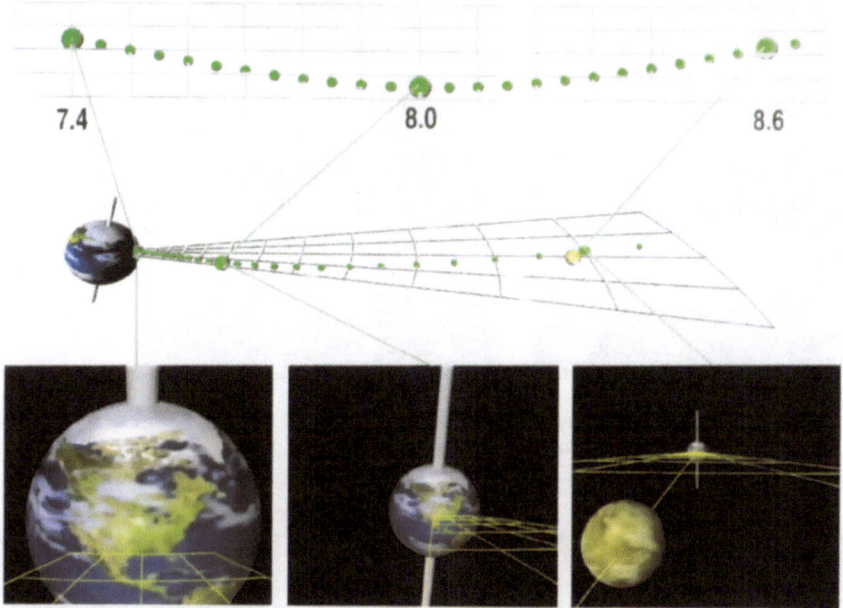

Fig. 4. Images corresponding to three viewpoints on the logarithmically scaled "wedge" path from the Earth to the Moon

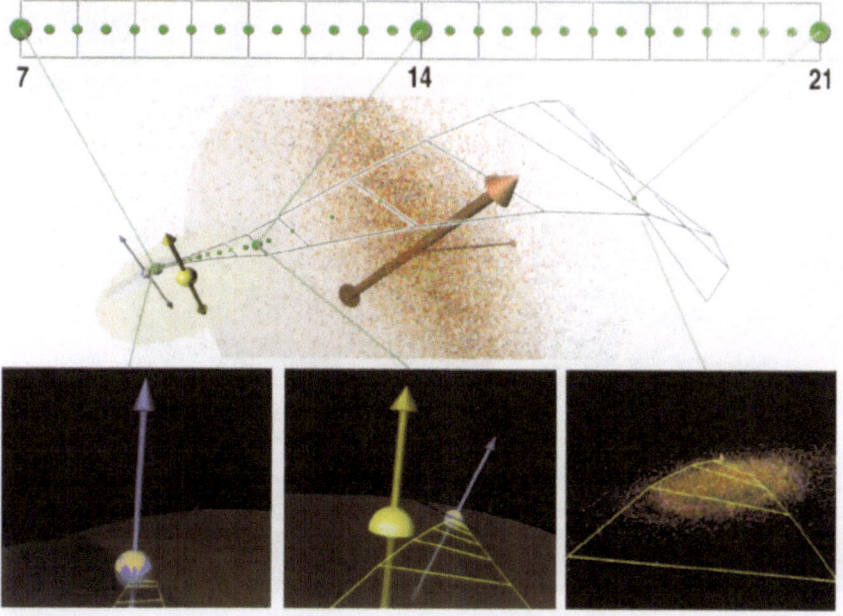

Fig. 5. Images corresponding to three viewpoints on the adaptive twisted-orientation constraint manifold encompassing the Earth, solar system, and the Milky Way galaxy

Pury et al. (pp. 125–135)

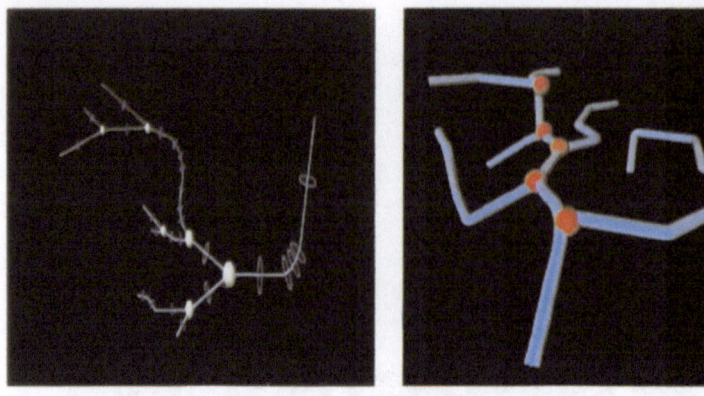

Fig. 5 (*left*). Symbolic model visualization: polylines
Fig. 6 (*right*). Symbolic model visualization: pipes and spheres

Fig. 7 (*left, middle*). Surface visualization. **Fig. 8** (*right*). Inner surface navigation

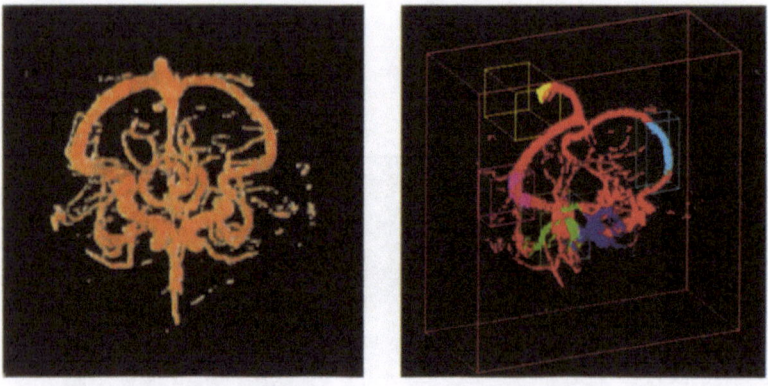

Fig. 9 (*left*). Shaded volume visualization with a BTF-method
Fig. 10 (*right*). Volume features visualization with a BTF-method

Ebert et al. (pp. 137–146)

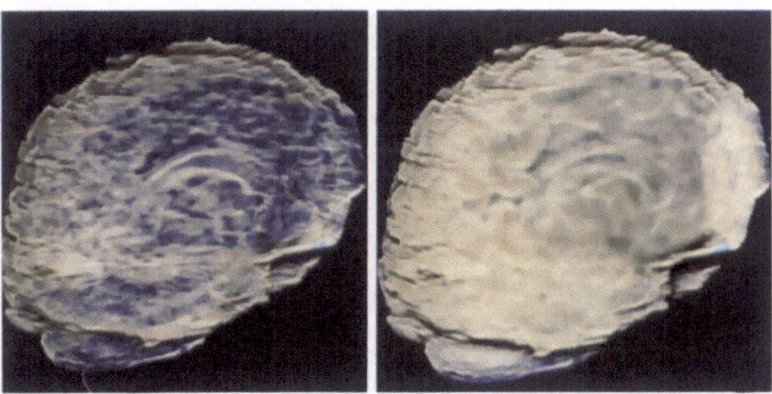

Fig. 2. Volume rendering of segmented photographie visible male brain, without and with enhancement of low gradient areas

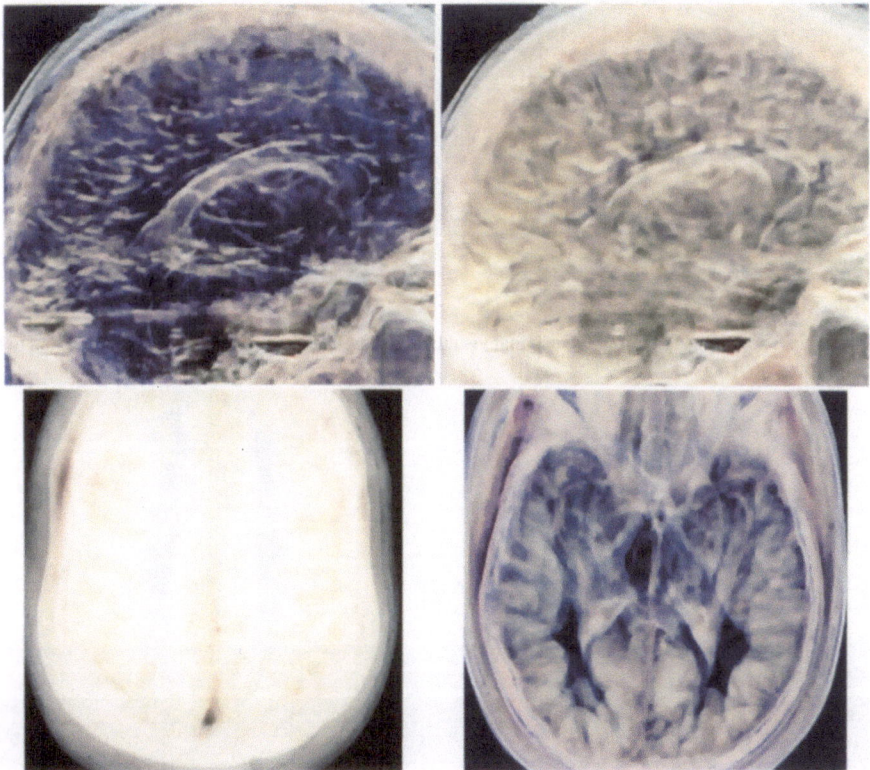

Fig. 3. Volume rendering of photographic visible male head, without and with enhancement of low gradient areas from side and top views

Engel et al. (pp. 167–177)

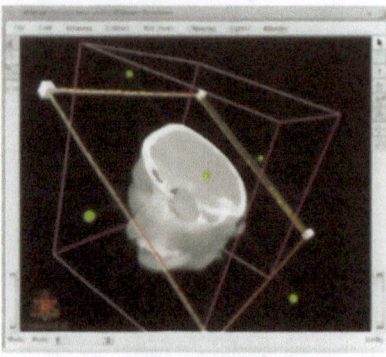 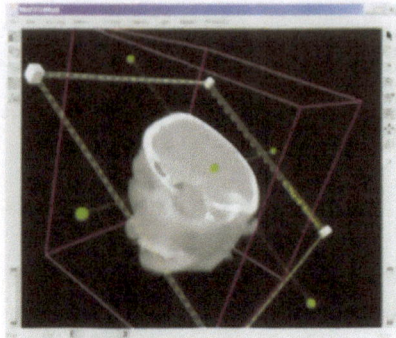

Fig. 3. Transparent remote access to 3D texture mapping hardware of a C++ server application (*left*) from a Java client application (*right*). Note that the client provides the same look-and-feel as the stand-alone application by providing an Open Inventor-like decoration. The displayed manipulators can be picked and dragged remotely

Fig. 4. Transfer of image data from the original visualization application (*left*) via a socket connection to a HTML browser (*right*)

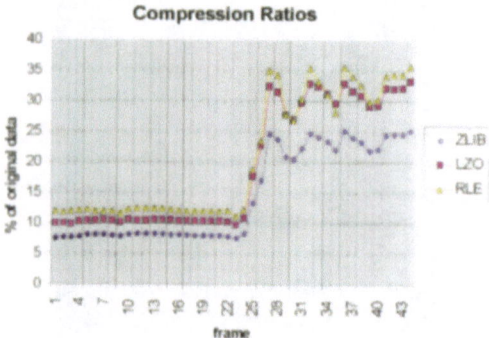

Fig. 5. Compression ratios of ZLIB, LZO and RLE compression for a typical image sequence. A rotation of the volume followed by a magnification was performed

Telea and van Wijk (pp. 189–198)

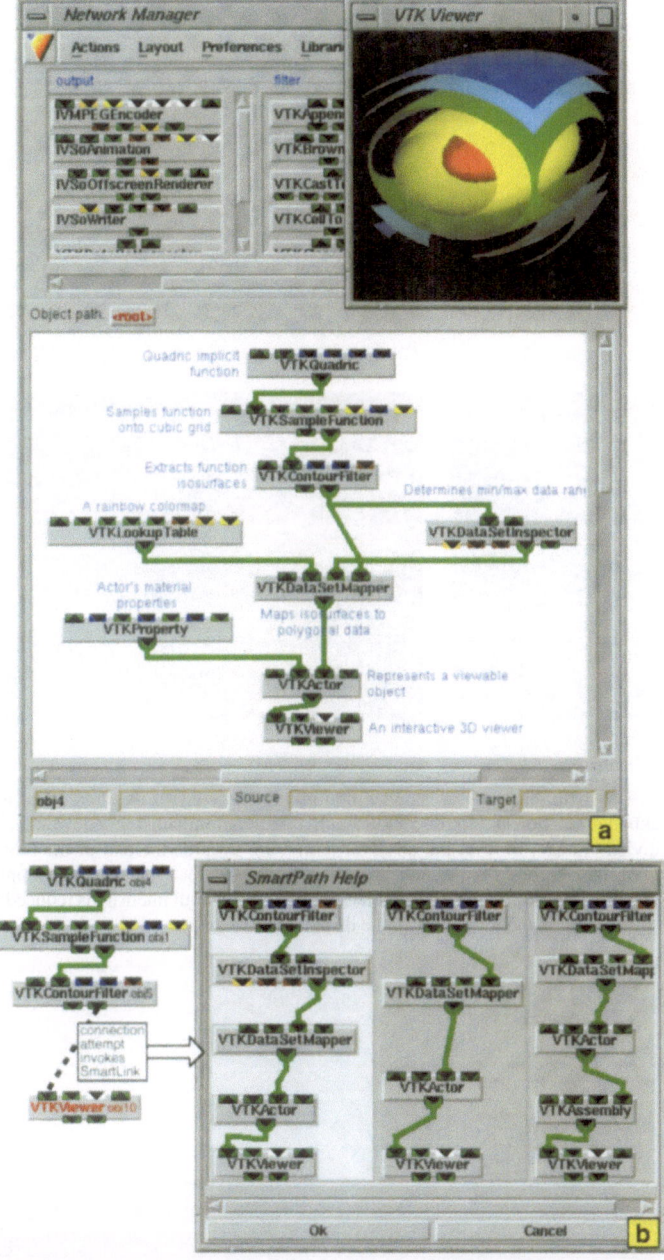

Fig. 8. VTK visualization example of quadric function isosurfaces (**a**) and SMARTLINK help agent invoca-tion (**b**)

Weiskopf (pp. 219–228)

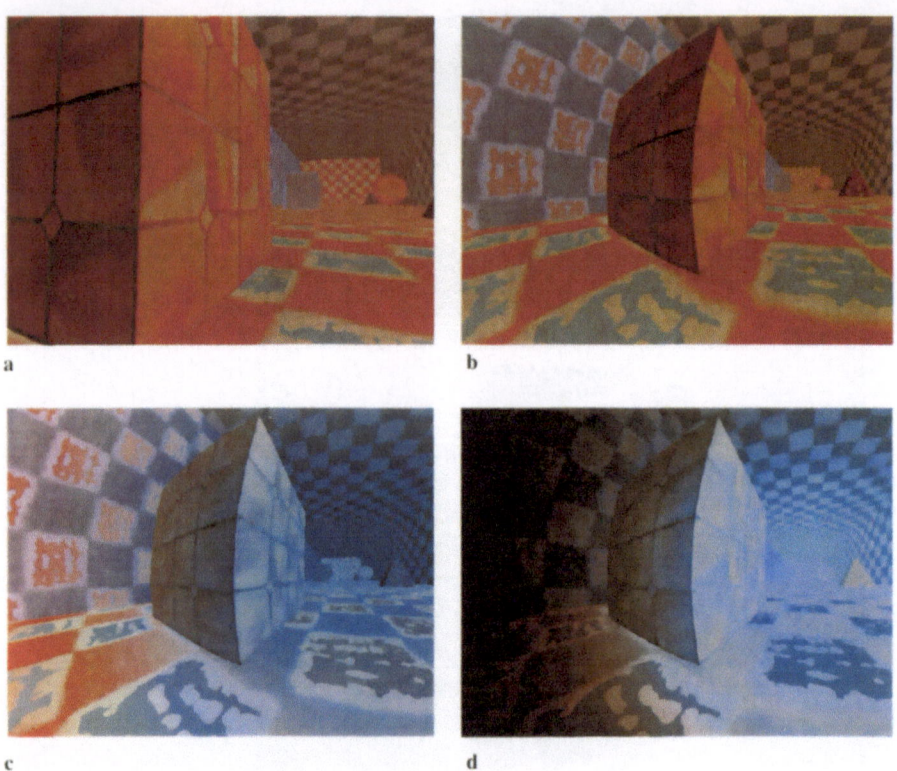

a b

c d

Fig. 3. Example of special relativistic rendering. **a** shows a non-relativistic view of the tunnel-like test scene. The scene emits blackbody radiation at 3500K, 5900K, and 10000K, producing the red, white, and blue colors in the non-relativistic image. In **b–d**, the observer is moving towards the end of the tunnel with 60 percent of the speed of light. **b** shows the visualization of apparent geometry, **c** adds the Doppler effect, and **d** illustrates the complete transformation of illumination. In **d**, the overall intensity is reduced to 10 percent of that in **a–c** in order to keep the intensities in the displayable range

König et al. (pp. 229–238)

Fig. 5. *Left*: sample with cut-away part. *Right*: distance measuring metaphor. The current length of the rubber band is 0.6 mm

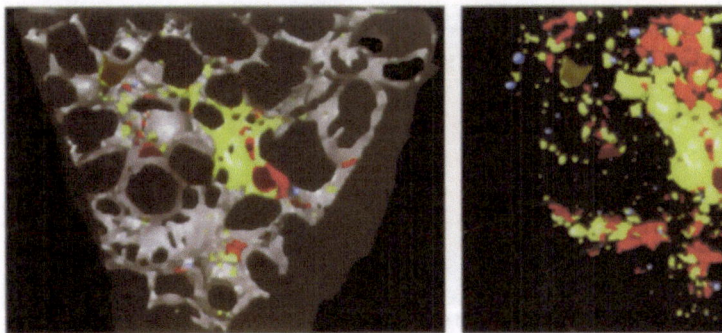

Fig. 6. Application of a cutting plane (*left*). Focusing (*right*) renders only cells, which are not cut open by the faces of the sample. Colors indicate the shape factor of cells

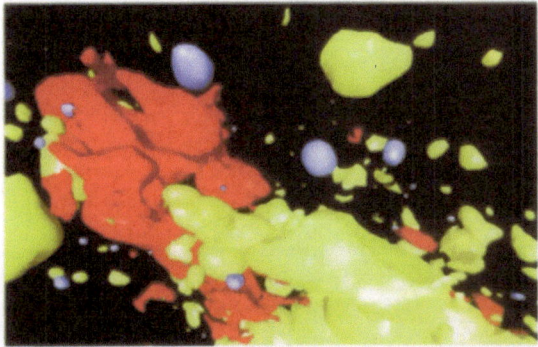

Fig. 7. Focused view based on shape of cells. Blue cells: close to the shape of a sphere, small value of f; green: medium value of f; red: large value of f

Maurits et al. (pp. 239–248)

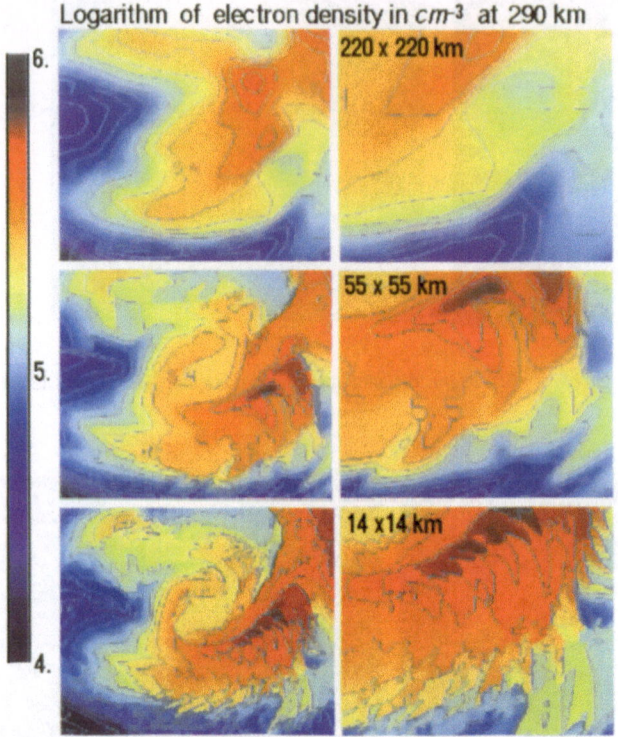

Color Plate 1. Electron density at 290 km, simulated with EPPIM at specified resolutions. Isocontours are drawn at every 0.1 in the log-scale, correspondingly to a factor 1.26 for density

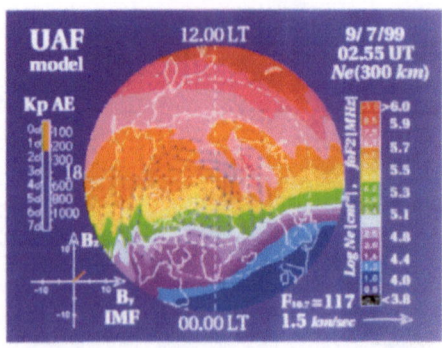

Color Plate 2. Example of ionospheric map from the EPPIM WWW-site dac3.gi.alaska.edu/~sergei

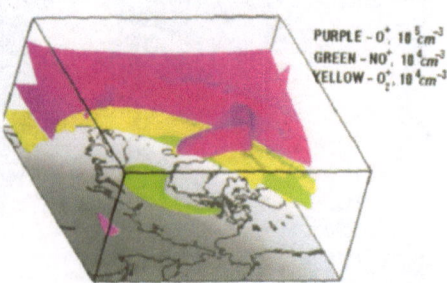

Color Plate 3. Vis5D representation of the EPPIM volumetric output

Sutton et al. (pp. 259–268)

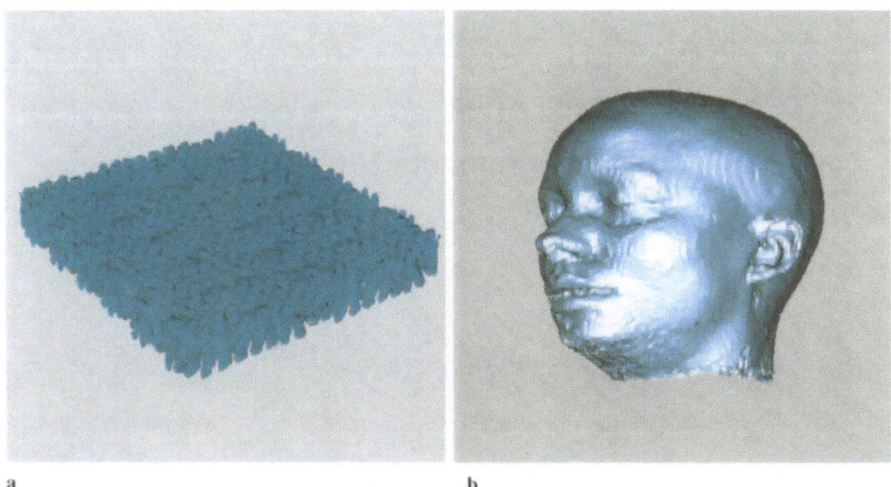

a   b

Fig. 4. Isosurfaces from simulation and measurement sources. **a** depicts an isosurface from the RAGE computational fluid dynamics simulation, showing the bubbles formed by Rayleigh-Taylor instability. **b** shows the skin isosurface from the volumetric data set produced by a CT scan

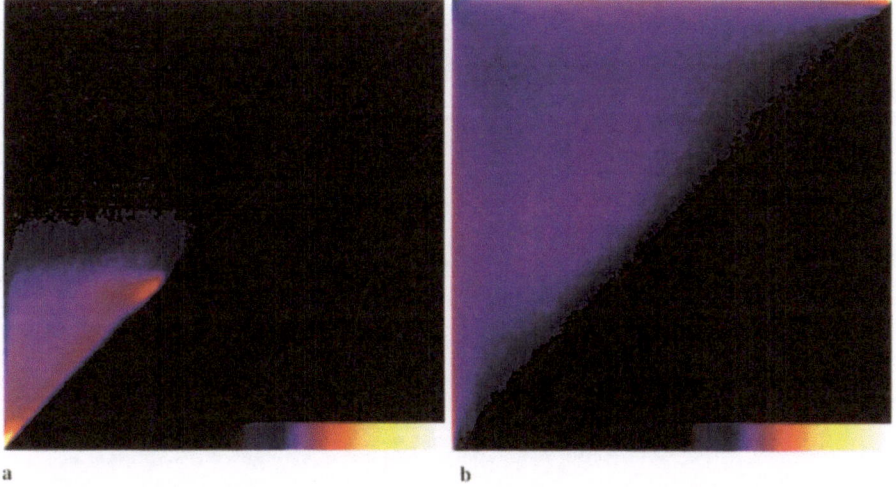

a   b

Fig. 5. Span space representations of the **a** Head128 and **b** Rage256 data sets. Blue represents low density of points, white represents high density

SpringerEurographics

Jurriaan Mulder,

Robert van Liere (eds.)

Virtual Environments 2000

Proceedings of the Eurographics Workshop,
Amsterdam, The Netherlands, June 1–2, 2000

2000. X, 217 pages. 95 partly coloured figures.
Softcover DM 98,–, öS 686,–, (recommended retail price)
ISBN 3-211-83516-4. Eurographics

From the Contents
- Practical Calibration Procedures for Augmented Reality
- Evaluation of Rotation Correction Techniques for Electromagnetic
  Position Tracking Systems
- A 'Plug and Play' Approach to Testing Virtual Environment
  Interaction Techniques
- Developing Effective Navigation Techniques in Virtual Environments
- The Effects of Group Collaboration on Presence in a Collaborative
  Virtual Environment
- An Asynchronous Architecture to Manage Communication, Display,
  and User Interaction in Distributed Virtual Environments
- Time Critical Computing and Rendering of Molecular Surfaces Using
  a Zonal Map
- A Volumetric Virtual Environment for Catheter Insertion Simulation
- Interacting with Simulation Data in an Immersive Environment
- ERGONAUT: A Tool for Ergonomic Analyses in Virtual Environments
- Accelerometer-Based Motion Tracking for Orchestra Conductor
  Following

For further information please visit our homepage: **www.springer.at**

 SpringerWienNewYork

A-1201 Wien, Sachsenplatz 4–6, P.O.Box 89, Fax +43.1.330 24 26, e-mail: books@springer.at, Internet: www.springer.at
D-69126 Heidelberg, Haberstraße 7, Fax +49.6221.345-229, e-mail: orders@springer.de
USA, Secaucus, NJ 07096-2485, P.O. Box 2485, Fax +1.201.348-4505, e-mail: orders@springer-ny.com
Eastern Book Service, Japan, Tokyo 113, 3–13, Hongo 3-chome, Bunkyo-ku, Fax +81.3.38 18 08 64, e-mail: orders@svt-ebs.co.jp

SpringerEurographics

Eduard Gröller, Helwig Löffelmann,

William Ribarsky (eds.)

Data Visualization '99

Proceedings of the Joint EUROGRAPHICS and IEEE TCVG Symposium
on Visualization in Vienna, Austria, May 26–28, 1999

1999. XII, 340 pages. 230 partly coloured figures.

Softcover DM 118,–, öS 826,–

(recommended retail price)

ISBN 3-211-83344-7

Eurographics

In the past decade visualization established its importance both in
scientific research and in real-world applications.

In this book 21 research papers and 9 case studies report on the latest
results in volume and flow visualization and information visualization.
Thus it is a valuable source of information not only for researchers but
also for practitioners developing or using visualization applications.

Contents
- Information Visualization
- Flow Visualization
- Volume Visualization
- Visualization of Medical Data & Molecules
- Geometry, Grids, and Systems
- Information Visualization and Systems
- Volume, Medical, & Molecular Visualization
- Authors Index
- Color Plates

SpringerWienNewYork

A-1201 Wien, Sachsenplatz 4–6, P.O.Box 89, Fax +43.1.330 24 26, e-mail: books@springer.at, Internet: www.springer.at
D-69126 Heidelberg, Haberstraße 7, Fax +49.6221.345-229, e-mail: orders@springer.de
USA, Secaucus, NJ 07096-2485, P.O. Box 2485, Fax +1.201.348-4505, e-mail: orders@springer-ny.com
Eastern Book Service, Japan, Tokyo 113, 3–13, Hongo 3-chome, Bunkyo-ku, Fax +81.3.38 18 08 64, e-mail: orders@svt-ebs.co.jp

SpringerEurographics

Nadia Magnenat-Thalmann,

Daniel Thalmann (eds.)

Computer Animation and Simulation '99

Proceedings of the Eurographics Workshop in Milano, Italy,
September 7–8, 1999

1999. X, 230 pages. 148 partly coloured figures.

Softcover DM 89,–, öS 625,–

(recommended retail price)

ISBN 3-211-83392-7

Eurographics

The 20 research papers in this volume demonstrate novel models and concepts in animation and graphics simulation. Special emphasis is given on innovative approaches to Modelling Human Motion, Models of Collision Detection and Perception, Facial Animation and Communication, Specific Animation Models, Realistic Rendering for Animation, and Behavioral Animation.

Contents
- Virtual Humans
- Collision Techniques
- Facial Animation and Communication
- Animation Models
- Realism
- Behavioral Animation
- Appendix

SpringerWienNewYork

A-1201 Wien, Sachsenplatz 4–6, P.O.Box 89, Fax +43.1.330 24 26, e-mail: books@springer.at, Internet: www.springer.at
D-69126 Heidelberg, Haberstraße 7, Fax +49.6221.345-229, e-mail: orders@springer.de
USA, Secaucus, NJ 07096-2485, P.O. Box 2485, Fax +1.201.348-4505, e-mail: orders@springer-ny.com
Eastern Book Service, Japan, Tokyo 113, 3–13, Hongo 3-chome, Bunkyo-ku, Fax +81.3.38 18 08 64, e-mail: orders@svt-ebs.co.jp

SpringerEurographics

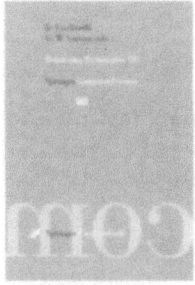

Michael Gervautz,
Axel Hildebrand,
Dieter Schmalstieg (eds.)

Virtual Environments '99

Proceedings of the Eurographics
Workshop in Vienna, Austria,
May 31–June 1, 1999

Dani Lischinski,
Ward Larson (eds.)

1999. X, 191 pages. 78 figures.
Softcover DM 85,–, öS 595,–
ISBN 3-211-83347-1. Eurographics

Rendering Techniques '99

Proceedings of the Eurographics
Workshop in Granada, Spain,
June 21–23, 1999

The special focus of this volume lies on augmented reality. Problems like real-time rendering, tracking, registration and occlusion of real and virtual objects, shading and lighting interaction and interaction techniques in augmented environments are addressed. The papers collected in this book also address levels of detail, distributed environments, systems and applications and interaction techniques.

1999. XI, 382 pages. 212 partly coloured figures.
Softcover DM 118,–, öS 826,–
ISBN 3-211-83382-X. Eurographics

The papers in this volume present new research activities in the "classical" rendering workshop topics: radiosity and Monte Carlo global illumination algorithms and illumination models, alongside papers on near-interactive ray tracing, hardware-assisted rendering algorithms, techniques for acquisition and modeling from images, image-based rendering, novel shadow algorithms, and inverse lighting and design.

Contents
- Levels of Detail
- Tracking
- Rendering of Virtual Environments
- Distributed Environments
- Systems and Applications
- Interaction.

All prices are recommended retail prices

 SpringerWienNewYork

A-1201 Wien, Sachsenplatz 4–6, P.O. Box 89, Fax +43.1.330 24 26, e-mail: books@springer.at, Internet: www.springer.at
D-69126 Heidelberg, Haberstraße 7, Fax +49.6221.345-229, e-mail: orders@springer.de
USA, Secaucus, NJ 07096-2485, P.O. Box 2485, Fax +1.201.348-4505, e-mail: orders@springer-ny.com
Eastern Book Service, Japan, Tokyo 113, 3–13, Hongo 3-chome, Bunkyo-ku, Fax +81.3.38 18 08 64, e-mail: orders@svt-ebs.co.jp

SpringerEurographics

Nuno Correia,
Teresa Chambel,
Glorianna Davenport (eds.)

Multimedia '99

Proceedings of the Eurographics
Workshop in Milano, Italy,
September 7–8, 1999

2000. IX, 222 pages. 85 figures.
Softcover DM 85,–, öS 595,–
ISBN 3-211-83437-0
Eurographics

Multimedia '99 covers technological and scientific areas of media production, processing and delivery. 24 contributions from research laboratories and universities worldwide give a broad perspective on multimedia research with a special focus on media convergence. The topics treated in this volume: image and sound content analysis and processing, paradigms and metaphors for multimedia authoring and display, applications such as education or entertainment, and multimedia content authentication and security.

David J. Duke,
Angel Puerta (eds.)

Design, Specification and Verification of Interactive Systems '99

Proceedings of the Eurographics
Workshop in Braga, Portugal,
June 2–4, 1999

1999. IX, 280 pages. 89 figures.
Softcover DM 118,–, öS 826,–
ISBN 3-211-83405-2
Eurographics

The collection of papers in this volume covers specification methods and their use in design, model-based tool support, task and dialogue models, distributed collaboration, and models for virtual reality input. Strong emphasis is laid on formal representations and modelling techniques and their use in understanding interaction and informing the design of artefacts.

All prices are recommended retail prices

 SpringerWienNewYork

A-1201 Wien, Sachsenplatz 4–6, P.O.Box 89, Fax +43.1.330 24 26, e-mail: books@springer.at, Internet: www.springer.at
D-69126 Heidelberg, Haberstraße 7, Fax +49.6221.345-229, e-mail: orders@springer.de
USA, Secaucus, NJ 07096-2485, P.O. Box 2485, Fax +1.201.348-4505, e-mail: orders@springer-ny.com
Eastern Book Service, Japan, Tokyo 113, 3–13, Hongo 3-chome, Bunkyo-ku, Fax +81.3.38 18 08 64, e-mail: orders@svt-ebs.co.jp